THE BIG BANG

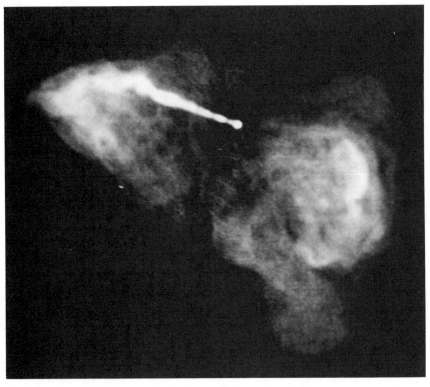

Frontispiece A radio image taken with the Very Large Array in Socorro, New Mexico, of the giant elliptical galaxy Messier 87. The complex filamentary structure of the radio lobes is shown, as well as the radio jet that powers the radio emission. The jet emanates from the galaxy nucleus, where a supermassive black hole is believed to be lurking.

THE BIG BANG
REVISED AND UPDATED EDITION

JOSEPH SILK
University of California, Berkeley

W. H. FREEMAN AND COMPANY
New York

Library of Congress Cataloging-in-Publication Data

Silk, Joseph.
    The big bang / Joseph Silk.—Revised and updated ed.
        p.   cm.
    Bibliography: p.
    Includes index.
    ISBN 0-7167-1812-X
    1. Cosmology.   2. Big bang theory.   I. Title.
QB981.S55 1989   89-16497
523.1'8—dc 19   CIP

Printed in the United States of America

1 2 3 4 5 6 7 8 9 0 VB 6 5 4 3 2 1 0 8 9 8

CONTENTS

PREFACE

The big bang is the modern version of creation, a topic of fascination since the dawn of humanity. Science has supplanted mysticism as the source of inspiration about the beginning of the universe. My purpose in this book is to present an accessible description of the scientific approach to the origin of the structures around us, ranging in scale from planets and stars to galaxies and great clusters of galaxies to the entire observable universe. The big bang theory provides the framework for my discussion. It is a theory based on astonomical data, painstakingly gathered at observatories around the world, and on recent advances in particle physics toward an understanding of the ultimate nature of matter. The search is far from over, and the theory is still incomplete. Nevertheless, the moment is at hand to describe where we are and where we are going.

I have kept the text free of mathematics but tried to present many of the underlying ideas faithfully. My hope is that I communicate the sheer excitement of the cosmologist's pursuit of the ultimate truth, without grossly diluting or oversimplifying the key ideas. For the reader who requires more detail, I have supplied an appendix that utilizes a low level of mathematics (no calculus!) to support and amplify some of the important fundamental concepts.

To set the scene for this cosmic saga, the opening chapters of the book describe the history of cosmological thought from the Greek philosopher-scientists to modern times. The evidence for the big bang theory is explained and a simple description of the theory, including a discussion of alternative big bang models, is given. Although much of the astronomical data is recent and many details deserve further study, the big bang theory provides the best explanation of the available evidence. One of the most substantial areas of progress since the first edition of *The Big Bang* has been in our understanding of the moment of creation. During the earliest instants of expansion of the universe, conditions were so extreme that particle energies surpassed those attainable in the world's largest atom-smashing machines. The big bang provided an environment for studying the fabric of matter itself. Particle physicists have used this cosmological laboratory to design the latest models of elementary particle interactions. Their search for the ultimate symmetry from which all properties of matter ultimately developed involves attempts to unify the quantum theory of the nuclear and electromagnetic forces with the theory of gravitation. It has led to new insights into the beginning of the universe; these are described in Chapter 6.

Since galaxies are the building blocks of the universe, we must understand their origin and evolution in order to understand the universe itself. Thus, we shall explore how galaxies formed and aggregated into groups and clusters. We shall unravel the mysteries of radio galaxies and quasars. We shall also see how the first stars formed and trace subsequent generations of star birth and death through eons of time. Our cosmic journey culminates with the formation of the heavy elements and the solar system. The final chapters assess the future course of the universe and discuss possible alternatives to the big bang theory.

Some of the topics covered in this book are a source of vigorous debate among active researchers. Where controversy exists, I have usually adopted the majority viewpoint, but I have not ignored alternative theories. Where there is no accepted dogma, I have allowed my own personal prejudices to intrude. Thus, the discussions of galaxy and star birth, for example, are necessarily highly subjective and speculative.

The bulk of the big bang theory, however, rests more on fact than on speculation. In addition, the relative simplicity of the big bang theory favors it over more exotic cosmologies. The ultimate cosmological tests can be performed only behind a telescope and not by means of philosophical debates. At present, the preponderance of cosmological data favors the big bang theory, as the reader may now judge.

The first edition of this book was begun when I held a Guggen-heim Fellowship and was on leave from Berkeley as a member of the Institute for Advanced Study, Princeton, in 1976. It was com-pleted while I was a visiting professor in the Physics Department of the University of California at Davis, whose hospitality I grate-fully acknowledge. This revised and updated edition was prepared at Berkeley during 1987. John Barrow, J. Richard Bond, William Burke, Bernard Carr, Larry Marschall, P. James Peebles, Bradley Peterson, Martin Rees, Michael Rowan-Robinson, and Robert Wa-goner read drafts of all or part of this book, and provided critical comments and suggestions. Thanks also to Pat Crowder, who so patiently typed the many revisions.

Joseph Silk

THE BIG BANG

· 1 ·

INTRODUCTION TO COSMOLOGY

Je m'en vais chercher un grand peut-etre.
—*François Rabelais*

C*osmology** is the study of the large-scale structure and evolution of the universe. When we survey the distant reaches of space, we are looking back in time, viewing the most remote galaxies as they were many eons ago, when their light began its long journey through space. Even the great galaxies were once young, so the issue of how cosmic structure arose is inextricably linked with cosmology. The study of the origin of observable structures in the universe, ranging from the huge clusters of galaxies down to the solar system, falls in the realm of *cosmogony*. Among the outstanding problems to be confronted are how and when the universe began, how galaxies formed and attained their observed array of shapes and sizes, how stars are born, and how the planets and life evolved.

Only thirty years ago it was not possible to address the central questions of cosmology and cosmogony with any degree of confidence. Our knowledge of the distant universe and the early universe was so scant that several widely varying cosmological the-

---

* Italicized terms are defined in the glossary.

ories appeared to explain the observable data. However, in recent years, astronomers have made exciting new discoveries about the nature of the universe, and these discoveries have provided overwhelming evidence to support one cosmological theory. Today the central questions of cosmology and cosmogony are being explored within the framework of the *big bang theory*.

Although we are not able to answer all the central questions of cosmology, the big bang theory provides us with a broad outline of the evolution of the universe. In the chapters that follow, we shall describe the discoveries that have provided evidence for a big bang, and we shall trace the evolution of the universe from the earliest moments. We shall find that as we attempt to answer some of the central questions of cosmology and cosmogony, new questions and issues inevitably arise. Many details of our theory remain uncertain, and in such cases, we can describe alternative hypotheses and point out the directions that further research may take to distinguish among them. Thus, our discussion encompasses the past and future of the universe as well as the past and future of humanity's attempts to understand it. To begin, we shall describe the principles that form the basis of any scientific cosmological theory.

COSMOLOGICAL PRINCIPLES

From earliest times, we humans have been reluctant to abandon the notion that we play a central role in the universe. First a geocentric universe was proposed and abandoned; then a heliocentric universe was proposed and abandoned. Not until the twentieth century was it realized that our sun is an unexceptional star located near the edge of an unexceptional galaxy. Our galaxy is part of a loose grouping of galaxies in the outer periphery of a great cluster of galaxies. Even this cluster, the Virgo cluster, is but a pallid imitation of the truly great clusters of galaxies that we observe elsewhere in the universe. Our location in the universe is insignificant.

This knowledge, acquired by observations with the largest optical telescopes, presents cosmologists with an awkward dilemma. Our observations are made from one specific location in the universe, yet construction of a cosmological theory requires a general knowledge of the distribution and properties of matter throughout the universe. To circumvent this unfortunate restriction, cosmologists postulate a universal principle or requirement: that our local sample of the universe be no different from more remote and inaccessible regions. There are strong philosophical reasons for advocating such a universal principle. For one, the laws of physics should be similar throughout the universe, for if they were not, experiments would not be repeatable and our physical laws would be irrelevant. A more compelling philosophical requirement is that the universe in the large should be as simple as possible. This is the principle of Occam's razor and the natural way that physics proceeds—by taking the simplest allowable model to explain the phenomena. There are, however, different formulations of cosmological principles.

In 1543, Copernicus proposed that the earth might not be at the center of the universe. A logical extension of the copernican theory is to displace our galaxy from any preferred spatial location. Thus, we are led to the key ingredient of modern cosmology, the *copernican cosmological principle*, which states that our perspective on the universe should not be from any preferred point in space. Inspection of the vast numbers of galaxies visible on photographic plates reveals their frequency of occurrence to be rather similar in different directions. This evidence demonstrates that the universe is locally *isotropic*—it appears the same in different directions, as viewed from the earth. (A sphere is isotropic when viewed from the center, whereas a hen's egg is not.) The copernican principle assumes that the universe is approximately isotropic about any point in space. A moment's reflection should suffice to verify that isotropy about any point requires that the universe must also be homogeneous in space, for if the universe were nonuniform, then it would only appear isotropic at special locations.

Some cosmologists have attempted to generalize the cosmological principle by including the concept of invariance in time. According to this principle, the universe is eternally unchanging, at least over the largest scales. Thus, the *perfect cosmological principle* states that the universe should present a similar aspect when viewed from any point in space *and* time. This assumption led to the *steady state theory*, which has been discredited on observational grounds. Consequently, cosmologists generally accept the less restrictive form of the cosmological principle, and we are content to accept that the universe is approximately homogeneous and isotropic in space but not necessarily unchanging in time.

We cannot complete our review of cosmological principles without describing the *anthropic cosmological principle*. This assumption takes precisely the opposing view to that of the perfect cosmological principle; it asserts that we are viewing the universe at a privileged time, although the present universe would appear the same when viewed from any point in space. This assumption of a privileged era follows from the necessity for special conditions to arise that favor the evolution of life. For example, if the universe were very much hotter or denser than it is now, galaxies could not have formed. If the force of gravity were very different from its observed strength, planetary systems either would not have formed or would not be congenial to life as we know it. It is, after all, a remarkable coincidence that the age of the earth turns out to be similar (within a factor of four) to the ages of the oldest stars or galaxies that astronomers have found. The anthropic cosmological principle explains this similarity by fiat. The universe could be vastly more irregular and disordered than it is. The anthropic cosmological principle asserts that if it were, conditions would be uncongenial to life. Thus, as observers, we inhabit a very special universe, and only this universe is isotropic and homogeneous. Of course, other universes could exist that are both isotropic and homogeneous but that nevertheless do not allow life. For example, the production of carbon requires nuclear fusion operating in the cores of massive stars and requires a period of at least several million years before the stars die and the carbon can be released. A younger universe, although isotropic and homogeneous, would not permit life as we know it. The anthropic argument is a fundamental one because it purports to explain the copernican cosmological principle, which is central to practically all viable cosmologies. Nevertheless, many physicists refuse to take it seriously, believing that the ultimate questions about the origin of the universe must be answerable by physics rather than by philosophy.

The anthropic approach has been further developed. In its strongest form, the anthropic principle asserts that life, even intelligent life, is inevitable. Not only was the universe conducive to the development of intelligent life forms, but given the inexorable nature of biological evolution, the strong anthropic principle actually requires them. One of the principal weaknesses in all anthropic arguments is their vagueness: we cannot distinguish a man from a mammoth or a dolphin from a dinosaur; all life forms are acceptable in the most favorable of cosmologies.

MODERN COSMOLOGIES

Each version of the cosmological principle leads to a radically different view of the universe. The copernican cosmological principle is the basis of the big bang theory. Indeed, the big bang theory actually preceded the discovery of the expansion of the universe. If the universe is required to be everywhere homogeneous and isotropic and general relativity is the correct theory of gravity, then we are uniquely led to a big bang cosmology. There are actually two distinct models of big bang cosmology. According to one, the universe is destined to expand forever. According to the other, the universe will eventually recollapse. In both models, the kinetic energy of the initial explosion, which makes the universe tend to expand, is balanced by the gravitational attraction, which we can quantify as potential energy and which makes the universe tend to recollapse. The difference between the two models is the greater extent of kinetic energy predicted by the first model and the greater force of gravitation predicted by the second.

According to the big bang theory, the universe must have been much denser and hotter in the past. This theory violates the perfect cosmological principle, which requires the universe to appear the same at all times. The steady state theory was constructed to comply with the perfect cosmological principle. Introduced by Hermann Bondi, Thomas Gold, and Fred Hoyle in 1948, this theory predicts a similar aspect for the universe at all times by postulating that matter is being continuously created at precisely the right rate to maintain the same mean matter density everywhere in the universe. Steady state cosmology was a very bold theory, at least in its initial form. The weakest element of the big bang theory, the initial instant of creation, was laid bare. If the big bang theory could assert that the universe was created at an instant of time in the remote but finite past, why was it not as reasonable to assert that creation occurs everywhere, in all space and at all times?

Observational evidence, the harsh and final arbiter of any rival theories, eventually caused the demise of steady state cosmology. The proponents of steady state cosmology gradually modified it during the 1950s, as more sophisticated and discriminating astronomical observations were performed. As the theory became progressively more ad hoc, only its most obdurate supporters remained loyal. Finally, steady state cosmology was overthrown in 1965 by the discovery of the *cosmic microwave background radiation,* which provided irrefutable evidence for an early, hot phase of the universe. The steady state theory is now little more than a footnote of considerable historical interest in the development of modern cosmology.

Although the proponents of the steady state theory have been persuaded to concur that the cosmic expansion was initiated at a single instant many eons ago, an infinite number of possible cosmological models for the behavior of the early universe remain viable. The copernican cosmological principle can be justified only in terms of the observable universe, for which the big bang theory provides an excellent description. However, for early eras, when the universe was young, we can conceive of a cosmology very different from the homogeneous and isotropic expansion of the *standard big bang model.* The expansion might have been *anisotropic,* occurring rapidly in some preferred direction and simultaneously allowing collapse in a different direction. Or the universe could have been highly inhomogeneous; in denser regions, local collapse could have formed *black holes.* There is no scientific reason to favor a simple, regular big bang model over a more exotic beginning for the universe. Either possibility is consistent with our physical laws, and astronomical observations do not distinguish between them.

Despite this infinite array of possible beginnings, we can attempt to appeal to the anthropic cosmological principle to select a unique past for the universe. There must be, according to this principle, a standard big bang model, for were the universe to evolve in a highly irregular fashion, we would not likely be present to bear witness. All these chaotic cosmologies, given enough time, are likely to develop in a way that is hostile to life. Only the standard big bang model, from an infinity of choices, is destined to provide an environment congenial to the evolution of life.

Cosmologists who deny the anthropic principle are reconciled to a chaotic origin for the universe. Admittedly, it could take an eternity for such universes to evolve adversely, and so we could consider this concern to be merely academic. Cosmologists who invoke the anthropic principle, however, select a universe that remains simple for all time, from the initial instant to eternity. This choice of a homogeneous versus a chaotic, or anisotropic, early universe constitutes one of the central options of modern cosmology. It permeates space and time, to be reflected today in the large-scale structure of the universe. Initial conditions, eons ago, manifest themselves in the processes that give birth to galaxies and to the clustering of galaxies. The goal of understanding the nature of galaxies and their distribution in space will be shown in later chapters to strongly constrain our speculations about the origin of the universe.

The anthropic principle, however, is not a fully satisfactory answer. It does not explain *why* the universe is so regular and so simple. For a more complete response, we require a physical theory of the beginning of the universe. In later chapters, we shall see how far modern physics has taken us in probing the origin of the big bang.

THE BIG BANG

The big bang theory reveals an immense vista of cosmic evolution since the cosmic expansion was initiated about 15 billion years ago. Conditions at this initial instant and before this instant are matters for speculation that the conventional theory does not address, although we will confront them in subsequent chapters. The early universe was very hot, very dense, and perhaps also very irregular. The irregularity and anisotropy gradually decayed. Within minutes after the big bang, some nuclear reactions occurred; essentially all the helium in the universe was synthesized at that time. As the universe expanded, it cooled, much as hot air expands and cools. The cosmic microwave background radiation is a vestige of this early era; it has been aptly christened the relic radiation of the *primeval fireball*. As the matter in the universe cooled, it eventually condensed into galaxies, according to one scenario for the evolution of the universe. The galaxies fragmented into stars and clustered together to form great aggregations over vast regions of space. As the first generations of stars were born and died, the heavy elements, such as carbon, oxygen, silicon, and iron, were gradually synthesized. As stars evolved into red giants, they ejected matter that condensed into dust grains. New stars formed from clouds of gas and dust. In at least one such nebula, the cold dust collapsed into a thin disk surrounding the star. Dust grains adhered to one another by coalescence and accumulated into larger bodies that grew in size by their gravitational attraction, forming the diverse array of bodies, from tiny asteroids to giant planets, that constitutes the solar system.

Our big bang theory leads us through the evolution of the entire universe—from the first microseconds of time to the formation of the earth and the development of life and into the possibly infinite future. Before we examine the details of this evolution, we shall discuss in Chapter 2 the historical origins of scientific cosmology.

· 2 ·

ORIGINS OF MODERN
COSMOLOGY

The Universe is not bounded in any direction. If it were, it
would necessarily have a limit somewhere. But clearly a thing
cannot have a limit unless there is something outside to limit it.
. . . In all dimensions alike, on this side or that, upward or
downward through the universe, there is no end.
—*Lucretius*

The origins of cosmology are unknown, but we can imagine that the ancients observed the starry heavens and asked the
obvious questions: What is the nature of the heavenly bodies? What
holds them up? Does the earth move? We can distinguish two types
of responses to these early cosmological questions. The mythological responses can be traced to the earliest writings of the Baby-
lonian, Egyptian, Greek, Indian, and Chinese civilizations. In rare
instances, such mythological cosmologies may have approached
truth, but they inevitably stagnated because of their inflexibility
and dogmatism.

The scientific approach to cosmology contrasts sharply with the
mystical viewpoint. Science seeks to explain the universe in terms
of observable and measurable phenomena. Let us begin our historical survey with the astonishing transition from mythology to
the birth of scientific inquiry that abruptly occurred in the middle
of the sixth century B.C. on the shores of Asia Minor. Profound and

lasting concepts emerged from the early attempts of the Greek philosophers to explain observed phenomena by natural causes. Many of their notions may seem absurd to us, but the underlying ideas exerted an important influence on the subsequent development of cosmology.

Consider, for example, the philosophy of Epicurus, who argued that the universe was initially in a continually changing state of primordial chaos, from which order may eventually have developed. One of the principal opponents of this viewpoint was Aristotle, who maintained that the extraterrestrial universe was perfectly ordered and unchanging. This theme of chaos versus order is reflected throughout the ages in theories of the origin of the solar system; this theme has also played an important role in our modern understanding of the early universe.

The ancients held a diverse array of opinions about the nature of the heavenly bodies and their motions. A few great thinkers revealed tremendous foresight in discussing such concepts as atoms, action at a distance, and the stellar structure of the Milky Way. Philosophical reasoning led Lucretius toward the atomicity of matter, yet millennia elapsed before experiments established the scientific merit of such notions. Cosmology has always attracted the most bizarre array of mystics, theologians, philosophers, and scientists into its domain. Some of their speculations have survived over the centuries to play a role in modern cosmology. Although these ideas were buried amid many erroneous conceptions and had to be rediscovered thousands of years later, we must nevertheless search for the origins of cosmology in the surviving fragmentary writings of these early natural scientists and in the spirit of inquiry that they launched.

THE GIANTS OF CLASSICAL COSMOLOGY

The early attempts at cosmology were naturally limited by the primitive state of astronomy. Early cosmologies focused on the stars, the sun, the earth, the moon, and the five known planets. The earliest surviving attempt at a rational cosmology is probably that of Pythagoras, who taught that the earth is round and rotates on its axis. The Pythagorean theory was a radical departure from the prevailing view that the earth was flat. Pythagoras based his ideas on an analogy between the harmony of the musical scale, expressed in terms of rational numbers, and the celestial harmony of the motions of the planets. Perhaps his most important contribution to cosmology was the idea that the celestial motions obey certain quantitative laws.

Pythagoras believed that whole numbers, especially those characterizing the ratios of the string lengths of a lyre, could account for the arrangement of the planets, the moon, and the sun. These bodies and the earth all turned around a hypothetical central fire (not the sun) on concentric spheres filled with air (Figure 2.1). The motion of the celestial bodies produced the harmony of the spheres, which his followers maintained, only their master was attuned to hear. Pythagoras put forth the revolutionary ideas that

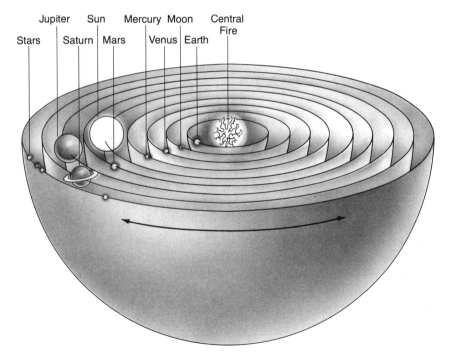

Figure 2.1 The Pythagorean Universe
The sun, moon, stars, and planets revolve on concentric spheres around a central fire. The "fixed" stars form the outermost sphere.

the earth moved and was not at the center of the universe. Although pythagorean philosophy prepared the way for a *heliocentric cosmology* and persisted for several centuries, its mystical emphasis on celestial harmony based on the musical scale made it eventually obsolete. Nevertheless, the pythagorean philosophy still prevails among some modern cosmologists who argue that improbable coincidences that lead to apparently "magic" numbers reflect an underlying, not yet understood, symmetry of nature.

The concept of a universal principle capable of accounting for the observed universe became firmly established during the fourth century B.C. with the ideas of Plato and Aristotle and dominated scientific thinking for more than nineteen centuries. Plato held that the circle, with no beginning or end, was a perfect form and therefore the celestial motions had to be in circles, since the universe was created by a perfect being, God. The earth also had to be spherical, as did the universe itself—"smooth and even and everywhere equidistant from the center, a body whole and perfect." Plato advocated the idea of daily rotation of the heavens around the immovable earth. The planets moved in circular orbits at different rates, with Venus and Mercury moving from west to east, but the other heavenly bodies moved in the same direction as the sun. Plato took little interest in the details of the heavenly motions; he did not notice, for example, that the apparent westward motions of Mercury and Venus occur only during part of their orbits (when the planets are brightest). His main contribution to classical cosmology was in spreading the pythagorean doctrine of a spherical earth and the circular motions of the planets.

In fact, the apparent motions of the planets were more complex than the explanation provided by Plato's simple scheme. The outer planets mostly move from east to west, but they occasionally appear to move backward. These *retrograde motions* are easy to explain if the sun is centrally located (Figure 2.2), but the notion of a heliocentric universe was alien to the early cosmologists. After all, they had little reason to believe that stars shine as brightly as the sun. And, in such a heliocentric universe, the relative positions of the stars would be expected to change during the course of the year as the earth moves around the sun; this most definitely did not occur. Of course, the modern resolution of these paradoxes is that the stars appear faint and fixed relative to one another because of their immense distances.

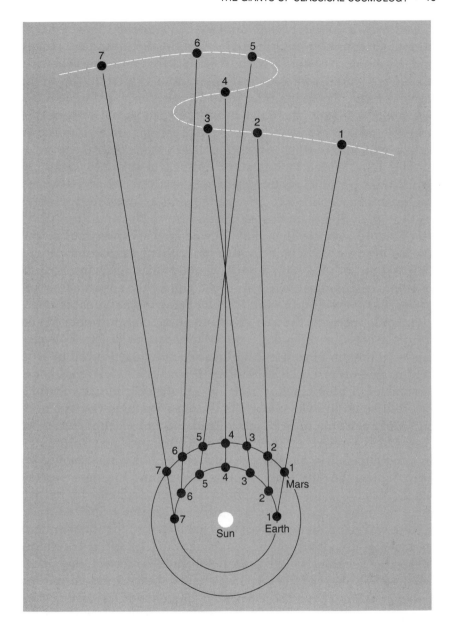

Figure 2.2   Retrograde Motions of the Planets

Planets may appear to have retrograde motion, because the earth orbits the sun in a shorter time than do planets farther from the sun. Earth completes half an orbit (in half a year) while Mars goes through little more than one-quarter of an orbit (in one-quarter of a Martian year). Consequently, as Mars is viewed from earth in positions 1 through 7, Mars appears temporarily to move backward.

Eudoxus, a younger contemporary of Plato, made a serious attempt to account for the irregularities of the motions of the planets and the moon. His work was the first really scientific astronomy—no longer would philosophical speculations without any observational basis play a significant role in astronomy. In the ingenious scheme of Eudoxus, the circular orbit of each planet was fixed on a sphere, which was allowed to rotate. Each sphere carrying a particular planet was attached at its poles to a concentric secondary outer sphere, which rotated about a different axis. This sphere in turn would be attached to a third sphere, as necessary. Thus, for each planetary sphere, there were several nonplanetary spheres, and the path of a planet as seen from the earth would be considerably more complex than that allowed by a simple circular orbit. By including a sufficient number of concentric rotating spheres (thirty-three in all), each simultaneously rotating about an independent axis, Eudoxus was able to account for the apparent motions of the planets within the limits of accuracy then attainable.

Aristotle, perhaps the last great speculative philosopher to contribute to classical cosmology, set out to construct an actual physical model of the universe. He devised a scheme that allowed the concentric spheres of Eudoxus to rotate in practice as well as in theory. Around each nonplanetary rotating sphere of Eudoxus, Aristotle inserted an additional sphere. This device enabled the axis of rotation of each sphere to be prolonged and to be connected to the two adjacent spheres. The earth, at the center, was now surrounded by nine concentric transparent spheres on which the moon, Mercury, Venus, the sun, Mars, Jupiter, Saturn, and the stars were situated. Aristotle's cosmology required fifty-five spheres for its elaborate machinery to operate. According to Aristotle, the outermost sphere was fixed, being that of God, who caused the inner spheres to rotate. Humankind was far removed from God and was assigned to the inner region, which was temporal and imperfect; everything outside the sphere of the moon was held to be perfect, eternal, and permanent. Such transient celestial phenomena as meteors and comets were of terrestrial origin and were carried along by the rotation of the upper atmosphere.

One genius of ancient Greece actually argued that the planets, including the earth, revolved in circular orbits around the sun. In about 280 B.C., Aristarchus offered a theory vastly simpler than that of Aristotle, but the other philosophers were reluctant to accept such a radical alternative. It seems that Aristarchus did not exploit the potential simplifications of the theory of planetary motions implicit in a heliocentric theory, and as more irregularities in the planetary motions were discovered, the scheme of Aristarchus was abandoned. It was not to be revived until more than eighteen hundred years had elapsed.

However, Aristotle's theory could not account for the variations in apparent brightness of planets such as Mars. The obvious interpretation was that planets appeared to approach and recede from the earth rather than remain on a sphere of fixed radius. This interpretation appeared to be true also for the moon, which causes an eclipse of the sun by passing between the sun and the earth— some eclipses of the sun are total, and some are annular, because the earth–moon distance varies.

Some five centuries later, Ptolemy, who lived in Alexandria during the second century A.D., introduced a *geocentric cosmology* that was not seriously challenged for fourteen hundred years. His goal was "to show that the phenomena of the heavens are reproduced by circular, uniform motions." Ptolemy achieved this aim by imagining that the sun moved around the earth on a giant wheel (Figure 2.3). Each planet was fixed to a smaller wheel, whose axis was perpendicular to the rim of the giant wheel. The big wheel rotates slowly, the smaller wheels rotate rapidly, and each planet performs an epicyclic motion in space.

The *epicycle theory* had been introduced several centuries before, but Ptolemy greatly improved it. He was able to account for both the apparent motions of the planets in the sky and the varying distances of the planets from the earth. To explain various irregularities in the observed motions, more wheels were added; Ptolemy himself required thirty-nine wheels for his theory (see Table 2.1 for a summary of the important early cosmologies). The epicyclic system was a geometrical model for the celestial motions that provided a good physical description of the universe. As such, it was extremely successful. The actual formation of the world was thought to approximate Aristotle's scheme of crystal spheres, but Aristotle's machinery no longer worked.

The theologians moved to fill the obvious gap in knowledge. The early leaders of the Church insisted on a literal interpretation

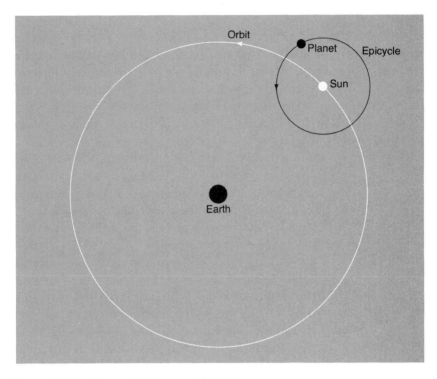

Figure 2.3 The Geocentric Theory of Ptolemy
Each planet moves in a small circle, or epicycle, the center of which moves in a circular path around the earth.

Table 2.1   The Great Cosmologists of Antiquity

Name	Date	Contribution
Pythagoras	ca. 580–500 B.C.	Spherical rotating earth revolving around a central fire
Plato	427–367 B.C.	Planets in circular orbits around stationary Earth
Eudoxus	408–355 B.C.	Mathematical model of motions of heavenly bodies with 33 concentric spheres rotating around stationary Earth
Aristotle	384–322 B.C.	An elaborate working model with 55 concentric spheres; immutability of the heavens
Aristarchus	ca. 280 B.C.	Heliocentric system—rotating Earth and planets revolve around central Sun
Ptolemy	ca. A.D. 140	Geocentric system; perfected epicycle theory

of the relevant biblical passages, and the earth became flat again. The destruction wrought by the barbarian hordes in the sixth century devastated the Roman Empire, and the fruits of Greek learning were swept aside. The long dark night of the middle ages commenced, and the progress of science was set back a thousand years or more.

THE RENAISSANCE OF COSMOLOGY

Not until the thirteenth century did the works of Aristotle, preserved by Arabic translations, finally become read in the Western world. The Ptolemaic system became widely known in the course of the following two centuries, and it was even systematically improved with additional epicycles. Once the writings of the Greek astronomers and cosmologists became available, the development of a new and more satisfactory theory was inevitable. Learned men had once speculated that the earth could move, and this idea led Nicolaus Copernicus (Figure 2.4) to consider whether allowance for the motion of the earth could reduce the complexity of the ptolemaic system.

Copernicus was able to show that the motion of the planets around the sun, with the moon orbiting around a rotating earth that was no longer the center of the universe, provided a far simpler and more elegant explanation of the planetary motions. Copernicus did not discard all the prejudices of his time, for he retained the notion of circular orbits in his cosmology, with the sun at the center of the earth's orbit (Figure 2.5). Reliance on circular orbits forced Copernicus to retain some of Ptolemy's epicycles. Nevertheless, the heliocentric system accounted for the most obvious aspects of the motions of the planets, and Copernicus' system received widespread acceptance after its publication in 1543.

The next great advance came as a result of numerous new observations of the planets by the Danish astronomer Tycho Brahe. Tycho refused to accept the copernican system but advocated a geocentric solar system in which the planets revolved around the sun, which itself orbited the stationary earth. However, his data laid the foundation for the final step in synthesizing the copernican model with planetary observations. This work fell to his assistant,

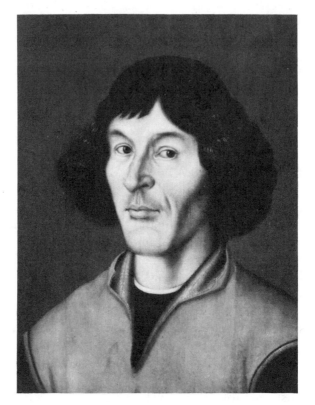

Figure 2.4   Nicolaus Copernicus (1473–1543)
A drawing of Copernicus as a student.

Johannes Kepler. Tycho's principal contribution to cosmology was to demonstrate that comets were much more distant than the moon and had highly elongated orbits. This discovery greatly discredited the Aristotelian notion of heavenly spheres that were fixed, permanent, and solid.

Tycho's legacy of accurate astronomical observations led to Kepler's discovery that the planets actually move in elliptical orbits, with the sun at one focus of the ellipse. Moreover, the planets speed up as they approach the sun and slow down as they recede from it. Kepler was able to show that this change in velocity corresponds to an imaginary line joining planet to sun that covers equal areas within the ellipse in equal times. His final great achievement, sum-

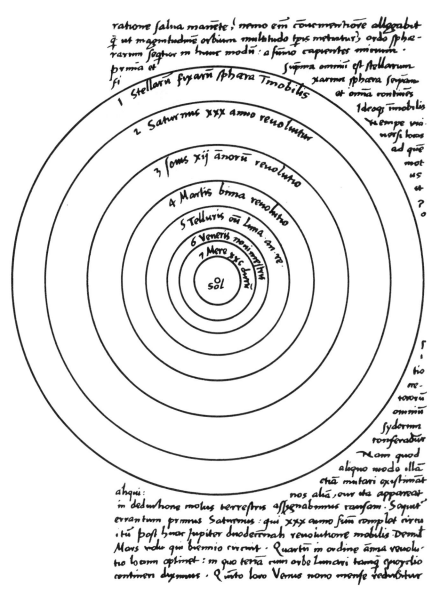

Figure 2.5   The Copernican Theory
This drawing from Copernicus' original manuscript placed the sun at the center of the universe.

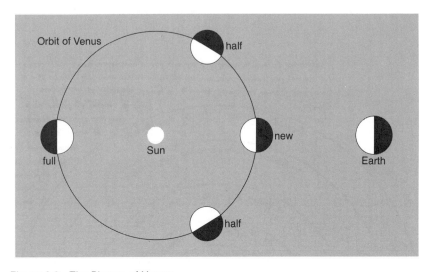

Figure 2.6 The Phases of Venus
Galileo pointed out that the different phases of Venus can be explained if Venus moves around the sun.

marized in his third law of planetary motion, was to show that the square of the orbital period of each planet is proportional to the cube of its mean distance from the sun.

Kepler also retained strong links with the ancients, and he indulged in wild and mystical speculations. He firmly believed that the five regular solids provided the key to understanding planetary orbits. These solids, which are the only bodies that can be constructed from straight lines with perfect symmetry, consist of the cube, tetrahedron, dodecahedron, icosahedron, and octahedron. The five regular solids were to be identified with the five spaces between the six known planets. Kepler placed the sun at the center, and the planets moved on each of six concentric spheres that fitted symmetrically inside and outside each of the regular solids. To account for the ellipticity of the planetary orbits that he inferred from Tycho's observations, Kepler gave each sphere a finite thickness. We must acknowledge Kepler's persistence in pursuing these mystical notions that nevertheless led to his discovery of the laws of planetary motion. He had finally freed the solar system from the grasp of epicyclic theory.

At this stage, the heliocentric theory was regarded as only a working model. The earth had not yet been dethroned from the center of the physical universe. However, during the course of the sev-

Figure 2.7   Isaac Newton (1642–1727)

enteenth century, the heliocentric cosmology was to gain complete acceptance. Galileo pioneered this scientific advance by innovating systematic methods of observation and experiment. He used the newly developed telescope to discover the four large satellites of Jupiter, which provided an analogy between the earth-moon system and other celestial bodies. The discovery of the phases of Venus helped to demonstrate that Venus revolves around the sun (Figure 2.6), and the discovery of sunspots helped finally to demolish the aristotelian doctrine of the immutability of the heavens. The fact that the stars remained unresolved points of light when viewed through the telescope reinforced the notion that they were exceedingly distant. Galileo himself did not contribute significantly to cosmological theory, but his discoveries made a path for others to follow.

Isaac Newton (Figure 2.7), who was born within a year of Galileo's death in 1642, provided the insight into planetary motions that enabled cosmology to become a modern science. No longer were mysterious movers invoked to account for the motions of the celestial bodies. By considering the force necessary to maintain the moon in orbit around the earth, Newton demonstrated that the gravitational force of attraction between two bodies varies inversely with the square of their distance of separation and is proportional to their masses. The force acting on an apple that falls to the ground is less than the force acting to hold the moon in its orbit, by the square of the ratio of the moon's distance to the radius of the earth times the ratio of the mass of the apple to that of the moon. Put another way, the force on each gram of terrestrial apple is greater by the inverse square of the distances. To demonstrate that the gravitational force between two bodies could be calculated as though the entire mass of each body were located at a central point required Newton to develop an entirely new branch of mathematics, calculus. Newton generalized the hypothesis of gravitational force to explain not only Kepler's laws of planetary motion, but also the interaction between any pair of particles in the universe.

The universality of the law of gravitation was eventually demonstrated by the eighteenth-century astronomer William Herschel, who found that binary stars in orbit around one another obey the same law of gravitation as the planets in the solar system. The discovery of the planet Uranus made Herschel famous, but cosmology received an enormous boost when he observed diffuse patches of light, or *nebulae,* through his telescope. To other astronomers, such as Charles Messier, the nebulae were of little interest, complicating the search for comets. However, Herschel considered these nebulae to be "island universes." Thomas Wright and Immanuel Kant, among others, had previously speculated about such nebulae, but Herschel's observations established extragalactic astronomy as an independent branch of astronomy. Herschel argued that the Milky Way was a separate disk-shaped island universe with the sun in a central position. He considered all "milky nebulosities" to consist of stellar systems when adequately resolved, and by so doing, he indiscriminately included gaseous nebulae, such as planetary nebulae and supernova remnants, as island universes of stars. Despite these errors, by realizing that the Milky Way might be similar in structure and absolute scale to other faint nebulosities, Herschel took a major step toward placing the

Table 2.2 Pioneers of Modern Cosmology

Name	Dates	Contribution
Nicolaus Copernicus	1473–1543	Heliocentric cosmology
Tycho Brahe	1546–1601	Accurate astronomical observations
Johannes Kepler	1571–1630	Laws of planetary motion
Galileo Galilei	1564–1642	Discovery of satellites of Jupiter, sunspots, phases of Venus
Isaac Newton	1643–1726	Universal law of gravitation
William Herschel	1738–1822	Observation of nebulae, discovery of Uranus

earth in its proper perspective with respect to the rest of the universe (see Table 2.2 for a summary of the contributions made by the pioneers of modern cosmology).

Herschel was able to resolve the globular star clusters in our own galaxy into stars, but the 72-inch reflector of the Earl of Rosse was needed to discover the spiral structure of nearby galaxies in 1850. The nature of the spiral nebulae remained a source of speculation until 1924, however, when the most prominent of our extragalactic neighbors, the Andromeda galaxy, was first resolved into stars.

ISSUES IN MODERN COSMOLOGY

Following Herschel's discovery of the nebulae, one of the most urgent issues in cosmology was the distance scale. Were these objects really island universes, comparable in size to the Milky Way? Or were they mere satellites of our galaxy? The pioneering work of American astronomer Edwin Hubble resolved this issue in the 1920s. Hubble isolated in the nebulae certain types of variable stars that seemed identical to other relatively nearby stars. If the distances to the nearby stars could be established, he could calculate the distances to the nebulae. With this key, Hubble opened what he aptly labeled the "realm of the nebulae." We know now that Hubble's distance scale was wrong. He thought the separations of the galaxies to be 10 times smaller than modern measurements reveal them to be. This error was not corrected until the middle of the twentieth century, and it led to a debate concerning the age of the universe that is now mostly of historical interest.

Even before Hubble, astronomers had discovered that the nebulae were systematically receding from the Milky Way but considered that perhaps this was only a local effect. The new calibration of the distance scale implied that the entire universe was expanding and that the expansion began some 2 billion years ago. The age of the earth was known to exceed 4 billion years, and this discrepancy provoked considerable discussion. Steady state cosmology, in which there is no beginning to the expansion and the age of the universe is infinite, was one by-product of the debate. Other attempts to explain the spectroscopic evidence that the galaxies were receding included the idea that the light from these galaxies was systematically affected in the vast intervening spaces so as to mimic the appearance of recession. As a result of the modern revision of the distance scale, we no longer have any discrepancy in the cosmological time scale, and the need for such alternative approaches to cosmology has largely disappeared.

The big bang theory provides the simplest interpretation of the astronomical data. Controversial issues remain, however. The initial instant of time, the *singularity,* lies beyond the frontiers of classical gravitation physics. A completely satisfactory theory of galaxy formation remains to be formulated. Was the initial state irregular and chaotic, or was it regular and orderly? Was the early expansion anisotropic or isotropic, and was the matter distribution initially inhomogeneous or homogeneous? One of the oldest questions remains the most controversial of all: Is the universe finite or infinite? After tracing the evolution of the universe, we shall suggest some possible solutions to these issues in Chapter 18.

One final topic deserves brief mention, if only because of the passions it arouses among cosmologists—the *cosmological constant.* Albert Einstein introduced this concept into his field equations of gravitation; it is equivalent to a repulsion force that is significant only over very large distance scales. The force is not present in Newton's theory of gravitation. Einstein introduced the cosmological repulsion force in order to devise a cosmological model, the *Einstein static universe,* in which the repulsion force is in balance with the gravitational force of attraction. When observations eventually revealed that the universe is not static, Einstein publicly regretted his introduction of this term. It is never-theless true that Einstein's equations in their most general form do contain this term. The advantage of the cosmological constant is that it permits a wider choice of cosmological models, including cases that avoid the initial singularity. However, it is a somewhat arbitrary addition because of the lack of any Newtonian analogue to the repulsion force, and cosmologists have felt ambivalent about incorporating it. The current view is that the cosmological constant is one additional parameter that we should introduce only if astro-nomical observations so require. So far, we do not seem to need it.

THE GIANTS OF MODERN COSMOLOGY

Edwin Hubble was the first and greatest modern observational cosmologist, but he was not working in a theoretical vacuum. After 1916, when Einstein introduced his theory of general relativity to supplant Newton's theory of gravity, theoretical cosmology flour-ished; the following decade proved to be a prolific period for new universes. Only a few of these models, however, have withstood the harsh test of astronomical observations.

The first cosmological model of the new theory was the Einstein static universe. Although the days of a static universe were num-bered, the notion seemed almost unchallengeable while it held sway. The earth, the sun, and indeed our Milky Way galaxy are systems in equilibrium, neither expanding nor collapsing. It was only natural to extrapolate this reasoning to the universe itself. But the logic was flawed. The greatest achievement of modern cos-mology is the Big Bang theory, which was a giant step forward at the time. Formulation of the big bang cosmology actually preceded the discovery of any firm observational evidence for a universal expansion. Prediction of the expansion of the universe constitutes one of the major successes of physics and the theory of relativity. This success was due primarily to two rather unlikely candidates, a Russian meteorologist and mathematician, Alexander Friedmann (Figure 2.8), and a Belgian cleric, Abbé Georges Lemaître (Figure 2.9).

Friedmann in 1922 and Lemaître in 1927 independently dis-

Figure 2.8   Alexander Friedmann
(1888–1925)

Figure 2.9   Georges Lemaître
(1894–1966)

covered the simplest family of solutions to Einstein's equations of gravitation that describe the expanding universe. Thus, they both may be said to have fathered big bang cosmology. Their stroke of genius was to abandon the notion of a static universe that Einstein (Figure 2.10) had been advocating. Following Einstein, they applied the cosmological principle, that the universe must be homogeneous and isotropic, but they boldly went a step further than Einstein had ventured by allowing the possibility of expansion. This resulted in a considerable simplification of Einstein's gravitational field equations and enabled them to derive a cosmological model. The static structure of our local environment has no parallel in the vast, unbounded domain of cosmology, where new concepts and indeed new physics play dominant roles. Einstein's theory of general relativity corrects Newton's theory of gravity on the earth by an infinitesimal amount, about one part in a billion, but it gave birth to the big bang theory of the universe.

Einstein's theory is noted for its generalization of our everyday concepts of ordinary space and time. The presence of matter is thought to cause the curvature of space, which is equivalent to gravity. In a curved space, Euclid's geometry is no longer valid. Parallel lines are no longer parallel, and the circumference of a circle is no longer given by 2π times its radius. However, the deviations from euclidean geometry are exceedingly small, unless the gravitational field is very intense. We generally are concerned about space curvature in two regimes. Near a black hole, or collapsed star, the enormous gravitational fields can cause a strong curvature of space. And, over the vast distances that light can travel through intergalactic space, space curvature becomes significant.

Einstein's gravitational field equations, when applied to cosmology, were found to possess at least three distinct solutions, each characterized by a unique curvature of space. The Friedmann-Lemaître universes include models in which parallel lines diverge, models in which they converge, and a model in which the lines stay parallel. These spaces are physically distinct, in that an object in one type of space would have to be grossly distorted by stretching or squeezing before it could fit in another type of space. In 1932, Einstein and a Dutch astronomer, Willem de Sitter (Figure 2.11), developed the expanding cosmology in which the spatial geometry resembles that of ordinary euclidean or flat space.

In 1929, when Hubble announced his discovery of the simple law that described the recession of the nebulae, he apparently was unaware of the expanding-universe cosmologies of Friedmann and Lemaître. However, one cosmological model that did undoubtedly influence Hubble was discovered as early as 1917 by de Sitter. This universe possessed the peculiar property that the light from the most distant regions became progressively reddened as the distance increased. Unlike the selective extinction by small atmospheric particles that scatters blue light more than red light, thereby producing red sunsets, or the interstellar dust that similarly reddens distant stars, the light from remote galaxies was entirely shifted to longer, or redder, wavelengths at great distances. Over the next decade, astronomers gradually became aware of this model, which was challenged principally by the Einstein static universe, but no attempt was made to relate the predicted redshift

Figure 2.10 Albert Einstein (1879–1955)
In 1930, Einstein visited Hubble at the 100-inch telescope on Mount Wilson to learn firsthand that the galaxies were receding.

Figure 2.11 Willem de Sitter (1872–1934)

to any systematic motion. Then, in 1928, the American cosmologist Howard Percy Robertson showed that, by a simple mathematical artifice, the de Sitter universe could be transformed into an expanding universe. Unfortunately, Robertson's transformation resulted in an empty universe, devoid of any matter, and this aspect of the solution lessened the relevance of this model to the real universe. Robertson, who was a colleague of Hubble's at the California Institute of Technology in Pasadena, noted that there was indeed a systematic relation between expansion velocity and distance of the same type that Hubble announced the following year, but this result was overshadowed by the earlier work of Friedmann and Lemaître. It is likely, however, that Robertson's work had a strong influence on Hubble's result.

The curved Friedmann-Lemaître universes and the flat Einstein-de Sitter universe form the core of the big bang theory. If we dispense with the cosmological repulsion force, then in each of these models, the universe expands from an arbitrarily high density to its present diffuse state. Recollapse occurs in the closed model, but the others predict indefinite expansion. Which of these models provides the best representation of the universe remains one of the central issues of modern cosmology.

· 3 ·

OBSERVATIONAL
COSMOLOGY

We find them smaller and fainter, in constantly increasing numbers, and we
know that we are reaching out into space, farther and ever farther, until, with
the faintest nebulae that can be detected with the greatest telescope, we
arrive at the frontiers of the known universe.
–Edwin Hubble

Once Herschel surmised that extragalactic nebulae might
be island universes of stars, cosmology became an observational
science. A century after Herschel's death, the distances to the spiral
nebulae were finally established, and the nebulae were recognized
as galaxies comparable in size to our own Milky Way galaxy.

Estimating the distances to remote galaxies was not a simple task.
We can try to condition ourselves to the vast distance scales in-
volved by examining the techniques that had been used to estimate
the distances to planets and stars. The key to understanding these
methods lies in the means by which human eyes judge distances.
Examine your thumb at arm's length, alternatively closing each
eye. The thumb's position appears to change relative to any background object when viewed with first one eye and then the other.
The shift in angle is roughly the thickness of three thumbs at arm's
length. One thumb at arm's length equals approximately 1 degree
of a circle; this is also roughly twice the angular extent of the sun
and the moon. The apparent angular shift of the thumb is 3 degrees;

31

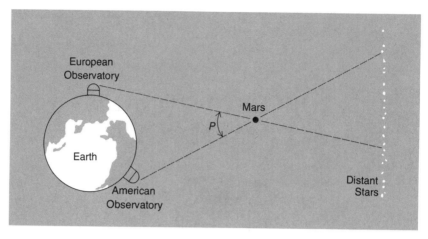

Figure 3.1 Parallax of a Planet

Two observatories on the earth can measure the distance to Mars by triangulation, or parallax. The distance is given by the separation between the observatories divided by the angle of parallax, *P*, expressed in radians.

simple geometry tells us that its distance, *D*, is about 20 times the baseline separation, *L*, between the left and right eyes. In other words, we infer that *D* is equal to *L* divided by the angular shift, provided that we express the angular shift as a fraction of the angle subtended by the amount of the circumference of a circle that is equal·in length to the radius of the circle. This unit of angle is known as a *radian;* there are precisely 2π radians in 360 degrees, and 1 radian equals about 57.3 degrees.

The apparent angular displacement that enables us to judge distances is known as *parallax.*[1]* A similar effect might have been used by primitive surveyors to measure the heights of mountains, using much longer baselines. On a grander scale, the parallax method was applied long ago to infer the distances to the planets (Figure 3.1). Observations are carried out at two observatories that are, say, 10,000 kilometers apart. Apparent shifts of much less than 1 degree are detectable, since the apparent size of a planet is quite small. The images of the nearest planets are several arc-seconds in diameter, and it is possible to detect the parallax of a planet to better than an arc-second. An arc-second is one-sixtieth of an arc-

* Numbered footnotes refer to the mathematical notes at the end of the book.

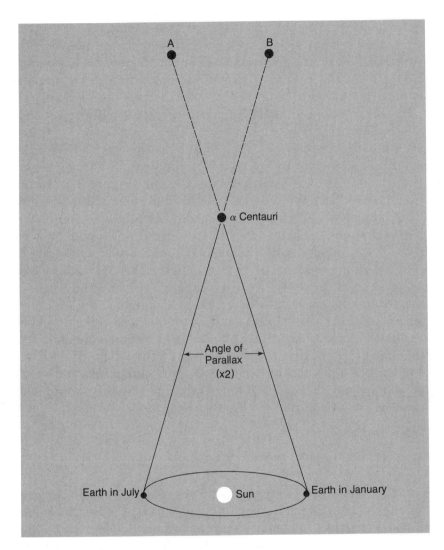

Figure 3.2 Parallax of a Nearby Star
A nearby star is observed at 6-month intervals. The shift in its apparent position from *A*
to *B* between January and July gives twice the star's parallax. The distance to the star
is equal to the mean radius of the earth's orbit divided by the angle of parallax. The star
Alpha Centauri has a parallax of 0.75 arc-second, and its distance is therefore 1.3
parsecs (or 4 light-years).

minute, and an arc-minute is one-sixtieth of a degree. So, 1 arc-second is one two-hundred-thousandth of a radian, and distances up to 200,000 times the baseline of 10,000 kilometers, or 2×10^9 kilometers, can be easily measured. The distance between the earth and the sun is, on the average, about 150 million kilometers, and the outer planets are up to 40 times more distant. The parallax method can therefore be applied to infer distances to all the planets in the solar system. In practice, more accurate techniques, such as radar ranging, are used.

A similar technique was applied to a nearby star in 1838 by an ingenious extension. Instead of using the earth as a baseline, scientists used the earth's orbit around the sun (Figure 3.2). In this case, the parallax p is defined as half the total angular displacement P. Measurements of the star were made 6 months apart, and a parallax p of slightly less than half an arc-second was found. Now we have a baseline of 3×10^8 kilometers, and the star is found to be 4×10^{13} kilometers, or 4×10^{18} centimeters, distant. Light would take 4 years to travel between the nearest star and the earth. So vast are the distances considered that we often use the *light-year* as a unit of distance—Alpha Centauri, our nearest stellar neighbor, is actually 4 light-years distant. Sometimes a more convenient astronomical unit of distance is a measure of parallax itself—a star that exhibits a parallax of 1 second of arc is said to be at a distance of 1 *parsec* (an abbreviation for *par*allax of one arc-*sec*ond). One parsec is roughly equal to 3 light-years.

THE EXTRAGALACTIC DISTANCE SCALE

The method of simple trigonometric parallax fails at distances of more than about 30 parsecs, or 100 light-years, because the angular shifts become too small to be measured accurately. However, other methods have enabled us to extend the distance scale. The sun is rushing through space at a velocity of about 20 kilometers per second relative to the nearby stars. Comparison of images taken over many years reveals small shifts in position for these stars because our perspective has changed. Because many stars in the same region of space will share the same parallax, relatively small parallaxes can be measured. This technique, known as *statistical parallax*, extends the distance scale to more than 100 parsecs.

One complication is that a cluster of stars with common motion will also exhibit a progressive angular shift relative to background stars. This shift is known as a *proper motion.* Measurements of proper motion can be used to estimate distances in much the same way as trigonometric parallax, but observations must be carried out for many years rather than for half a year. If the star cluster is moving away from the sun, the motions of the individual stars exhibit an apparent convergence of their proper motions to a point on one side of the cluster. An analogous effect is that of a train moving away into the distance—by a trick of perspective, the train appears to approach the tracks (Figure 3.3). This convergence effect contributes to the proper motion. If the star cluster is approaching us, the effect, like that of a train rushing toward us, is that of divergence.

To clarify matters, we must separate the effects of our relative motion toward or away from the star cluster from the relative motion of the sun perpendicular to the line of sight to the cluster that has given us a baseline for the measurement of proper motion. It is necessary to derive directly the relative motion of the stars away from us by measuring the Doppler shifts in their spectral lines. (We shall discuss this technique later in this chapter.) By analyzing stellar spectra and inferring the motion in our direction, it is possible to determine the cluster's distance from the proper motions of the member stars.

This "moving cluster" method has been applied systematically to the Hyades star cluster, which is at a distance of 40 parsecs. Once we know the distance to the Hyades, we can obtain the luminosity of each star from its observed brightness by applying the *inverse square law:* brightness = luminosity/[4π (distance)^2]: the brightness of a star declines as the square of its distance from us. The Hyades cluster therefore serves as a standard of distance, and we can obtain the distance to stars (not in the cluster) with similar spectra (and presumably similar luminosity) by again applying this inverse square law.

In the early years of the twentieth century, astronomers realized that our galaxy must be at least some thousands of light-years in

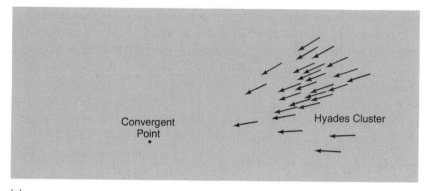

(a)

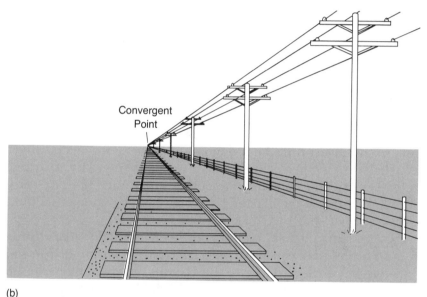

(b)

Figure 3.3 Distances to Star Clusters

A combination of proper motions and Doppler shifts of individual stars allows us to
determine distances to one or two nearby star clusters that are too far away for
parallaxes to be obtained easily. The proper motions of the stars in the Hyades cluster,
obtained from plates taken over a period of several years, appear to converge to a point
to one side of the cluster (a). The effect is analogous to that of a train moving into the
distance (b). Observations of the convergence point and proper motions of the cluster
stars, together with measurement of the Doppler shifts, enable us to calculate the
distance to the cluster and its true motion in space.

extent. Its diameter is now known to be about 30,000 parsecs, or 100,000 light-years. Confirmation of the distances to the spiral nebulae, however, required the development of a new type of distance calibrator. Parallax, of course, is negligibly small for such distant objects, and ordinary stars cannot be individually resolved. If we could identify in our own galaxy a specific type of star that was intrinsically bright enough to be seen in other spiral nebulae, such a star could be used to estimate extragalactic distances. *Cepheid variable stars* provided the link. Edwin Hubble first succeeded in resolving these stars in the Andromeda galaxy in 1923, and he used them to establish the distances to several nearby galaxies.

Cepheid variables pulsate regularly (Figure 3.4). First the star expands and brightens, and then it subsequently shrinks and dims. The periodic brightening and dimming occur over a time interval that can range from a few days to a year for different stars. During the decade preceding Hubble's discovery, Henrietta Leavitt and Harlow Shapley had found that the light variation of the Cepheids had a unique, periodic character, compared with other variable stars. Leavitt determined periods for dozens of Cepheid variables in the Magellanic Clouds, and she noted that the brightest Cepheids systematically exhibited the longest periods. There seemed to be a unique relation between period and observed brightness— by measuring one, the other could be inferred (Figure 3.5). Leavitt's work was based on relative brightnesses between stars at an unknown but essentially invariant distance, while Harlow Shapley in 1917 first used the *period-luminosity relation* to obtain *absolute brightnesses* for the Cepheids in the Milky Way. To obtain the distances to Cepheids in nearby star clusters was not an easy task. Such clusters are several hundred parsecs away, distances at which parallax methods are inaccurate. Astronomers have developed a technique that involves measuring the colors and magnitudes of many stars in a cluster and plotting the data in a color-magnitude diagram. In this diagram most stars fall near a characteristic band called the main sequence. In particular, stars of a given color or spectral type have a characteristic magnitude. Thus, the color-magnitude diagram can be calibrated on the Hyades cluster and then applied to infer distances to more distant star clusters that contain Cepheids.

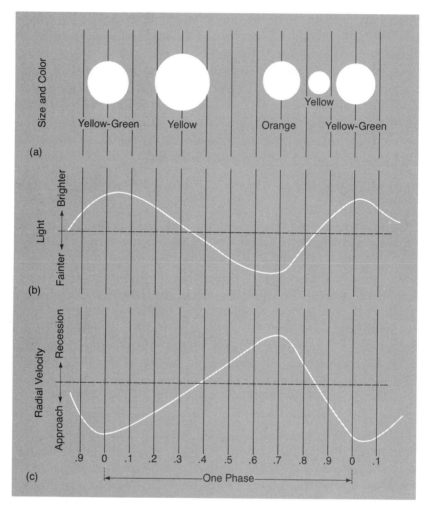

Figure 3.4 Light Curve of a Cepheid Variable Star

Cepheids vary in color (a) and radial velocity of the stellar atmosphere (c) as well as in light intensity (b). The star's atmosphere expands to maximum size midway between maximum and minimum light; the star's atmosphere shrinks to its smallest size in the opposite part of the cycle. The relative sizes have been exaggerated for clarity; the actual changes are less than 20 percent.

In this way, the distances to Cepheids in nearby clusters of stars in the Milky Way could be derived, and with this information, the period-luminosity relation could be calibrated to determine the intrinsic brightnesses of nearby Cepheids. Once the intrinsic

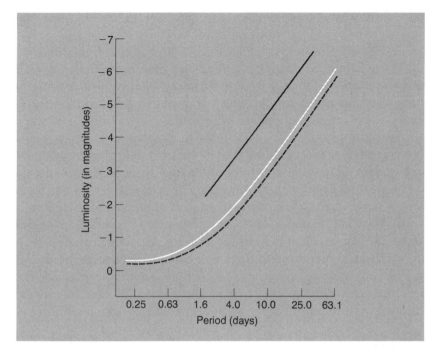

Figure 3.5 Period-Luminosity Relation for Cepheids

Harlow Shapley's estimates fitted all Cepheids onto one curve (*solid white line*), with luminosity increasing for the longer-period variables. Walter Baade divided Cepheids into two classes, Population I (*solid black line*) and Population II (*broken black line*), the latter type being fainter than Population I stars with the same period. The magnitude scale signifies that brightness increases by a factor of 2.5 for each smaller number (the smaller, or more negative, the magnitude, the brighter the star).

brightnesses were known, distances to galaxies containing Cepheids could be estimated.

With these calibrations we could see that the Magellanic Clouds are very far away; indeed, we now know them to be separate companion galaxies to the Milky Way, about 150,000 light-years distant but mere dwarfs by comparison with our galaxy. Our nearest neighbor of comparable size to the Milky Way is the Andromeda galaxy. Barely visible to the naked eye as a fuzzy patch in the sky, the Andromeda galaxy is some 2 million light-years distant. Both Andromeda and the Milky Way contain about 300 billion stars. Many of these stars are similar in brightness to the sun. A few rare, massive stars are a million times brighter. Most are less massive and much fainter, and some perhaps are not much greater in mass than a giant planet such as Jupiter.

Galaxies are the building blocks by which astronomers map the universe. Edwin Hubble continued the search for distance indicators to map the distant regions of space, and he found what he thought to be exceedingly bright stars that could apparently be resolved in more remote galaxies. Although these brightest "stars" are now identified as gaseous nebulae (known to astronomers as *HII regions,* in the conventional notation for ionized hydrogen) that surround massive stars, the largest of these are indeed relatively constant in diameter from galaxy to galaxy, and they can serve as distance indicators. Hubble used only their brightness to infer the distances to galaxies up to 10 times more distant than Andromeda. This first step out of our local region of space provided Hubble with the means of making a revolutionary discovery: the distant galaxies are moving away from each other. To fully appreciate this new finding, we must first consider the kind of information we can extract from the feeble light we receive from a remote galaxy.

THE MOST DISTANT OBJECTS

In recent years, the measurement of the most distant objects has been limited by what can be seen on photographic plates taken with the world's largest telescopes. For many years, the largest telescope in the world was the 200-inch telescope atop Mount Palomar in southern California. Now the world's largest optical telescope is the 240-inch telescope in the Caucasus Mountains in the Soviet Union. In development, however, are several 300-inch telescopes and one 400-inch telescope, due to see first light in the early 1990s.

Astronomers can identify many faint, fuzzy images as remote galaxies on a long exposure of a photographic plate at the focus of one of these great telescopes. In fact, modern techniques dispense with old-fashioned photography and use CCD detectors, similar to those in video cameras, to produce electronic images. Although these images are too small to reveal any detailed structure, we know that galaxies with similar shapes share certain general properties, and we classify galaxies according to their appearance. Galaxies with a prominent spiral structure are classified as *spiral;* those with a smooth, rounded shape, *elliptical.* These types are further subdivided according to the relative prominence of the spiral pattern or the degree of flattening. Many other galaxies do not fit this classification scheme, and these are known as *irregular* systems. The classification of galaxies is further discussed and illustrated in Chapter 10.

One key characteristic of a galaxy is the amount of radiation, or total luminosity, that it emits. Just as there are bright stars and dim stars, there are also bright galaxies and dim galaxies, but members of any particular class of a given type will generally vary little in total luminosity (Figure 3.6). For example, astronomers can measure the apparent brightness of a particular type of spiral galaxy. The intrinsic brightness of such galaxies can be inferred from study of the Cepheid variable stars or the brightest gaseous nebulae in a nearby system of the same type. Comparison of the apparent brightness of the photographic image of a more distant galaxy with that of a nearby galaxy of known distance then yields the distance to the more remote galaxy. This method of distance estimation is fairly reliable to distances of hundreds of millions of light-years.

As we probe even farther than these enormous distances, it becomes increasingly difficult to recognize different types of spiral structure in very distant galaxies, and astronomers use another characteristic property of galaxies to categorize them. Galaxies often occur in rich clusters (Figure 3.7), and rich clusters are much more prominent than single galaxies at great distances. It happens that the brightest or second- or third-brightest member of a cluster is almost invariably a giant elliptical galaxy, and these brightest cluster members differ in brightness from one another by only a small amount. Furthermore, 90 percent of these brightest-cluster members span a range of less than a factor of 2 in intrinsic brightness. This is really a very fortuitous circumstance, since individual galaxies vary greatly in intrinsic brightness. The most luminous known galaxies are about 100 times brighter than the Milky Way, which itself is about as bright as 10 billion suns. The least luminous dwarf galaxies are no brighter than a few million suns. The use of prominent galaxies in rich clusters as distance indicators has opened the distant universe to our examination. We can actually measure light emitted by galaxies at a distance of 10 billion light-

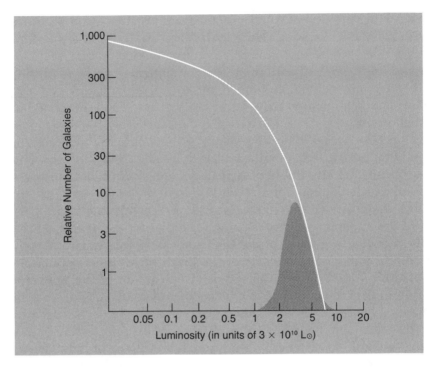

Figure 3.6   Galaxies as Distance Indicators
Galaxies of differing intrinsic brightness, or luminosity, are found in any large volume of space. If we include all galaxies, there is a very broad range of brightnesses (*continuous line*). Astronomers refer to this distribution as the *galaxy luminosity function*. If we restrict ourselves only to the brightest, or, say, third-brightest, galaxy in a cluster of galaxies, we find that these brightest galaxies are remarkably alike and that they all have the same luminosity to within a factor of 2 (*shaded region*). This class of galaxies can therefore be used as a distance indicator: once we have inferred the intrinsic brightness of this class from observing several nearby clusters of galaxies, we can then infer the distances to more remote clusters containing such galaxies.

years. Stars in these galaxies were emitting the light that we observe billions of years before the earth was formed (see Table 3.1).

The most distant galaxies seem to differ little from nearby galaxies. Their stars also consist of hydrogen and heavier elements such as helium, oxygen, and iron. Some recent evidence suggests that more blue galaxies may occur in remote clusters of galaxies than in nearby clusters. In our galaxy, the bluest stars are often very hot, luminous, massive, and young. Any excessive blueness

Figure 3.7   A Remote Cluster of Galaxies

The Coma cluster of galaxies is at a distance of roughly 300 million light-years. Many elliptical and spiral galaxies are visible as diffuse images on this photograph, which covers the central regions of the cluster, several millions of light-years across.

Table 3.1 The Extragalactic Distance Scale

Method	Astronomical objects	Distance surveyed
Parallax (terrestrial baseline)	Planets	10^{14} cm, or 1 light-hour
Parallax (earth-orbit baseline)	Nearby stars	50 light-years
Moving cluster	Hyades star cluster	120 light-years
Statistical parallax Color-magnitude diagram	Galactic star clusters	10^3 light-years 3×10^5 light-years
Period-luminosity relation	Cepheid variable stars	10^7 light-years
Diameters of HII regions Rotation velocity–infrared luminosity correlation	Spiral galaxies	10^8 light-years
Brightest galaxy in a cluster	Remote clusters of galaxies	10^{10} light-years

of the distant galaxy clusters must be due to active star formation. If evidence for a greater degree of star formation is observed in the galaxies of more distant clusters than in nearby clusters of similar type, then, because the light from more distant clusters takes longer to reach us, we conclude that the more distant galaxies must be younger and more vigorous than nearby galaxies. Thus, observations of distant galaxies lend support to the big bang theory, in which cosmic evolution is occurring on a grand scale.

The composition of distant galaxies puzzled astronomers until the early years of the twentieth century, when the science of astronomical *spectroscopy* was developed. In spectroscopy, light (from any source) is split into its constituent colors. A rainbow is a common example of a natural spectroscope—the tiny water drops in the earth's atmosphere disperse sunlight into the colors of the spectrum. This effect occurs because the water droplets scatter the different wavelengths that compose white light at different angles; rays of red light, for example, are bent less than blue light. In this way, white light is dispersed to reveal the colors that compose it.

The *spectrum* of light from a distant star or galaxy can be obtained by passing the light through a prism at the focus of a large telescope. More commonly, a diffraction grating is used in place of a prism. The diffraction grating consists of a glass plate etched with many closely spaced parallel lines, whose spacing is comparable to the wavelength of light. The grating works by the principle of diffraction—a beam of light passes through what is in effect a slit whose width is comparable to the wavelength of the light. This results in the bending of the light at an angle dependent on its wavelength, so the light is dispersed into its constituent colors. The astronomical instrument that splits light into its constituent colors is known as a spectrograph (see Figure 3.8).

If a low-dispersion grating (one which does not spread the relative wavelengths very much) is used, only a few colors are obtained. The effect is similar to that of a prism, which spreads light into multiple overlapping images of different colors, ranging from red to blue. As the dispersion of the grating (or the light-spreading power, which depends on the number of lines) is increased, more and more subtle gradations of color are obtained. Astronomers insert a narrow slit in front of the grating to produced narrow images of different colors that will not overlap. The resulting enhanced emission or absorption at certain wavelengths is easily measured provided that there are some preferred colors or wavelengths of enhanced (or reduced) emission in the incident light.

Indeed, at sufficiently high dispersion (when the grating is densely etched), starlight does not display a continuous sequence of colors like a rainbow but disperses into many narrow *spectral lines*, each of which is characteristic of emission by a specific type of atom. These spectral lines are superimposed on a weaker continuous light distribution. Specific spectral lines correspond to a specific species of atom. The lines always occur at precisely the same wavelength, or color, if the atom is at rest. For example, common salt turns a flame yellow-orange; if this light is examined with a dispersion grating, intense emission at wavelengths of 5889 and 5896 Ångstroms is found (1 *Ångstrom* $= 10^{-8}$ cm). Ångstroms provide a precise measure of color; for example, the human eye recognizes light as red between 6500 Ångstroms and 7000 Ångstroms and as blue between 4000 Ångstroms and 4500 Ångstroms. Sodium emission appears yellow-orange, because any light between 5000 Ångstroms and 6000 Ångstroms is perceived as yellow-orange by the eye. This characteristic light from sodium occurs at a precise wavelength that can be measured to within one-billionth of a centimeter.

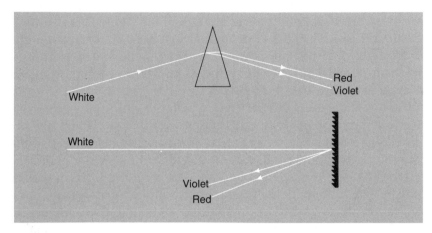

(a)

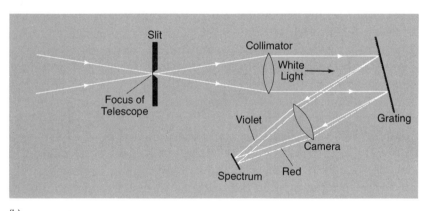

(b)

Figure 3.8 How a Spectrograph Works

A diffraction grating contains many finely spaced grooves, the separations of which indicate wavelengths of the light. The light is refracted through an angle that depends on its wavelength and is thereby split into its constituent colors. A prism does the same thing (a) by bending the light rays through different angles, depending on their color. A spectrograph (b) consists of a slit in the focal plane of the telescope (which isolates the star under study), a collimating lens to make the light parallel, a grating, and a camera to focus the light into a spectrum.

The light from a star contains thousands of these spectral lines (Figure 3.9), each identified with different elements. Sunlight also contains spectral lines. The spectrum of the sun has been studied since the early years of the nineteenth century, when Joseph Fraunhofer described the spectral lines that he discovered. In 1859, the physical interpretation of these lines in terms of emission and absorption was developed, and the first spectrum of a star was obtained shortly thereafter. In sunlight, there are many narrow dark lines as well as some bright lines. The bright lines are produced

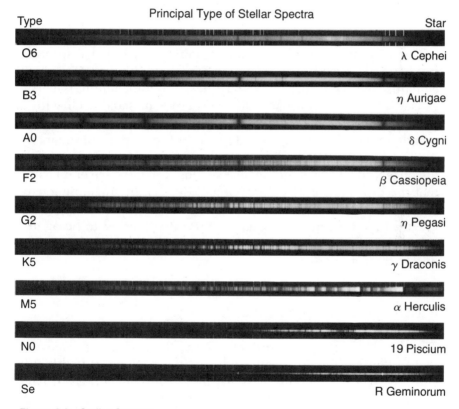

Figure 3.9 Stellar Spectra

A sequence of the principal types of stellar spectra for several stars, spanning types O (hot), B, A, F, G, K, M, N, and S (cool). The bright lines are absorption features in these negative photographs and the dark lines are emission features. The spectra of hot stars are dominated by lines of hydrogen and helium; spectra of cooler stars show prominent lines of calcium and iron.

in exactly the same way as the yellow-orange light radiated by the flame when sodium burns. The process is that of *emission* of radiation by a hot gas. The dark lines are *absorption* lines. These are produced when continuous bright radiation passes through a cooler gas. The cool gas absorbs the incident light at the same characteristic wavelengths the gas would be emitting if it were sufficiently hot. The cool gas layer in the sun is known as the *photosphere;* it surrounds the hotter interior regions, where the light originates. Study of the relative strengths of the spectral lines reveals that the sun and the nearby stars consist predominantly of hydrogen (70 percent by mass) and helium (28 percent). There is a small (but crucial) admixture of heavier elements, such as nitrogen, carbon, oxygen, and iron.

The spectrum of a spiral galaxy contains many broad absorption lines and is quite unlike the spectrum of an individual star, which contains relatively sharp spectral lines. The broadening is partly caused by the mix of many different types of stars, each with a slightly different spectrum, but the predominant cause of line broadening in the spectrum of a galaxy is the motions of the many individual stars. We shall describe how motions of stars can affect the observed spectrum in the next section. For now, we emphasize that the broad absorption lines of galaxies can be unambiguously identified as originating in many stars of mass similar to or slightly greater than the sun. The galaxies are thus inferred to be aggregations of many billions of stars that are individually unresolvable due to their great distances.

RECESSION OF THE GALAXIES

The development of spectroscopy led to the surprising discovery that the universe is in a state of dynamic expansion. It is difficult to appreciate the revolutionary nature of this concept. Even Einstein discarded the possibility of an expanding universe in his early papers on the theory of gravity. To understand how this dramatic result was obtained, let us examine the way the velocity of a source of light can be measured. The speed of light is finite, about 300,000 kilometers per second, and is constant in a vacuum. Light emitted by a moving star travels at the same velocity as light from a star at rest. These characteristics of light are a key theme of the theory of special relativity, which concluded that no material object can exceed the speed of light. The constancy of the speed of light has been directly measured by astronomers studying *binary stars*, pairs of stars in close orbit. Some of these double stars pass behind, or eclipse, one another at regular intervals. One can measure the time at which the eclipse begins, when the eclipsed star is moving away from the earth, and the time when it terminates, when the star is emerging from behind its companion and moving toward the earth. By also measuring the duration of the eclipse (from the time the star begins to the time it ends its eclipse), we are able to determine whether light traveled more slowly at the onset than at the termination of eclipse. No effect of this sort is found: the onset and termination of the eclipse are perfectly symmetrical in duration, and we conclude that the velocity of light does not depend on the motion of the emitting star.

Nevertheless, we can determine the motion of the star toward or away from the earth by studying the spectrum of the light emitted by the star. When light from a star that is moving away from the earth arrives at the astronomer's telescope, the star will be slightly farther away from earth than it was when the light left the star. Of course, light is continuously leaving the star and arriving at the telescope. We can think of light as traveling in waves—the waves actually register the infinitesimal electromagnetic impulses that constitute light. Let us suppose that during a certain time, the star produces N waves. The frequency is so great (about 10^{15} waves per second are emitted by a source of visible light) that the eye perceives the emission as being continuous. The distance between successive waves, or *wavelength,* of a high frequency wave is small and corresponds to blue light; longer wavelengths of lower frequency correspond to red light. A star emits N waves at a certain wavelength L, but these waves are spread over a longer distance by the time they arrive at the earth, since the star is moving away. In other words, the wavelength of the light measured by the astronomer is greater than the emitted wavelength $L!$ The light has shifted to a longer wavelength; we say it has become redder, or has acquired a *redshift.* Thus, the lines of the spectrum of a star moving away from earth will have increased in wavelength, compared with the spectrum of a star at rest (Figure 3.10).

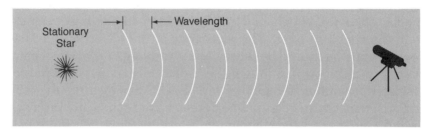

(a)

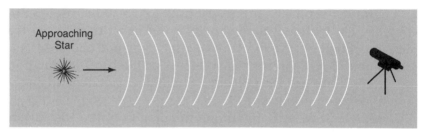

(b)

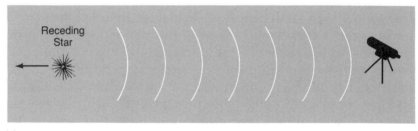

(c)

Figure 3.10 The Doppler Effect

The Doppler effect applies to any moving source of radiation, including sound waves and light waves. In (a) the star is stationary, and the distance between successive wavefronts (or wavelength of the light) is measured at the telescope to be the same as the wavelength of the light emitted by the star. In (b) the star is approaching the earth; consequently, the light waves appear to occupy a smaller region by the time they reach the telescope, and the wavelength is reduced, or shifted toward the blue region of the spectrum, relative to that of a stationary star. Conversely, rapid motion away from the earth (c) would result in a lengthening of the wavelength, or a shift of the spectral lines toward the red.

because we know exactly the expected wavelength from a star at rest, we can determine whether distant stars are moving away from us and at what velocity. The redshift (the fractional increase in wavelength) is directly proportional to the speed of the star. We can express this relationship quantitatively: the fractional increase in wavelength (relative to the wavelength of the light when emitted) is equal to the velocity of recession divided by the velocity of light. If the observed wavelength is reduced rather than increased, we then would observe a *blueshift*. The blueshift must be the symmetrical fractional decrease in wavelength observed when the light is emitted from an object moving toward us, compared with a star at rest. The amount of blueshift is equal to the velocity of approach divided by the velocity of light. Because the blueshift or redshift of an emitter that is approaching the speed of light will increase without limit, we have to modify these relationships when dealing with very large velocities. However, the simple proportionality of redshift (blueshift) to velocity of recession (approach) is generally adequate for light from stars and galaxies.

The change in wavelength of light produced by the motion of the light source is known as the *Doppler effect*.[2] A similar effect occurs when the pitch of a train whistle rises and falls as the train approaches and then recedes. If the wavelength of the light can be precisely measured, we can infer both the direction of motion and the velocity of the star. A photographic or digital spectrum can yield wavelengths to an accuracy of better than 0.01 Ångstrom for nearby stars. Astronomers can therefore detect stellar velocities that amount, in order of magnitude, to the speed of light multiplied by the ratio of 0.01 Ångstrom to the wavelength of visible light (around 5000 Ångstroms), or about 0.6 kilometer per second. With special techniques, we can measure radial velocities as low as 100 meters per second for bright stars; of course, for faint galaxies, the accuracy is considerably less.

We can now understand why spectral lines are broadened. The motions of the atoms in a star shift the wavelength of each line to the blue or to the red, and if the motions are random, as we would expect for a hot gas, the net effect is a broadening of each spectral line. Absorption lines are produced in relatively cool gas, and they are usually much narrower than emission lines.

The spectrum of a typical star contains numerous lines produced by various elements in the star, including hydrogen, helium, sodium, calcium, carbon, nitrogen, oxygen, and iron. Measurement of the shifts of these lines for many stars led to the discovery that our galaxy is spinning in space like a gigantic pinwheel. The sun and many of the stars in the Milky Way are orbiting the center of our galaxy at a speed of more than 200 kilometers per second. In our local solar neighborhood, we do not notice the effects of such a high velocity for most of the nearby stars, because they are moving along with the sun. These stars, the low-velocity stars, are part of the galactic disk that constitutes the luminous regions of the Milky Way. A number of high velocity stars in the solar neighborhood form part of the more slowly rotating galactic halo population. Relative to our stellar neighbors, the sun is moving at the modest rate of 20 kilometers per second toward a region in the constellation Vega. (By comparison, an escape velocity of 11 kilometers per second is required for a spaceprobe to leave the earth.) The earth orbits the sun with a mean velocity of 30 kilometers per second.

The light from galaxies is decomposed by the spectrograph into a composite of the spectra of many stars. It is rarely possible to resolve individual stars, either photographically or spectroscopically, except for the brightest ones in a few nearby galaxies. In the Andromeda galaxy, more than 300 billion stars contribute to the light. Although these stars comprise a variety of stellar types with different spectra, we can recognize in Andromeda distinct spectral lines and can use these to measure the velocity of the galaxy. The motions of the stars in a distant galaxy lead to a composite spectrum, in which the spectral lines are broadened still further. If the stellar motions have a systematic component that results from galactic rotation, it will show up as an asymmetry in the shape of the spectral lines. If the bulk of the stars on one side of the galaxy are moving toward us, the spectral lines emitted by these stars will be preferentially blueshifted. In light from stars on the opposite side of the galaxy, the spectral lines will be redshifted, because the stars are mostly moving away from us as the galaxy spins around. The net effect is that blueshifted lines come from one side and redshifted lines from the opposite side of the galaxy, so that if the slit of the spectrograph is placed along the major axis of the galaxy, the composite spectral lines are asymmetrical. We can then infer the rotational velocity of the galaxy from the asymmetry of the spec-

Relation Between Red-Shift and Distance for Distant Galaxies

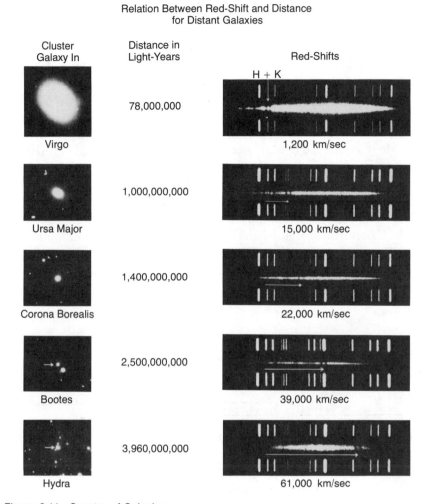

Cluster Galaxy In	Distance in Light-Years	Red-Shifts
Virgo	78,000,000	H + K / 1,200 km/sec
Ursa Major	1,000,000,000	15,000 km/sec
Corona Borealis	1,400,000,000	22,000 km/sec
Bootes	2,500,000,000	39,000 km/sec
Hydra	3,960,000,000	61,000 km/sec

Figure 3.11   Spectra of Galaxies

Photographs and spectra of galaxies in several rich clusters. The clusters range in distance from nearby (Virgo) to remote (Hydra). Prominent absorption features are indicated by arrows; these lines, known as the *H* and *K lines*, are due to calcium absorption. The lines in cluster spectra are very broad compared with lines in individual stellar spectra because of the motions of the stars in each galaxy that are contributing to the spectrum. The absorption lines are progressively redshifted as the distance to the cluster increases. This redshift is due to the Doppler effect (Figure 3.10) and can be expressed in terms of the relative velocity of recession of the cluster. The recession velocity is found to increase linearly with the distance to these clusters.

tral lines. In this way, we have learned that the Andromeda galaxy and many other spiral galaxies are rotating at much the same rate as our own Milky Way.

During the second decade of the twentieth century, Vesto Melvin Slipher and others recognized that the more remote galaxies were almost all receding from the Milky Way. Edwin Hubble then demonstrated that the rate of recession was directly proportional to the distance of the galaxy—the greater the distance to a galaxy, the greater is its apparent velocity (Figure 3.11). The spectra of remote galaxies are usually redshifted; from the amount of redshift, we can determine the velocities at which they are receding from us.

One exception is the Andromeda galaxy, which is moving toward the Milky Way at a velocity of about 50 kilometers per second. Only a handful of nearby systems, including Andromeda, are not receding from the Milky Way, and these possess relatively small velocities. Hubble depicted the observable galaxies as constituting

Figure 3.12 Our Cosmological Frame of Reference

According to the astronomer Gerard de Vaucouleurs, the galaxies in and around the Virgo cluster constitute a system called the Local Supercluster. This system extends to a distance of about 50 million light-years and has a flattened distribution of galaxies. Our Local Group of galaxies is falling toward Virgo, relative to the Hubble flow, at a velocity of about 250 km s^{-1}. This velocity is about 20 percent of the Hubble recessional velocity at the distance of the Virgo cluster. Even beyond the Local Supercluster, there may be still larger-scale inhomogeneities in the galaxy distribution. According to the American astronomers Vera Rubin and Kent Ford, such an inhomogeneity extends over about 400 million light-years. It is detectable by measurement of different recession velocities for distant galaxies in different directions. The amplitude of this deviation from large-scale uniformity, however, amounts at most to about 10 or 15 percent of the Hubble recession velocity. Over larger scales, the astronomical evidence suggests isotropy, and the microwave background measurement shows that the net velocity of the Milky Way galaxy relative to the cosmological reference frame is about 600 km s^{-1}. The actual measured velocity of the earth relative to the background radiation is only 390 km s^{-1}; the higher velocity for our galaxy results when account is taken of the earth's motion around the sun (30 km s^{-1}), the solar motion around the galactic center (220 km s^{-1}), and the motion of the Milky Way galaxy toward the Andromeda galaxy in the Local Group of galaxies (100 km s^{-1}). The net direction of the Local Group's center-of-mass velocity is about 45° away from the Virgo cluster; consequently, when account is taken of the Local Group infall, the entire Virgo Supercluster is inferred to be moving at a velocity of about 400 km s^{-1} toward a region in the southern sky where the Hydra and Centaurus clusters are located.

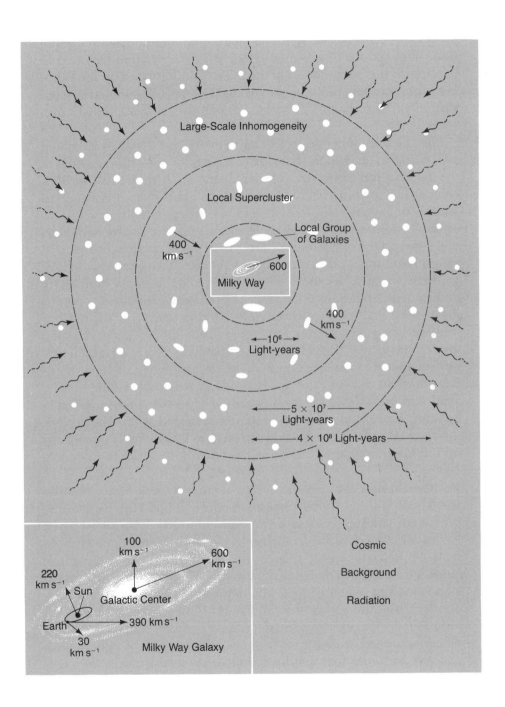

Large-Scale Inhomogeneity

Local Supercluster

Local Group
of Galaxies

400
km s⁻¹

600

Milky Way

400
km s⁻¹

←10⁶→
Light-years

←5 × 10⁷→
Light-years

←4 × 10⁸ Light-years→

100
km s⁻¹

600
km s⁻¹

220
km s⁻¹

Sun

Galactic Center

Earth

390 km s⁻¹

30
km s⁻¹

Milky Way Galaxy

Cosmic

Background

Radiation

the expanding *metagalaxy,* or universe. Yet superimposed on this universal cosmic expansion is a small component of random motion. The motions of galaxies are similar to the motion of waves in the ocean: the bulk motion of the ocean is dominated by the tides, but other, random motions also occur. Figure 3.12 summarizes the various types of motion that characterize the earth's orientation in space.

The relation between recession velocity and distance is known as *Hubble's law,* which states that velocity of recession is equal to the distance multiplied by a certain quantity, H, known as *Hubble's constant.* The value of H is found by measuring the recession velocities of galaxies whose distances are independently known (for example, from the dimensions of the brightest gaseous nebula in the galaxy). One of the most reliable distance indicators is obtained by applying an empirical correlation between the observed rotation velocity (which is distance-independent) and the galaxy luminosity (a distance-dependent quantity). Astronomers estimate the resulting value of H to be approximately 20 kilometers per second per million light-years. That is, for every million light-years of distance to a remote galaxy, the galaxy is receding at the rate of 20 kilometers per second. A substantial (and vocal) group of astronomers maintains that the correct value for H is nearly twice this value, but we shall adopt the lower determination of H in our subsequent discussion, for reasons that will become apparent when we discuss the age of the universe. This controversy will not likely be resolved until a large optical telescope is launched into space in the early 1990s.

The value for the expansion rate originally derived by Hubble was about 10 times larger than its modern value. Hubble's velocities were not greatly in error, but his inferred distances were far too small; the observable universe is much larger than Hubble realized. Galaxies in the nearest rich cluster to us, in the constellation Virgo, are at a distance of 50 million light-years. Their recession velocity is roughly 1000 kilometers per second, or one three-hundredth of the speed of light. The Hubble expansion has been observed to exceedingly great distances, where galaxies have a recession speed of one-third or more of the speed of light. These galaxies are more than 5 billion light-years away.

THE HOMOGENEITY OF THE UNIVERSE

Edwin Hubble made another contribution to cosmology that is perhaps as fundamental as his discovery of the expansion of the universe. Herschel, a century before, had counted many thousands of stars, and he concluded that the Milky Way was of finite extent. When Hubble counted galaxies, probing deeper into space to fainter and fainter limiting magnitudes, he found that the number of

galaxies increased proportionately, just as would be expected for a uniform distribution of galaxies in ordinary euclidean space.

Astronomers use the scale of magnitudes[3] (a logarithmic scale of brightness) to measure the relative brightness of stars. The naked eye can see down to sixth magnitude; the faintest objects (mostly remote galaxies, if we look out of the Milky Way) photographed, on a long exposure taken with one of the larger telescopes, are about twenty-fourth magnitude. Modern CCD detectors have probed the galaxy distribution to about twenty-seventh magnitude: at this level, the image of a galaxy is only a 1-per cent enhancement over the night sky background. Hubble's result has now been confirmed nearly to this limiting magnitude.

This result demonstrates that the universe is approximately homogeneous on the largest scales. The brightness of the Milky Way indicates that our galaxy is a gross local inhomogeneity, as are the stars within it, and that the great clusters of galaxies are lesser inhomogeneities, in the sense that the local mean density of matter is increased relative to the mean background level to a lesser extent within the volumes occupied by these clusters. Even on larger scales of tens of millions of light-years or more, irregularities in the distribution of galaxies are found. Superclusters of galaxies have been identified, some extending over perhaps 100 million light-years. Our own galaxy appears to be on the outskirts of one such supercluster, the Virgo supercluster, which is centered on the nearest cluster of galaxies in the constellation Virgo. There appears to be an even larger supercluster, dubbed the Great Attractor, just beyond Virgo, some 300 million light-years distant. However, evi-

Figure 3.13   The Large-Scale Structure of the Universe

A map of galaxy counts in the northern galactic hemisphere. All galaxies down to nineteenth magnitude were included, and nearly a million galaxies were counted. The north galactic pole is at the center, and the galactic equator is at the edge. The degree of whiteness of each spot corresponds to the number of galaxies counted in square cells that are one-sixth of a degree across (about one-third the apparent diameter of the sun); blackness corresponds to an absence of galaxies. The prominent feature near the center is due to the Coma cluster of galaxies. Although many clusters and aggregations of galaxies stand out, the uniformity of the galaxy distribution is very apparent over the largest scales except near the edge of the picture, where our view is obscured by the Milky Way. The large blank area in the lower-right region corresponds to the portion of the sky that can be seen only from the southern hemisphere.

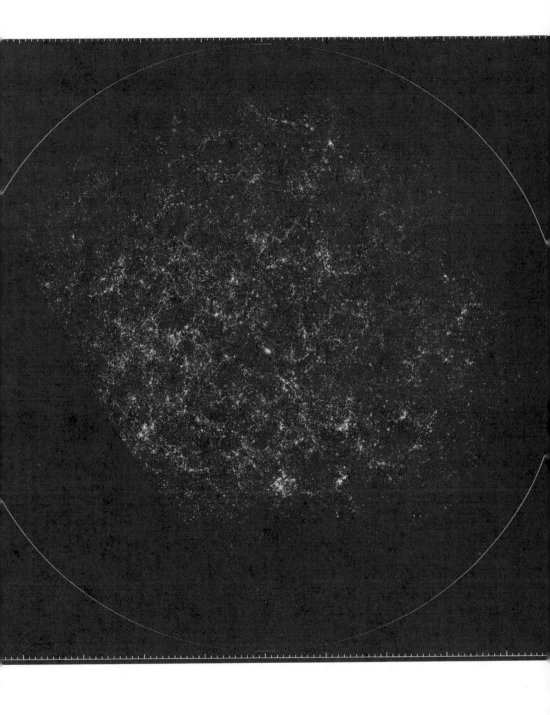

dence obtained from counting galaxies beyond our own supercluster generally confirms their homogeneous distribution on the very largest scales (Figure 3.13). The density of luminous matter in the form of stars and galaxies appears to be the same in the vicinity of the Virgo supercluster as it is in more remote regions of the universe.

No deviations from homogeneity on large scales are apparent in any direction. If the universe had a preferred center or a boundary, we might expect to see evidence for this from galaxy counts—for example, more galaxies should be counted toward the center than away from the center. No such effect has been found.

Perhaps the most striking confirmation of the large-scale homogeneity of the universe has come from recent studies of the cosmic background radiation. These studies have confirmed the isotropy, or complete uniformity in all directions, of the radiation. If the universe possesses a center, we must be very close to it, within less than 0.1 percent of the radius; otherwise, excessive observable anisotropy in the radiation intensity would be produced, and we would detect more radiation from one direction than from the opposite direction. Moreover, no small-scale angular variations in the radiation intensity are detected. Evidently, the very early universe, where the radiation was produced, must have been highly homogeneous. We will further discuss the nature of the cosmic background radiation in Chapter 4; the important point here is that it provides further observational evidence for the copernican cosmological principle, which serves as a basis for the big bang theory.

OLBERS' PARADOX

Away from the Milky Way, the night sky is surprisingly dark. This apparently superficial observation has deep significance for cosmology. Johannes Kepler and Edmund Halley speculated on the apparent contradiction with stars shining forever in an infinite universe. The paradox was clearly posed by the nineteenth-century astronomer Heinrich Olbers (and a century earlier, by Jean Philippe de Cheseaux), who had made a number of simple assumptions about the universe—it was static, it was comprised of stars of similar luminosity, and the stars were uniformly distributed when sufficiently large volumes of space were considered.

A remarkable paradox emerges from these assumptions. Consider any large spherical shell enclosing the earth at its center. Within this shell, the amount of light produced by stars can be calculated. Then consider a shell of twice the radius. Within this shell, the stars are on the average only one-quarter as bright, but there are 4 times as many of them, and so they would make a similar contribution to the light of the night sky. For each doubling of the radius, the amount of light received on the earth is doubled, and so the night sky must double in brightness. Continuing this argument indefinitely, we find that as we consider larger and larger shells, the night sky continues to increase in brightness without limit. Yet the night sky is very dark, apart from the Milky Way, which leads to an apparent contradiction. (This argument is somewhat misleading, because we have not allowed for the fact that intervening stars may intercept radiation from more distant stars. This effect reduces the inferred sky brightness to the average brightness at the surface of a star. The night sky should therefore be about as bright per unit area as the sun's surface, if we accepted this reasoning! This translates to a total brightness for the night sky of nearly 100,000 suns.)

Olbers attempted to resolve this paradox by suggesting that space was filled with a tenuous absorbing medium. This explanation is invalid, however, because the intervening gas would be heated by the radiation it absorbed until it attained a temperature at which it radiated as much energy as it received, so that no reduction in the average radiation field would result.

Olbers' paradox can be resolved by modern theories of how the radiation originates. The most fundamental limitation on the total radiation is the finite age of the universe. Stars do not live forever; their finite lifetimes for producing radiation limit the resulting diffuse radiation density that stars can produce. Of course, it is unlikely that most of the matter in the universe has already been processed in stars; the sun, for example, which is still mostly hydrogen, seems to be rather typical of most stars in the Milky Way. Thus, stars cannot contribute more than a fraction of the Milky Way's brightness to the night sky.

We can also appeal to the redshift of the light from distant galaxies as an alternative means of resolving Olbers' paradox. The redshift amounts to a loss of energy, and the light from distant galaxies is highly redshifted. Its effective contribution to the night sky brightness is correspondingly diminished. However, this only accounts for visible light. Modern observations at longer wavelengths tell us that the sky is also dark even in the radio part of the spectrum. Therefore, the redshift effect is not an explanation of Olbers' paradox. Nevertheless, the expansion of the universe plays an important role by limiting the volume of the universe within which stars are visible to us. It is precisely this fact, which restricts the observable universe to a distance of about 10 billion light-years, that results in a dark night sky: stars are inadequate radiators over this time scale. In this way, we can again reconcile the observed darkness of the night sky with theoretical expectations.

We may consider an interesting footnote to Olbers' paradox: the concept of night sky brightness has provided a significant constraint on the luminosity that can be attained by very remote galaxies. These remote galaxies are also very young and vigorous stellar systems, which, according to theoretical models of galactic evolution, should be extremely luminous. It seems likely that, even in the infrared region of the spectrum, the night sky is significantly darker than the Milky Way, as is the case at optical wavelengths. When one looks out of the Milky Way toward the poles of our galaxy, the sky is obviously very much darker than in the direction of the Milky Way. We do not know precisely how great the extragalactic contribution to the brightness of the night sky is, largely because of the zodiacal light and terrestrial airglow, but it is undoubtedly small, perhaps only 1 per cent or so of the Milky Way brightness. To avoid predicting an excessive night-sky brightness, astronomers have been compelled to conclude that most of the luminosity from young galaxies must have been greatly redshifted. Although the modern resolution of Olbers' paradox lies in recognizing that luminous stars are not perpetual energy machines, but have a finite lifetime of billions of energy-years, the redshift of starlight from distant galaxies plays an important role in quantitatively explaining the darkness of the night sky. A successful cosmology must be able to explain Olbers' paradox, and the big bang theory satisfies this fundamental observational requirement.

MACH'S PRINCIPLE: THE CONCEPT OF INERTIA

Do the distant stars exert any influence over the local properties of matter? Although this question seems to invite astrological speculation, it played an important role in Einstein's conception of cosmology. To find the answer, we must compare Newton's conception of absolute space with the ideas of the nineteenth-century physicist Ernst Mach. Consider how Newton might have measured the velocity of rotation of the earth. He would have used an entirely terrestrial experiment, such as watching the precession of the plane of motion of a pendulum, and he would thereby have inferred the rotation of the earth relative to a *local* frame of reference. This reference frame is said to be an *inertial frame,* because the apparent motion of a body is governed by its own inertia (that is, unless acted on by a force, it continues to move in a straight line at constant velocity). In a modern version of this experiment, we might launch a communications satellite into synchronous orbit around the earth, where the satellite appears to be stationary with respect to the earth's surface. Its orbital speed must be equal to the rotation speed of the earth for this to occur; if the earth were not rotating, the satellite could not remain stationary relative to the earth.

Mach realized that Newton's measurement was entirely local and that no reference to the rest of the universe was necessary. Alternatively, to measure the earth's rotation, Mach might have gazed up at the night sky and observed the apparent motion of the stars. He then would have determined the rotation rate of the earth by a *global* (or astronomical) measurement.

Mach was profoundly struck by the coincidence between these two results, which could be obtained by such different techniques of measurement. He argued that Newton's law had nothing to say about the relation between local and global inertial frames and referred only to local ones. To understand the coincidence, Mach argued that there had to be a causal relationship between the motion of the distant stars and the local inertial frame of reference. Now, it is obvious that the local inertial frame does not affect the motion of the distant stars. According to Mach, however, the converse must be true and this implies *Mach's principle:* the inertia of any body is determined by the distribution of the distant matter in the universe.

Einstein was strongly influenced by Mach's argument. However, Einstein's theory of general relativity does not satisfy Mach's principle, and many cosmologists, including Einstein himself, have tried in vain to incorporate the principle into the theory. The acceptable cosmological models of general relativity do, however, satisfy a limited form of Mach's principle in the following sense: there is a preferred local frame of reference in which the distant galaxies of the universe are receding isotropically.

Astronomers have been able to perform an experiment that enables us to measure this preferred reference frame. The cosmic microwave background radiation was effectively emitted from very distant regions of the universe. These regions coincide with the reference frame of the isotropic universal expansion. Behind the experiment is the idea that in the direction we are moving, the average wavelength of the radiation will appear slightly blueshifted. In the opposite direction, it will appear slightly redshifted (Figure 3.14). Since the radiation is blackbody radiation, the small variation in wavelength is equivalent to a small change in temperature. Such an effect has indeed been detected. The temperature of the background radiation has been measured to an accuracy of about one-thousandth of a degree kelvin. There is a slight deviation from uniformity in the radiation of about 0.1 percent, which is what we would expect if the earth is moving relative to the cosmic reference frame of the background radiation.

The motion of the earth with respect to the distant regions of the universe is not well understood from theoretical considerations. As Figure 3.12 illustrates, the earth revolves around the sun, the sun orbits the galaxy, and the galaxy moves through the Local Group of galaxies. After allowance for these motions, we find that our net measured velocity relative to the cosmic frame amounts to about 600 kilometers per second in the general direction of the Virgo cluster. The entire Local Group may be moving toward the Virgo cluster, or at least toward the *supergalactic plane*, which defines the greatest concentration of galaxies in our galactic neighborhood.

Such a conclusion can, in principle, be tested by optical measurements of the local distribution of galaxy redshifts. It is not yet clear whether these yield concordant results. However, such measurements are always relative to a more local frame than that defined by the cosmic background radiation. Inhomogeneities in the

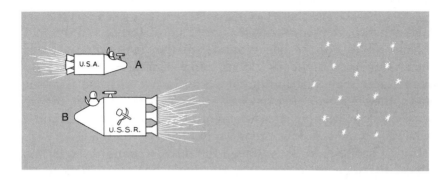

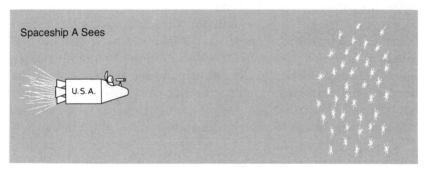

Spaceship A Sees

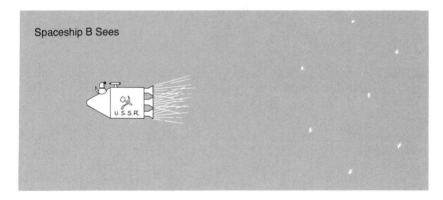

Spaceship B Sees

Figure 3.14 Anisotropy of the Background Radiation

Imagine two spaceships, one traveling rapidly toward a cluster of stars (A) and the other traveling rapidly away from the cluster (B). Spaceship A observes the stars to be more closely spaced, and their light is blueshifted; B sees the stars to be more separated and redshifted. Of course, these effects are only this significant if the spaceships are traveling near the speed of light. However, a similar effect occurs in the case of the background radiation. From a moving body, the background radiation appears more intense in the direction of motion. The blackbody radiation remains blackbody radiation, but it becomes hotter. Conversely, in the opposite direction, there is an apparent cooling. Just as for the Doppler shift, the magnitude of the relative temperature change is equal to the velocity of the observer divided by the speed of light.

galaxy distribution could cause significant distortions in the local velocity field. But the background radiation is unique, in that it anchors us to a cosmic, or global, frame of reference determined by the large-scale distribution of matter in the universe. The success of the *microwave background anisotropy experiment* provides us with a quantitative measurement of a preferred local inertial frame and a modern interpretation of Mach's principle.

In this chapter, we have traced the development of the cosmological distance scale from the relatively nearby planets to the deepest regions of space that can be probed by earth-based telescopes. In so doing, we have seen how the earth's location in space is determined by these observations. We have established our cosmological frame of reference. The distribution of velocities of the distant galaxies clearly implies an expanding universe; thus, the observational evidence supports the big bang theory. In the next chapter, we shall trace the development of the cosmological time scale and present additional evidence to support the big bang.

· 4 ·

EVIDENCE FOR
THE BIG BANG

Measurements of the effective zenith noise temperature of the 20-foot horn-reflector antenna at the Crawford Hill Laboratory, Holmdel, New Jersey, at 4080 Mc/s have yielded a value about 3.5 K higher than expected. This excess temperature is, within the limits of our observations, isotropic, unpolarized, and free from seasonal variations.

—*Arno Penzias and Robert Wilson*

The central thesis of big bang cosmology is that about 20 billion years ago, any two points in the observable universe were arbitrarily close together. The density of matter at this moment was infinite. We have previously referred to this initial instant of the big bang as the singularity. How do we know when it occurred? Did the universe exist prior to this moment? Clearly, if it did, the universe might have existed for an infinite time.

THE AGE OF THE UNIVERSE

Strangely enough, big bang cosmology can answer the first question approximately but cannot answer the second question at all. By the *age of the universe*, we shall therefore mean the time elapsed since the big bang. We do not exclude the possibility of a prior phase of existence, but we can say essentially nothing about it.

Remote galaxies are receding from one another at high velocities. The farther away the galaxy, the greater is its velocity of recession. The system of galaxies is in a state of expansion. Whether it is exploding like a bomb, never to implode, or whether the galaxies will eventually fall back together again is still an unresolved question, which we shall discuss further in Chapters 5 and 17. However, this issue is largely irrelevant for early eras, when elementary particles and radiation exploded from a state of infinite density.

We can try to visualize the initial expansion by imagining an immense swarm of bees crammed into a tiny hive. Suddenly the beekeeper removes the hive, and the bees rush off in all directions. Any given bee will observe all its neighbors to be moving away from each other. Suppose all the bees fly in straight lines but in random directions. The swarm of bees will steadily spread out, covering an ever-increasing volume, and the fastest bees will be farthest away. A simple relation connects the velocity of any bee with the distance traveled—velocity equals distance divided by time elapsed.

Galaxies, or at least their constituent atoms, must have begun to expand in a similar way. Immensely concentrated in density, matter rushed away in all directions. As the expansion proceeded, the matter must have condensed into galaxy-sized aggregations, which eventually fragmented into stars. Much of our discussion in Chapters 6 through 13 will concern these evolutionary processes in the early universe.

We can date the initial instant by simply calculating the time that has elapsed since any two galaxies that are now observed to be receding were in contact. Most of these galaxies' lifetimes have clearly been spent apart. Let us ignore for the moment the details of galactic evolution during the time they would have been in intimate contact, because this phase can occupy only a small fraction of their present age. The age of the universe is therefore roughly given by the ratio of *relative distance* to relative *velocity* for the pair of galaxies. According to Hubble's law for the recession of the galaxies, the recession velocity equals Hubble's constant, H_0 (which is H at the present time), multiplied by the distance of their separation. Just as in our analogy of a swarm of bees, the time elapsed since the beginning of the expansion is simply distance divided by velocity, or $1/H_0$, if the velocities have not changed with time.

Contemporary astronomers have measured H_0 more accurately than Hubble could, by introducing certain technical improvements. The various indicators of distance (Cepheid variable stars, luminous stars, bright nebulae, spiral galaxy rotation velocities, and elliptical galaxies) have been remeasured with considerable accuracy in different types of galaxies. The distance scale has been extended to far more remote galaxies. Even with the most recent data, the simple proportional relationship of velocity and time is approximately satisfied: our preferred modern value of $1/H_0$ is about 15 billion years, although the Hubble time scale could be as short as 10 billion years or as long as 20 billion years.

Because H_0 is so low, the recession velocities for the galaxies nearest to us are small, and they are masked by the galaxies' local

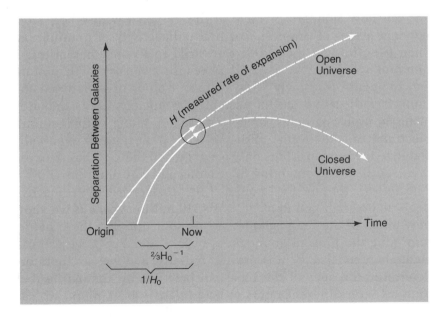

Figure 4.1   Age of the Universe
Measurement of the Hubble constant H_0 provides us with an estimate of the time elapsed since the big bang. In a closed universe, the age is less than $\frac{2}{3} H_0^{-1}$; in an open universe, the age is between $\frac{2}{3} H_0^{-1}$ and $1/H_0$. Current observations indicate that $1/H_0$ is equal to about 15 billion years, with an uncertainty of $\pm$ 5 billion years.

motions in the Local Group of galaxies. This system consists of the Milky Way galaxy, a few satellite galaxies dominated by the Milky Way, and the Andromeda galaxy and its companions. These galaxies move in orbit around one another; therefore, they do not take part in the overall expansion of the more distant galaxies.

Galaxy velocities can also change with time, because the cosmic expansion is decelerating. This change in velocity could affect our calculation of $1/H_0$ as the time elapsed since the big bang. In fact, $1/H_0$ is a good approximation to the age of an open universe. However, in a closed universe that is destined to recollapse, the age must actually be less than $1/H_0$. If the universe is closed, the age of the universe is somewhat less than $\frac{2}{3}H_0^{-1}$, or about 10 billion years, if we adopt our standard value for H (see Figure 4.1).

THE COSMIC TIME SCALE

We can now compare this age of the universe with other, independent scales of time. The natural radioactivity of uranium has been used to date the oldest terrestrial rocks and meteorites. A uranium isotope, U^{238}, so labeled because it consists of 92 protons and 146 neutrons, slowly decays into an isotope of lead. Since uranium is radioactive and is unstable, its nucleus emits radiation (actually alpha particles, or helium nuclei, along with neutrons, electrons and neutrinos) until the nuclear forces can eventually stabilize the nucleus. This only occurs when the U^{238} has decayed to a lead isotope containing 82 protons and 124 neutrons (Pb^{206}). The *half-life* (or time for 1 pound of U^{238} to decrease by radioactive decay to one-half pound of U^{238}) is 4510 million years. If we know how much of the lead isotope was present initially, then, by determining the present abundances of U^{238} and Pb^{206}, we can calculate how much U^{238} was present initially, and we can therefore determine the ages of the rocks. In practice, by comparing rock samples with different ratios of lead relative to uranium, we can circumvent the need to know the initial Pb^{206} abundance. By this method, the oldest known earth rocks (discovered in Greenland) are found to have ages of about 3.9 billion years.

The age of the solar system is inferred to be about 4.6 billion years from dating of the oldest meteorites (Figure 4.2). The oldest

(a)

(b)

(c)

Figure 4.2 Meteorites

There are three main types of meteorities: iron (*a*), stony (*b*), and a hybrid form, the stony-iron meteorities (*c*). The oldest meteorities are the stony variety; ages of about 4.6 billion years have been estimated for some of these. They are the oldest known objects in the solar system.

moon rocks found by the Apollo astronauts have also yielded a similar age. From these studies, we infer that all the planets formed within a period of less than 100 million years about 4.6 billion years ago (Table 4.1).

Table 4.1   The Cosmic Time Scale

Cosmic time	Era	Redshift
0	Singularity	Infinite
10^{-43} second	Planck time	10^{32}
10^{-36} second	Inflation	10^{28}
10^{-4} second	Hadronic Era	10^{12}
1 second	Leptonic Era	10^{10}
1 minute	Radiation Era	10^{9}
1 week		10^{7}
10,000 years	Matter Era	10^{4}
300,000 years	Decoupling Era	10^{3}
1–2 billion years		10–30
2 billion years		5
3 billion years		
3.1 billion years		
4 billion years		3
7 billion years		1
10.2 billion years		
10.3 billion years		
10.4 billion years		
10.7 billion years		
11.1 billion years	Archeozoic Era	
12 billion years		
13 billion years	Proterozoic Era	
14 billion years	Paleozoic Era	
14.4 billion years		
14.55 billion years		
14.6 billion years		
14.7 billion years		
14.75 billion years	Mesozoic Era	
14.8 billion years		
14.85 billion years	Cenozoic Era	
14.95 billion years		
15 billion years		

We could apply the radioactive dating technique to infer the age of the uranium itself if we knew the ratio of uranium isotopes when the uranium was created or synthesized. This argument leads to an age not of the earth but of the epoch prior to the formation of

Event	Time from now
Big bang	15 billion years
Particle creation	15 billion years
Annihilation of proton-antiproton pairs	15 billion years
Annihilation of electron-positron pairs	15 billion years
Nucleosynthesis of helium and deuterium	15 billion years
Radiation thermalizes prior to this epoch	15 billion years
Universe becomes matter-dominated	15 billion years
Universe becomes transparent	14.9997 billion years
Galaxy formation begins	13–14 billion years
Galaxy clustering begins	13 billion years
Our protogalaxy collapses	12 billion years
The first stars form	11.9 billion years
Quasars are born; Population II stars form	11 billion years
Population I stars form	8 billion years
Our parent interstellar cloud forms	4.8 billion years
Collapse of protosolar nebula	4.7 billion years
Planets form; rock solidification	4.6 billion years
Intense cratering of planets	4.3 billion years
Oldest terrestrial rocks form	3.9 billion years
Microscopic life forms	3 billion years
Oxygen-rich atmosphere develops	2 billion years
Macroscopic life forms	1 billion years
Earliest fossil records	600 million years
Early land plants	450 million years
Fish	400 million years
Ferns	300 million years
Conifers; mountains formed	250 million years
Reptiles	200 million years
Dinosaurs; continental drift	150 million years
First mammals	50 million years
Homo sapiens	2 million years

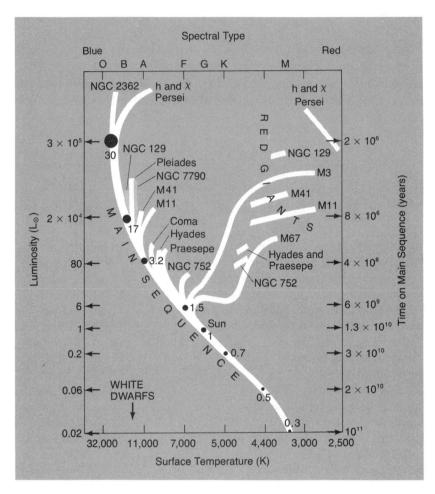

Figure 4.3   Ages of the Stars

We can estimate stellar ages from the Hertzsprung-Russell diagram, which plots stellar luminosity (in solar luminosities) against surface temperature, that is, against color or spectral type. The hydrogen-burning phase of a star (the main sequence) is indicated for stars from 0.3 solar mass to more than 30 solar masses. The more luminous a star is, the more rapidly it exhausts its supply of hydrogen and the shorter is its main-sequence lifetime. Many star clusters contain stars that have begun to burn helium and have evolved off the main sequence. For the older clusters, old stars of low mass have already left the main sequence, and the turn-off occurs well down the lower main sequence. In young clusters, however, only massive stars have had time to evolve off the main sequence. The location of the main-sequence turn-off provides the principal clue to the ages of many star clusters. Stars in the upper-right region are very luminous but cool; consequently, they must possess very large radii. These red-giant stars enter the post-main-sequence phase of stellar evolution when the stellar core becomes hot enough to burn helium. When the sun becomes a red giant, some 5 billion years hence, it will swell 1000 times in radius, enveloping the inner planets. Eventually, the nuclear fuel supply will be exhausted, and many of the red-giant stars will become white dwarfs, like those in the lower-left region of the diagram.

the earth when the element uranium was created. It is therefore a question of the rate of synthesis of heavy elements in the galaxy during prior generations of massive stars that exploded as supernovae and polluted the interstellar medium with uranium (and many other elements). Because this rate depends on detailed models of galactic evolution, all we can say is that the uranium must have been synthesized between 10 and 15 billion years ago.

A similar time scale can be independently estimated from the theory of stellar evolution applied to a star of known mass. The total supply of hydrogen in a stellar core determines the luminosity of a star of specified chemical composition. We can thus infer the age of a star from its mass (Figure 4.3), and we find that the oldest stars in the galaxy appear to be between 12 and 15 billion years old (Figure 4.4). This time scale has led us to prefer the lower value of H_0 that we adopted in Chapter 3. If we used the higher value, the oldest stars would appear to be slightly older than the big bang. Despite the uncertainty about the value of H_0, these independently determined time scales show a remarkable convergence (summarized in Table 4.2). Could this convergence be mere coincidence?

Most astronomers regard this convergence as evidence favoring a finite age for the universe. For many years, these time scales appeared to contradict each other. Because of various errors and uncertainties that were unknown to him, Hubble originally derived a value for H_0 that was 10 times larger than the modern value. One of these errors was a confusion between two different types of Cepheid variable stars that had similar periods but different luminosities (Figure 3.5). Using Hubble's original value for H_0, astron-

Table 4.2 Dating the Universe

Technique	Object	Age
Velocity-distance relation	Galaxies	10–20 billion years
Radioactive dating	Lunar rocks and the oldest meteorites	4.6 billion years
Radioactive dating and models of galactic evolution	Uranium and uranium isotopes	10–15 billion years
Models of stellar evolution	Oldest stars in Milky Way	12–15 billion years

omers calculated the age of the universe to be significantly less than the age of the earth. This inconsistency was one of the prime motivations for the emergence of the steady state theory. Once Walter Baade and Alan Sandage established the modern distance scale in the 1950s, the inconsistency in the time scale was removed, and steady state theory was no longer required to account for the known astronomical data. However, stronger evidence for the big bang was needed before it could be unreservedly accepted as a theory of origin for the universe. In the 1960s, this evidence was forthcoming from a very modern branch of astronomy—radio astronomy.

RADIO GALAXIES

Ancient astronomers were very limited, insofar as they could observe only the optical region of the radiation spectrum. The visible range from 4000 to 8000 Ångstroms is but a small fraction of the spectrum of possible wavelengths. Modern astronomers have developed a variety of technologies that have opened the nonvisible spectrum to our examination: gamma-ray astronomy (at wavelengths shorter than 0.01 Ångstrom), x-ray astronomy (0.01 Ångstrom to 100 Ångstroms), ultraviolet astronomy (100 Ångstroms to 3000 Ångstroms), infrared astronomy (8000 Ångstroms to 10^7 Ångstroms, or 0.1 cm), microwave astronomy (0.1 cm to 10 cm), and radio astronomy (10 cm to 100 meters or more). Many of these branches of astronomy require observatories in space, and so they were not developed until the 1970s. For example, the radiation between x-ray and ultraviolet wavelengths does not penetrate the earth's atmosphere (Figure 4.5). This is fortunate, because otherwise life would be destroyed by ultraviolet radiation from the sun. Before about 1970, astronomers could not examine these regions of the spectrum for cosmological information.

Restriction of astronomy to the visible region of the electromag-

Figure 4.4 A Globular Cluster, M3

The globular cluster M3 in the constellation Canes Venatici is one of the brightest visible clusters. It contains about 1 million stars. These stars are between 12 and 15 billion years old.

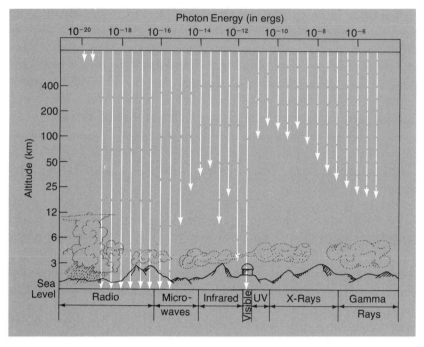

Figure 4.5 Transparency of the Earth's Atmosphere to Radiation
Radiation at wavelengths in the microwave, infrared, ultraviolet, and x-ray regions of the spectrum usually cannot reach the surface of the earth because of absorption in the upper atmosphere by such molecules as ozone, oxygen, and water vapor.

netic spectrum has a number of consequences. For example, only the nearby regions of the Milky Way can be studied at optical wavelengths. The Milky Way contains a great amount of interstellar gas and dust. These small, solid particles, known as *interstellar grains*, are believed to consist of graphite and rocklike materials similar to quartz, and their characteristic dimensions are comparable to the wavelength of light. Such grains strongly scatter and absorb electromagnetic radiation at wavelengths that are comparable to their sizes. Consequently, optical astronomers cannot see more than a few thousand light-years in the plane of the Milky Way, although there are occasional holes where a deeper glimpse can be obtained. Our galaxy is almost completely transparent, however, at radio, infrared, and x-ray wavelengths. Radio astronomers routinely study the central core of our galaxy, a region that will never be captured by an optical photograph from a terrestrial telescope.

When they look in directions away from the Milky Way, optical astronomers can observe to great distances and are able to study remote galaxies. However, the development of radio astronomy in the 1950s led to a major insight into the nature of the universe because many very remote extragalactic sources were detected, with our own galaxy providing little interference. Extragalactic radio sources generally do not emit spectral lines characteristic of emission from a hot, highly ionized gas; instead, they emit a continuous emission, like noise or static, with no preferred frequency. Although there are radio spectral lines, emitted by a cold atomic or molecular gas, one of the best-studied being the emission of hydrogen at a characteristic wavelength of 21 centimeters, the lines are relatively weak, and they cannot easily be measured in emission from very distant galaxies. Normal galaxies such as the Milky Way are weak emitters of continuous radio waves, but some, generally very distant, galaxies are strong emitters. These radio galaxies, as they are called, were initially discovered in radio surveys of the sky. Only subsequently was the source of radio waves identified with an optically observed galaxy, whose distance could then be determined in a measurement of its redshift. In the early days of radio astronomy, however, the locations of radio sources were very imprecise, and in many cases no optical counterparts were seen. The radio astronomer resorted to the ingenious approach of counting radio sources to extract cosmologically significant information from the radio sky.

We know that in our galaxy, the number of stars we can see increases as we observe fainter and fainter stars. Through even a small telescope, the Milky Way appears immeasurably richer than it appears to the naked eye. The increase in the number of stars with distance actually obeys a simple law that is a fundamental property of euclidean space: the number of stars will increase in direct proportion to the volume of the region being sampled. The greater the distance sampled, the fainter are the stars one can detect and the greater is the number of stars. Of course, at the boundary of our galaxy, this law breaks down; we see relatively few stars between the galaxies. Hubble applied this principle to the large-scale distribution of galaxies to demonstrate the approximate homogeneity of the universe.

Similarly, radio astronomers expect to detect more and more radio sources as they survey the sky at fainter and fainter levels of

Figure 4.6 Map of Radio Sources
This map of a small patch of the sky was obtained with the 1-mile radio telescope, which actually consists of several smaller radio telescopes that observe in phase with one another at Cambridge, England. Each set of peaks represents a radio galaxy or a quasar. There are far more numerous faint radio sources than we would expect in such charts of the radio sky. Since the faint sources are at great distances and are observed as they were in the remote past, we may infer that radio galaxies were much more numerous in the past than they are now.

radio *flux*, or intensity of signals. Suppose that the radio source distribution is homogeneous throughout space. According to the inverse square law, the flux from a source varies as the inverse square of its distance. However, the number of sources detectable down to some flux level varies as the volume of a sphere with radius equal to the distance to the faintest sources, or as the third power of this distance. In other words, the source number must vary as the inverse of the square root of the flux raised to the third power. This argument applies even when the sources differ in power; that is, a diverse distribution of intrinsic source luminosities will still produce a similar result.

However, radio astronomers in the 1950s discovered far more faint radio sources than were predicted on the assumption of a uniform distribution in space (Figure 4.6). The explanation appears to be that distant *radio galaxies* emitted more radiation, or were more abundant, or both, in the remote past, some tens of billions of years ago. Thus, many of the strongest radio sources are inferred to be the most remote, and we observe them as they were long ago, when their emission was at its peak. From this we infer that radio galaxies evolve from stronger to weaker sources on a cosmological time scale. This inference is in fact evidence for big bang cosmology, in which the evolution of the universe is a central theme. Rival theories, in particular the steady state theory, could not incorporate a scheme of cosmological evolution for radio galaxies, and nonevolutionary theories have lost credibility as a result.

THE COSMIC MICROWAVE BACKGROUND RADIATION

Probably the most persuasive evidence favoring big bang cosmology is the cosmic microwave background radiation, the cooled residue of the primeval fireball that constituted the early universe. *Microwave* is the radio astronomer's term for short wavelength radio waves (those with wavelengths less than several centimeters). Of course, optical radiation is much lower in wavelength than radio waves, but microwave radiation is not visible to the human eye, and microwaves normally do not produce much heat, unless their intensity is increased to a high level. The universe is a prolific source of microwaves. The intensity of cosmic microwaves is as great as the brightness of the Milky Way if we imagined the Milky Way to extend over the entire sky. Human beings are probably safe, however, from any cosmic contamination because the energy flux of cosmic microwaves absorbed by any individual is minuscule, amounting to about 10^{-5} watt, or only one ten-millionth of the power expended by a 100-watt lightbulb.

In fact, detection of the radiation required the construction of an elaborate but small radio horn, which was designed to make measurements with unprecedented accuracy. The original horn that was used was developed at Bell Laboratories in Holmdel, New Jersey, for satellite communications. In 1965, radio astronomers Arno Penzias and Robert Wilson made a series of measurements with this radio telescope (Figure 4.7). They found an excess radio

Figure 4.7 Discovery of the Background Radiation

This horn antenna was used by Nobel laureates Arno Penzias (*left*) and Robert Wilson at Bell Laboratories in New Jersey to detect the cosmic microwave background radiation in 1965. The antenna is small by radioastronomical standards, because, surveying a diffuse background requires relatively low resolution. However, an exceedingly precise method had to be developed for calibrating the antenna to an accuracy far greater than that to which radio astronomers had previously been accustomed.

noise that seemed to be independent of the direction the antenna was pointing. After trying very hard to calibrate their telescope and to eliminate the possibility of a terrestrial origin for the radio signal, they concluded that the radiation was uniform in all directions. The radiation was not more intense in the direction of the sun or of the Milky Way, for example, so it could not be of solar or of galactic origin.

The cosmological significance of the background radiation was immediately realized by a group of Princeton University physicists led by Robert Dicke. In one of the great detective stories of modern physics, Dicke realized that the background radiation provided a crucial clue to the origin of the universe. Dicke independently rediscovered a theory proposed by George Gamow and his students more than a decade earlier. Gamow had argued that some of the elements were produced in the first minutes of the big bang. As a consequence, the leftover primordial radiation should be omnipresent. As a result of the cosmic expansion, this original radiation should have cooled to a temperature of about 5 degrees above absolute zero.

Gamow's theory had fallen into obscurity when astrophysicists concluded that no significant amount of elements heavier than helium could be synthesized in the big bang. It was eventually realized, however, that the big bang provided a plausible and possibly essential environment for the synthesis of helium. This common element, second only to hydrogen in abundance, constitutes about one-third of the mass of the universe. It seems unlikely that ordinary stars would have synthesized just the right amount of helium in their cores, where thermonuclear fusion is presently occurring, to satisfy the predicted abundance. The universality of the helium abundance is thus strong additional evidence for its primordial origin in the big bang.

Dicke and his collaborators were building an antenna to detect this primordial radiation, which they had concluded should be detectable at radio wavelengths, when they heard of Penzias' and Wilson's remarkable findings in the nearby Bell Laboratories. The two groups of scientists published their discovery and the account of its significance simultaneously.

In subsequent studies, the cosmic microwave background radiation has been found to possess a very high degree of uniformity, to better than 1 part in 10,000 (Color plate 1). This uniformity, or isotropy, attests to its origin in the farthest reaches of the universe, as the following simple argument shows. Any radiation produced near the sun, in our galaxy, or even in nearby galaxies would undoubtedly be unevenly distributed. Therefore, we assume that the sources of the radiation are evenly distributed throughout space. Suppose we divide the universe into a large number of concentric and equally spaced spherical shells, all centered on and enclosing the earth. In this case, the amount of radiation coming from sources within any pair of adjacent shells is the same, because the area of a sphere increases with distance in just the same way as the intensity of the radiation decreases. A uniform background radiation must come mostly from the distant parts of the universe, where the majority of the sources are found. Very little of the radiation could originate in our local region of space, and any isotropic background radiation must be produced at cosmological distances.

The intensity of the background radiation has now been measured at many wavelengths. It has been found to possess a spectrum characteristic of radiation that originates from a state of perfect equilibrium: when matter and radiation are in equilibrium with one another, the temperatures of both must be identical. Imagine an enclosure with walls that are so dense and opaque that no heat or radiation can penetrate them. There will be a radiation field inside this box that is characterized by the temperature of the walls. We refer to this radiation field as *blackbody radiation* at this characteristic temperature.[4] A *blackbody* is thus a hypothetical, ideal radiator, or absorber, of radiation.

The cosmic microwave background radiation appears to possess an almost perfect blackbody spectrum (Figure 4.8). Determination of this result required measurements from balloon-borne telescopes, because the blackbody spectrum of the background radiation peaks at a wavelength of about 1 millimeter. At millimeter wavelengths, strong water vapor absorption occurs in the earth's atmosphere. Radio astronomers mostly observe at much longer wavelengths, where absorption by the earth's atmosphere is insignificant. Because the cosmic background radiation is very intense only in the millimeter region of the spectrum, its peak wavelength corresponds to a characteristic temperature of only 3 degrees above absolute zero. This is cold indeed. Such a low temperature is in accord with the notion that the observed radiation is the pale rem-

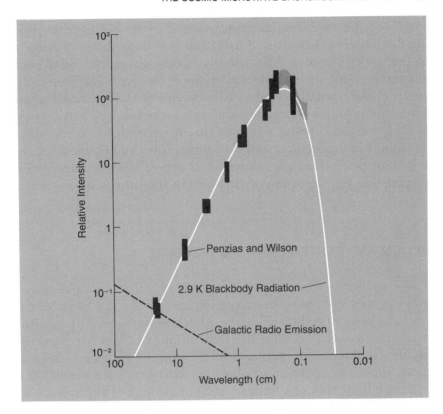

Figure 4.8 Blackbody Radiation
An ideal, or blackbody, radiator emits a characteristic spectrum that depends only on its temperature. The peak of the radiation occurs at a wavelength that is inversely proportional to the temperature of the radiation. The vertical lines and shaded region represent results obtained by many different astronomers at radio, microwave, and far-infrared wavelengths. The background radiation reasonably approximates the spectrum of radiation that is characteristic of a blackbody at 2.7 degrees Kelvin.

nant of the extremely hot primeval fireball that once pervaded the very early universe.

As we trace the history of the universe back in time, the temperature of the cosmic blackbody radiation rises. At earlier and earlier times, the radiation becomes hotter and hotter, and the universe denser and denser, until an instant is reached when blackbody radiation can be created. At this time, conditions are such that perfect equilibrium exists between matter and radiation; this is how the blackbody nature of the radiation gets established. Discovery of the radiation and confirmation of its blackbody nature (no deviations are measured to within less than about 10 percent of the spectrum of a 3-degree blackbody spectrum) have been among the most successful predictions of the big bang theory. Recently (1988), tentative evidence has been reported for small (about 5 per cent in temperature) deviations from the blackbody spectrum. If confirmed, this result will provide important information about the minor imperfections and deviations from homogeneity in the big bang that ultimately, as we shall see in later chapters, were responsible for the origin of structure in the universe.

HELIUM AND DEUTERIUM IN THE UNIVERSE

There are persuasive arguments for believing that certain elements and isotopes may have been synthesized in the big bang, just as big bang theory predicts. First, the high temperatures and densities predicted to occur during the first minutes of the big bang are highly conducive to synthesis of the lighter elements. Second, there appear to be no other plausible astrophysical sources for at least one light element, helium, and one isotope of hydrogen, *deuterium*. The fusion of hydrogen into helium is the source of energy by which stars radiate for most of their lives. We know that only a small fraction (far less than 10 percent) of hydrogen has been converted into helium during the course of the evolution of our galaxy, and most of this helium is still locked up in the cores of stars like the sun. Moreover, in many other galaxies, as well as throughout the Milky Way, we invariably find 1 helium atom for every 10 hydrogen atoms. This uniform distribution of light elements contrasts markedly to the distribution of heavier elements, which often shows pronounced variations; for example, the abundance of heavy elements decreases with distance from the center of our galaxy. Production of heavy elements can occur in supernovae. The rate at which supernovae are found increases with galactic luminosity, that is, with the mass in stars. More supernovae of a particular type are found toward the spiral arms of galaxies. The amount of stars increases in the inner regions of galaxies, and both supernovae and the heavier elements synthesized by supernovae also increase in abundance in a corresponding manner. Helium, however does not show any such concentration: its abundance is essentially independent of location; diverse regions of galaxies, whether metal-poor or metal-rich, show similar abundances of helium. This observation provides indirect evidence for a primordial origin for helium—or at least a pregalactic origin.

Deuterium (heavy hydrogen) is a fragile isotope that cannot survive the high temperatures achieved at the centers of stars. Stars produce energy by thermonuclear fusion of hydrogen into helium; deuterium is only an intermediate step in this reaction chain. Stars do not make deuterium; they only destroy it. Deuterium is observed to be present in our galaxy in interstellar matter that has not yet condensed into stars. Most astronomers now believe that helium and probably also deuterium originated in the first minutes of the big bang. At that time, conditions were such that nucleosynthesis would inevitably occur. The case for the origin of deuterium in the big bang is not as strong as the case for the creation of helium, for deuterium is relatively rare—there is approximately 1 atom of deuterium for every 30,000 hydrogen atoms in the interstellar medium. It is possible, for example, to conceive of other, nonstellar sources for deuterium at early stages of the evolution of the galaxy. However, deuterium plays a critical role in the big bang because its fragility and low abundance imply that it is sensitive to cosmology in a way that helium is not. Confirmation of the universality of the initial deuterium abundance in other galaxies should become possible within the foreseeable future. Such confirmation would greatly strengthen our confidence in the big bang theory.

We shall discuss the abundances of helium and deuterium in Chapter 7, and we shall develop the theory of heavy-element formation in Chapter 15. All the evidence discussed in Chapters 3 and 4 will reemerge as we trace the evolution of the universe, beginning in Chapter 6. For now, we stress only that the observable evidence strongly favors a big bang cosmology—so much so that astronomers are posing the questions to be decided by future research in terms of alternative big bang models. Yet, as we have seen in this chapter, most of the evidence supporting the big bang

· 5 ·

COSMOLOGICAL
MODELS

There is a single general space, a single vast immensity which we may freely call
Void: in it are innumerable globes like this on which we live and grow; this space
we declare to be infinite, since neither reason, convenience, sense-perception nor
nature assign to it a limit.

—*Giordano Bruno*

It is fairly certain that our space is finite though unbounded.
Infinite space is simply a scandal to human thought.

—*Bishop Barnes*

The observational evidence presented in Chapters 3 and 4 strongly favors a big bang model, and most cosmologists today are big bang cosmologists. Establishing that the universe originated in this manner was an exciting, revolutionary development, which led to further research and discovery as astronomers tested the implications of the model. Yet many fundamental questions about the nature of the universe remain unanswered, and alternative big bang models have been proposed to describe the properties of the universe about which we are yet uncertain. In this chapter, we shall describe some simple analogies that help to clarify these questions, and we shall then examine the models that remain viable alternative descriptions of the real universe.

THE CURVATURE OF SPACE

One of the fundamental questions of cosmology concerns the nature of space. Two of the standard big bang models assume that space is curved. What does this mean? One way to visualize the concept of curved space is to use a two-dimensional analogue. Consider a map of some mythical territory we shall call the Land of Lilliput. On the map in Figure 5.1, distances are marked between various cities. How do we decide whether Lilliput exists in a flat or a curved space? Take a compass and choose two towns, A and B, as points of reference. Then draw a circle of radius 4 units of Lilliput distance, centered on A, and 5 units, centered on B; the intersection is town C. Repeat, but use circles of radius 12 and 9 units; this gives point D. Now, if Lilliput space is flat, the distance between C and D has a unique value that is determined by elementary euclidean geometry. If the actual distance differs from the euclidean value, we must bend the map to fit the correct distances. If we have to bend the map, Lilliput cannot be flat.

Map makers encounter such problems in trying to construct a

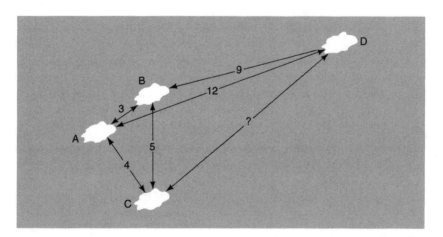

Figure 5.1 A Map of Lilliput
Imagine communicating with a Lilliputian, who tells us how to draw a map of his homeland. Towns A and B, at a distance of 3 units, are our base points. The four distances from A and B to C and D enable us to locate these towns on our map, and we can infer the distance between C and D. Does it agree with the actual Lilliputian distance? If not, we must conclude that the Lilliputian lives in a curved space.

two-dimensional map of the surface of the earth. They often use a Mercator projection, which distorts the shapes and sizes of geographic features, particularly near the poles; for example, generations of school children have grown up with the illusion that Greenland is a gigantic territory. Not only do the geometries of a spherical surface and a plane differ radically, but another geometry may also describe a two-dimensional surface. The third type describes a saddle-shaped surface (Figure 5.2). Imagine the task of inhabitants from these different surfaces who try to produce a two-

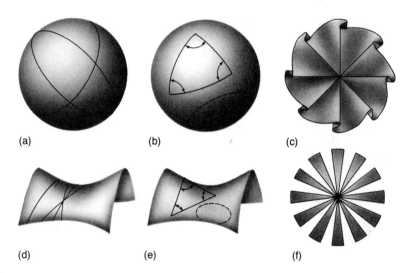

(a)   (b)   (c)

(d)   (e)   (f)

Figure 5.2   Curved Space

A two-dimensional analogy illustrates the concepts of curved space. The spherical surface (a) is positively curved. On such a surface, a line that is the shortest distance between any two points is a portion of a great circle, which will eventually intersect every other great circle. Therefore, no parallel lines can be drawn in this way. The circumference of a circle on the sphere is less than 2π times its radius, and the angles of a triangle add up to more than 180 degrees (b). If we try to construct a flat map of the surface, we have to crowd the map together at the edges (c); distances increase less rapidly toward the edges of a map of curved space than on a map of flat space. A saddle-shaped surface (d) is negatively curved. A line that is the shortest distance between two points is curved. Take a point not on that line; many other lines can be drawn through the point that intersect the original line. (On a plane, only parallel lines could be drawn that would not intersect the original line.) The circumference of a circle is more than 2π times its radius, and the angles of a triangle add up to less than 180 degrees (e). If we try to make a flat map, we have to stretch the map at its edges (f); distances increase more rapidly toward the edges than in a map of flat space.

dimensional map describing those surfaces. Suppose that on each surface are a number of randomly spaced islands. Each dweller makes a map and passes it to an observer from three-dimensional space (our space). What does the observer make of the maps?

The map of flat space reveals a number of randomly distributed islands, just as the observer expected. In the map of spherical space, however, the number of islands decreases toward the outer parts of the map. This is the same characteristic of spherical space that stretches out the northernmost countries on a projection of the earth's surface onto a two-dimensional map. To recover a random distribution, the map would have to be crinkled upward at the edges. The spherical surface is said to have a positive curvature. On such a curved surface, the circumference of a circle will be slightly less than the product of 2π times the radius.

The saddle dweller's map is of quite a contrary nature—the islands are crammed closer and closer together toward the edge of the map. This projection causes, in effect, a shrinking of all distances. The map would have to be stretched at the edges to recover a random distribution of islands. The saddle surface is said to have a negative curvature, and the circumference of a circle will exceed the product of 2π times the radius on this surface.

Let us now apply our models of two-dimensional space to the observed distribution of the galaxies in real, three-dimensional space. We must generalize our two-dimensional concepts to three dimensions. This is not as difficult as it might seem, given sufficient imagination. Consider for example the prophetic words of Giordano Bruno: "The center of the universe is everywhere, and the circumference nowhere." Of course, Bruno wanted an infinite universe, and here the spherical analogy is inadequate.

A spherical surface is a finite surface: a traveler on such a surface would eventually return to the starting point. Moreover, the curved surface has no boundary. A saddle surface does not provide a good analogy with real space, because a saddle surface has an edge, and the cosmological principle excludes any edge in the matter distribution of the universe. One could imagine extending the saddle shape indefinitely to avoid any edge, which suggests that, if the analogy of saddle geometry is somehow applicable to the universe, space could be infinite.

Mathematicians are indeed able to extend these concepts of two-dimensional surfaces to three-dimensional space. Three distinct types of space are found: *spherical space,* which corresponds in two dimensions to the surface of a sphere; *flat space,* which corresponds to a plane; and *hyperbolic space,* which corresponds to the saddle surface. The spherical space is finite and is said to be a *closed space.* However, the other spaces, like their two-dimensional analogs, are infinite and *open.*

HORIZONS

Imagine a rubber balloon with a large number of randomly distributed metal beads embedded within the rubber. Suppose the balloon is gradually inflated. If we imagine that the universe is confined to the surface of the balloon, we now have a two-dimensional model of a closed expanding universe (Figure 5.3). The metal beads represent clusters of galaxies. The beads are gradually receding from one another but do not themselves change shape.

Now imagine the earth as a point on the surface of the balloon. An observer on the surface could only survey a fraction of the area of the balloon. Similarly, we on earth will never be able to see very much more of the universe than we see at present, no matter how

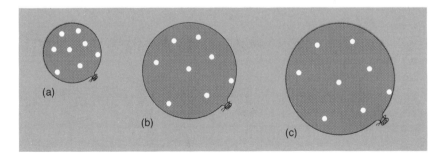

Figure 5.3 A Rubber Universe

Imagine a balloon in which metal beads are imbedded. As the balloon is inflated, the distances between the beads increase, but the intrinsic dimensions of the beads do not change. This model provides a crude analogy with the observed recession of the galaxies.

our telescopes improve, because we are limited by the observable _horizon_ (Figure 5.4). We can understand this concept by imagining that the balloon is expanding radially at a steady rate. For the moment, we ignore the additional complication that the universe is actually slowing from the self-gravity of the galaxies. Consider a galaxy separated from us by a distance of D light-years. Light takes D years to travel to us. Only after the universe has been expanding for a time in excess of D years will the galaxy become visible to us. At earlier eras, there would not have been time enough for its light to reach us. Hence, the galaxy would be completely inaccessible to any observations we might perform. We say that a galaxy first comes within our horizon after the universe has been expanding for a period equal to the time it takes for light to travel to us from the galaxy.

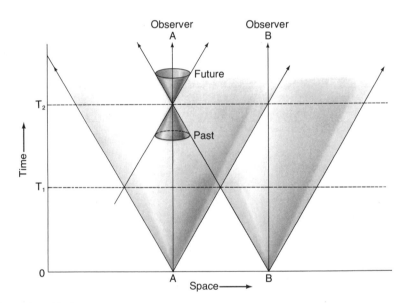

Figure 5.4 Horizons

In this space-time diagam, time is plotted vertically, and the three space dimensions are compressed into the horizontal axis. At the initial time of the big bang, two hypothetical observers A and B could not communicate with one another. The shaded regions denote their respective horizons—the distance traversed by light at any given time. Only about T_1 could A and B observe common parts of the universe, and only after T_2 do A and B enter one another's horizons.

The galaxies we observe just coming over the horizon are highly redshifted. Their recession velocities relative to us are close to the speed of light (otherwise, they would have been observable long ago). We can receive no light signals from the outlying regions of the universe; any other galaxies are completely inaccessible to us until they come within our horizon. The distance to the observable horizon can be most simply expressed as the distance that a light signal can travel in the available time since the big bang. This time determines the distance of the most remote observable objects in the universe. The distance to the horizon increases directly with the age of the universe.

In a more realistic model of the universe, the galaxies are slowed by the effects of gravity. Our horizon actually expands at a greater rate than the rate at which the galaxies are receding from one another. This means that in the early universe, very little matter would be contained within any hypothetical observer's horizon. As we trace the evolution back toward the big bang, the horizon eventually encompasses only the matter within a single galaxy, within a single star, and even, at incredibly short instants after the big bang, within a few atoms.

Our scheme of evolution requires us to consider regions of space that are bounded by the distance light can travel. Information cannot propagate beyond this ultimate boundary. In particular, applications of the laws of physics require careful scrutiny before we can discuss how a universe in which different sectors are not causally linked may evolve. Such a boundary in *space-time* may be one of the limiting factors in enabling us to trace the evolution of the universe back to the origin of time.

We can easily estimate the size of the section of the universe that we can see at present with Hubble's constant, which tells us that for every million light-years of distance, a galaxy recedes an additional 25 kilometers per second. The speed of light (300,000 kilometers per second) would be reached at a distance of 15 billion light-years. Of course, no galaxy actually attains light speed; a galaxy approaches this speed relative to us evermore closely as its distance increases. This distance thus constitutes the extent of the observable universe. Much of the universe may be outside our horizon. If the universe is spherical, a finite amount of space exists and will eventually, in the distant future, become accessible. This is not the case in a hyperbolic universe, whose space is infinite; we are indeed infinitesimal, and on such a scale, we always will be.

NEWTONIAN COSMOLOGY

Sophisticated mathematics is not required to understand big bang cosmology. The simple physics of gravity that Isaac Newton developed suffices for most purposes, because it enables us to describe the newtonian cosmological models, which are analogs of models derived in *relativistic cosmology*. The newtonian cosmological models are not perfect; if they were, we would have no need for more complicated theories. Although they provide a reasonable description of the evolution of the universe, they fail to describe how light from remote galaxies traverses space. Consequently, they are inadequate for the purposes of observational cosmology. However, they are very useful as a guide to the more esoteric aspects of relativistic cosmology, and we shall therefore examine them in some detail.

Early attempts to construct a *newtonian cosmology* failed because of a conceptual difficulty. To understand how it arose, we must introduce the concepts of *gravitational potential energy* and *kinetic energy*. Potential energy is the kinetic energy of motion that any particle is capable of acquiring when accelerated through a gravitational field. The sum of kinetic and potential energies is constant during the motion of the particle. Thus, a particle at rest has only potential energy. In bounded systems that are of finite extent, the gravitational potential energy of a particle is calculated by adding the contributions of a series of spherical shells. The presence of each shell contributes a certain amount to the potential energy. However, given an arbitrarily large number of such shells in an infinite system, one finds that the potential energy has no upper bound. Thus, the concept of potential energy is not meaningful in an infinite universe. An even more serious difficulty is that the universe cannot be spherical, finite, and static unless nongravitational forces are introduced. A similar result is true in relativity theory, and this led Einstein, in his earliest work in cosmology, to postulate a force of cosmical repulsion to balance the attractive force of gravity. As we have seen, Einstein abandoned this repulsion force once he adopted the expanding-universe theory.

These difficulties with newtonian theory were overcome only long after the relativity theory was developed, when the American mathematician George Birkhoff proved a general theorem that applies to any spherical distribution of matter. Consider a spherical volume of arbitrary but finite dimensions surrounding any given point (Figure 5.5). We can consider the gravitational potential energy of any particle within that volume to depend only on the matter within the spherical volume, provided the dimension of this region is small compared with that of the horizon. We can now interpret Big Bang cosmology in terms of simple newtonian gravity. The expansion of the universe enables the gravitational self-attraction to be overcome, thereby allowing a self-consistent cosmological model to be constructed.

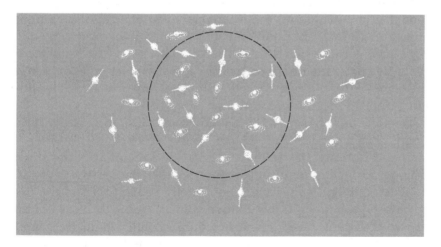

Figure 5.5 Newtonian Cosmology

An imaginary sphere of large but arbitrary size is drawn in the expanding universe, which we assume to satisfy the cosmological principle. An important theorem in general relativity tells us that only the matter within this sphere makes any contribution to the local gravitational field.

A RAISIN PUDDING MODEL OF THE UNIVERSE

We will next describe a simple model for the real three-dimensional universe. To avoid unnecessary complications, we shall use Newton's theory of gravitation; the curvature of space will not trouble us explicitly. Let us consider first the analogy of an ordinary raisin pudding. The raisins are scattered randomly through the pudding; they individually represent clusters of galaxies. The pudding is allowed to cook slowly. The pudding swells steadily, but the raisins do not expand. If the pudding always maintains a uniform consistency, the raisins will recede from one another at a relative velocity proportional to their distance. Of course, we cannot consider raisins near the edge of the pudding—the pudding must be imagined to be infinite. It is rather easy to make this model quantitative. We can derive Hubble's law, and by introducing the notion of gravitational attraction between the raisins, we can derive the relation between the expansion rate and the average density of matter. The gravitational attraction exerted by the raisins in the pudding tends to counter the expansion.

Starting from the cosmological principle, we can now apply this model to the evolution of the universe. Let us assume a uniform and expanding distribution of matter. At any time, observers moving with the expansion are required to see a similar appearance to the universe, which implies that the density must be everywhere the same. Also, the relative velocities that any observers measure must depend on distance and time.

To make this concept more quantitative, consider first any three points that form a triangle. As the universe expands, in order to maintain isotropy, this triangle must always retain a similar shape (Figure 5.6). This requirement implies that the relative velocity between any two points must be directly proportional to the distance between them. If the relative velocity depended, say, on the square of that distance, then an arbitrary triangle would grow more and more distorted as the expansion proceeded: the longest side would grow much longer, relative to the shortest side. In other words, we have inferred Hubble's law, that the relative velocity v between any two points is equal to their separation r multiplied by a universal constant H. Actually, this statement is more general than Hubble's law, which applies only to nearby regions of the

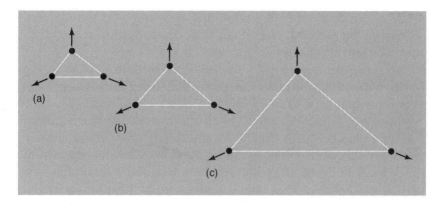

Figure 5.6   Isotropy of the Expanding Universe
Any three points define a triangle. As the universe expands, the cosmological principle requires the triangle always to maintain the same shape as it increases in size. It directly follows that the relative velocity between any two vertices of the triangle must be proportional to their separation; otherwise, the shape of the triangle would change.

universe. Since the travel time of light from nearby regions is short, compared with the age of the universe, we ordinarily interpret H in terms of the present era of the universe. Hubble's law is really far more fundamental, however, and we can evaluate Hubble's constant for any arbitrary time. H is not really a constant; the *Hubble parameter* would be a better term. For this reason we introduced H_0 to denote Hubble's constant, namely the value of the Hubble parameter at the present epoch.

It also follows from the relationship between distance and relative velocity that the distance between any two points must depend only on the galaxies' initial separation at some initial reference time and on the time elapsed subsequently. Thus, we can express the separation r between any two points as being equal to their initial separation, r_I, multiplied by a scale factor R that expresses how much the universe has expanded. When r is equal to r_I, no expansion will have occurred, and R must be unity. The scale factor R is evidently a function only of the time elapsed since r was equal to r_I. The meaning of H now becomes clearer, because we can interpret H as being the relative rate of change of R. H is really just an inverse measure of the age of the universe—at a time $1/H$ in the past, any two galaxies would have been in contact with one another. We thus infer that when the universe was young, H was very large. As the universe aged, H decreased, until it attained its present observed value, H_0.

Now, consider an arbitrary region of space bounded by a sphere that is expanding with the universe. We refer to this region as a *comoving sphere,* within which the matter is no longer expanding (because the volume expands with the matter, it always contains the same amount of mass). Let us ignore any possible processes of creation or destruction of matter. Let us also assume the absence of any large amounts of radiation, the introduction of which would cause complications that only the more sophisticated theory of relativistic cosmology can handle properly. It follows that the number of particles of matter in such a sphere must forever be constant. It further follows that the total energy of all the matter within the sphere must be forever constant.

Let us examine the nature of the energy content of this sphere more carefully. We divide the comoving sphere into a large number of concentric, thin, spherical shells, each expanding with the universe. The energy of the particles that constitute any one of these shells consists of two types—kinetic energy and gravitational potential energy. One of these can increase only at the expense of the other. Consider the example of a stone thrown into the air. At its greatest height above the ground, its kinetic energy of motion is zero (it is motionless), but its gravitational potential energy at this point is greatest. When it hits the ground, it has converted all its potential energy into kinetic energy. The total energy, equal to the sum of the kinetic energy plus the potential energy, is always constant.

From this analogy, we can infer that the sum of the kinetic energy of the expansion of a spherical shell and the gravitational potential energy of the shell does not change with time as the universe expands. We can express the kinetic energy as one-half of the product of the mass m of the shell and the square of its velocity of expansion v, or $\frac{1}{2} mv^2$. The gravitational potential energy at the center of the sphere is the energy a particle would acquire as it fell toward the center. Because gravitational potential energy is greatest when kinetic energy is least, we always consider gravitational potential energy to be negative energy. The total mass M contained within the shell contributes to its gravitational potential energy, which can be approximately written as a constant (Newton's constant of gravitation G) multiplied by the product of the mass of the sphere and the mass of the shell and divided by its radius r, or $-GMm/r$.

The analogy with the motion of the stone leads us to express the motion of any particle in the universe by a simple equation that expresses the *conservation of energy*. We now infer that the kinetic energy per unit of shell mass, which is equal to one-half of the square of the particle velocity (measured relative to another particle), plus the (negative) potential energy per unit of mass, which is equal to a constant divided by the distance between the particles, does not change with time. The constant is actually equal to the mass contained in a sphere between the two particles multiplied by Newton's constant of gravitation.

For a comoving sphere that expands with the universe, the mass will also be unchanging with time, because no particles are being created or destroyed. On the average, any particles that can escape from the sphere are replaced by ones that enter. If we denote the uniform density of the universe by d, we can now infer that because the number of particles within a comoving sphere never changes, the product of density and volume must be constant. In other words, d must be proportional to the inverse of the volume of the sphere, or R^{-3}. When R approaches zero, the density becomes arbitrarily large, and we reach the moment of the big bang, which we take as the origin of time. As time proceeds, the density decreases without limit if the universe continues to expand.

The energy conservation equation can be simplified by applying Hubble's relation to the expansion velocity. It then becomes a simple equation for the scale factor R. This equation is the *Friedmann equation* (Figure 5.7).[5] Relativistic cosmology, which utilizes Einstein's theory of gravitation, yields an identical equation. However, the interpretation of the constant that expresses the total amount of energy differs for the two equations. We can regard this constant, which we call k, as expressing the average amount of total energy possessed by an arbitrary gram of material in the universe. The value of k can be zero, positive, or negative. To understand the significance of k, we can consider the example of a rocket launched with a certain amount of kinetic energy; the rocket may possess enough energy to escape from the earth, or it may fall back to earth. This simple analogy describes the possible alternative fates for the

M = mass interior to shell
r = radius of shell
d = density within shell
v = velocity of expansion
H = Hubble's constant
G = Newton's gravitational constant
k = curvature constant (= +1, −1, or 0)
c = speed of light
R = scale effect = $r/r_{initial}$

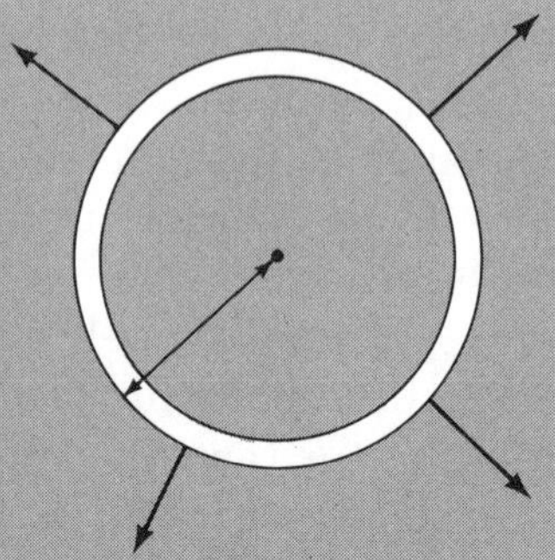

(a) Energy of expansion of shell + gravitational potential energy of shell = a constant

$$\tfrac{1}{2}v^2 - G\frac{M}{r} = -\tfrac{1}{2}kc^2r^2_{initial}$$

(b) Applying Hubble's law, $v = Hr$, this equation transforms to

$$\tfrac{1}{2}H^2 - \frac{4\pi}{3}Gd = -\frac{kc^2}{2R^2}$$

At early times, $\tfrac{1}{2}H^2 \approx \frac{4\pi}{3}Gd$

Figure 5.7 The Friedmann Equation
The Friedmann equation is the fundamental equation of big bang cosmology. It relates kinetic energy of expansion and gravitational potential energy for any spherical distribution of matter in the universe. The sum of these two types of energy must be constant with time.

matter in the universe: the galaxies may expand forever, or they may fall back together. Thus, our simple model has led us to the point where alternative big bang models emerge.

BIG BANG MODELS

We can visualize the early stages of the big bang as a gigantic explosion. The kinetic energy of a given amount of material was very great, and because energy must be conserved, the potential energy was also great. For such early times, the total energy, or k, which is the difference between these two large quantities, is relatively insignificant. The Friedmann equation reduces to a simple balance between the kinetic energy of a unit mass of material (proportional to one-half of the square of Hubble's constant H) and the gravitational potential energy (equal to the product of the density d times Newton's constant G times $4\pi/3$), or $\frac{1}{2}H^2 = (4\pi/3)Gd$. The radius of the spherical region under consideration enters identically in both the kinetic and potential energy terms. This leads to a surprising result: the Friedmann equation does not depend on the dimension of the region being considered. It is independent of scale, and it depends only on time.

In a particularly simple mathematical solution to this equation, R increases as the two-thirds power of the time elapsed since the big bang. This solution to the Friedmann equation is known as the *Einstein-de Sitter universe*, and it is the simplest of the big bang cosmologies (Figure 5.8).[6] The space in this universe is infinite and has the same properties as ordinary euclidean space. The universe expands forever, from a time when R was arbitrarily small. At this instant, considered to be the origin of time, the density of matter was infinite.

At early times, the Einstein-de Sitter universe is an excellent approximation to the open and closed Friedmann-Lemaître universes[7] (provided that the cosmological constant originally introduced by Einstein is set equal to zero). Later, however, the effect of k becomes important. Imagine a late stage in the expansion, when the density has dropped and become very small. The Friedmann equation now expresses a balance between a kinetic energy term, as before, and the energy constant k.

Suppose, for example, that k is negative, which corresponds to a universe having positive kinetic energy at an arbitrarily large radius. At any large value of the scale factor R, the universe will always be expanding at a certain rate determined by H. The universe expands without limit when k is negative. We call such a universe an *open universe*. Space is infinite; it has no edge, and we say it is unbounded. Solution of the Friedmann equation leads to the result that at late eras, R must vary proportionally with the time elapsed since the big bang.

However, if k is positive and we attempt to let R increase without limit, we find that there can be no real solution for R (we know

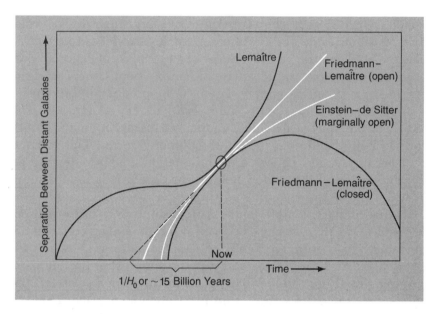

Figure 5.8 Big Bang Models

The viable alternative big bang models are the Friedmann-Lemaître open and closed models, the marginally open Einstein-de Sitter model, and the Lemaître universe. The separation between any two distant galaxies in different regions of the universe is plotted vertically; time is plotted horizontally. The small circle represents the present era. If the universe has always been expanding at its present rate, it would now be about 15 billion years old, as shown by the extrapolation (*dashed line at lower left*). If the expansion is decelerating, as shown by both the open and closed models, the universe is actually less than 15 billion years old. The open model is nearly 15 billion years old. The closed model has the lowest age because the deceleration must be greatest in this model for the expansion to reverse. The Lemaître universe is much older than 15 billion years because there is a long quiescent period (*the flat segment of the curve*) during which the expansion almost stops. Both the open Friedmann-Lemaître model and the Lemaître universe continue to expand forever.

that the square of any real solution for H must be positive, yet the Friedmann equation now equates a positive quantity with a negative quantity). In this case, the universe must stop expanding! We now find that the scale factor R increases from zero to a maximum value and decreases back to zero again. Such a universe is spatially finite, although unbounded. This again means that there is no edge to the universe. The geometry of such a universe is analogous to the geometry of the surface of a sphere; space is of positive curvature, and we say that such a universe is *closed*. One of the great ongoing debates of cosmology is whether the real universe is open or closed. In Chapter 17, we shall see how the controversy is presently faring.

The preceding discussion was wholly based on newtonian concepts of gravitation. Einstein's theory of relativity also leads to the identical Friedmann equation. As in the newtonian theory, the relativity theory arrives at this result by postulating that the universe is homogeneous and isotropic. Historically, the relativistic theory of cosmology actually preceded the newtonian theory—the newtonian theory was developed to provide a simple interpretation of the relativistic results. The most important difference between the two approaches to cosmology is the interpretation of the term k. The relativity theory shows that this term arises from the curvature of space. According to this theory, gravity is equivalent to a curvature of four-dimensional space–time: what we consider to be ordinary three-dimensional space may in fact be curved, or non-euclidean. In the strong gravitational field near a black hole, space becomes very strongly curved. Even in the relatively weak gravitational fields normally encountered in cosmology, parallel lines no longer need be parallel, and the three angles of a triangle no longer need add up to precisely 180 degrees. Many other consequences of euclidean geometry can similarly be violated. If parallel lines eventually meet, as would happen on the surface of a sphere, the space has a positive curvature. In cosmology, such a space is closed and finite. If parallel lines eventually diverge, space is of negative curvature. Negatively curved spaces in cosmology are open and infinite.

The Friedmann equation has a much deeper significance in relativistic cosmology. We no longer need to think of the energy of matter as defining the characteristics of the expansion. We call k the curvature of space and assign k three possible values: $+1$, 0, or -1. If $k = +1$, space is spherical and closed; if $k = -1$, space is hyperbolic and open. The case $k = 0$ also corresponds to an open, or infinite, universe, but the geometry is euclidean. Table 5.1 summarizes the characteristics of these alternative models.

If the universe were static, the expansion term H would not be present in the Friedmann equation. However, we know that the curvature term cannot be important at early times. Inspection of the Friedmann equation then suggests that the only static universe

Table 5.1 Alternative Cosmologies

Big bang models	k	Space	Extent	Destiny
Einstein-de Sitter	$k = 0$	Flat	Open and infinite	Expands forever
Friedmann-Lemaître	$k = -1$	Hyperbolic	Open and infinite	Expands forever
Friedmann-Lemaître	$k = +1$	Spherical	Closed and finite	Expands and will recollapse
Lemaître	$k = +1$	Spherical	Closed and finite	Expands forever with quasi-static phase
Models with no big bang				
Eddington-Lemaître	$k = +1$	Spherical	Closed and finite	Static initially, then expands forever
Steady State	$k = 0$	Flat	Open and infinite	Stationary (but not static)

capable of satisfying the cosmological principle is an empty universe. When the theory of general relativity was developed in 1916, the concept of an expanding universe was a revolutionary concept, which Einstein himself did not yet accept. He provided an ingenious solution to this apparent cosmological paradox by introducing, as we have already seen, the cosmological constant (independent of time and position) into the equations of gravity. According to Einstein, the gravitational force acting on a point near the boundary of a large sphere of matter consists of the ordinary newtonian attractive force with an additional repulsive force. We can write the attractive force as proportional to the mass divided by the square of the radius. Equivalently, if the density is everywhere uniform, we can express this force as proportional to the product of the density and the radius of the sphere. The additional repulsive force can be expressed as the product of the cosmological constant and the radius of the sphere (if we wished to be rigorous, we would also include a factor of one-third, to be in accord with Einstein's definition of the cosmological constant). Thus, cosmic repulsion and gravity both increase with distance. In the Einstein static universe, these forces exactly cancel one another, and a very precarious sort of equilibrium is achieved.

The equilibrium is precarious because the slightest disturbance or deviation from it could shift the balance in favor of one force over the other, and the universe would either collapse or expand. For example, any tiny deviations from an initial uniformity, such as those resulting from the random motions of the atoms, would eventually become larger. The gravitational force depends on density, but the repulsion does not; therefore, even infinitesimal enhancements in density generate an excess of gravity over repulsion. In the unlimited time available in a static universe, this excess is sufficient to result in the eventual collapse of large regions of matter—for galaxies to form. It was thought for some years that galaxy formation might cause the universe as a whole to expand, because the net gravitational attraction is slightly smaller when matter is gathered together into discrete regions than when it is spread out uniformly. This hope, now known to be a misconception, led to a cosmological model that the British astronomer Arthur Eddington was particularly active in promoting. The *Eddington-Lemaître universe* begins as a static universe and starts expanding only when galaxies begin to condense. Such a model is attractive, because it avoids the problem of an initial moment of time and yet might account for the observed expansion of galaxies.

More recently, it has been suggested that when galaxies condense, an Eddington-Lemaître universe would actually collapse rather than expand, because a great deal of radiation must be produced by newly forming galaxies. Naively, one might expect radiation to aid the expansion because of its pressure. It turns out, however, that the opposite would occur: the extra pressure has really disastrous results for this cosmology, because it tends to aid collapse. These results can be properly understood only in terms of relativistic cosmology. Crudely, we can say that pressure is associated with a certain density of energy, which is equivalent to a certain amount of mass, which enhances the effects of gravity. The extra gravity caused by the pressure speeds the collapse. A similar effect is found in studies of massive collapsing stars: the pressure of the collapsing material actually helps to increase the gravitational attraction and accelerate the formation of a black hole.

Despite these theoretical difficulties with the Eddington-Lemaître universe, there are viable cosmological models that incorporate the cosmological constant. In a cosmology known as *Lemaître's universe*, the expansion begins in a standard big bang. At some late stage, the force of cosmic repulsion exerts its influence. Like the curvature term, the cosmic repulsion is only important when the density of matter in the universe has fallen considerably. The repulsion causes the universe to enter a coasting phase. Eventually, the curvature term makes its presence felt, and the expansion accelerates. Because the expansion never completely ceases, this universe is not liable to the same fate that befell the Eddington-Lemaître universe; galaxies would ultimately drift apart, just as they did in the open Friedmann-Lemaître model. Lemaître's universe has often seemed attractive to cosmologists because the coasting period provides a plausible era when condensation of galaxies could occur. One other noteworthy feature of the Lemaître universe is that it provides sufficient time for light to propagate around the universe (because this model is spatially closed). This characteristic could lead to odd, counterintuitive results. In principle, by using a large enough telescope, one could see the back of one's head without using a mirror! One might also expect the formation of ghost images; for example, an overabundance of remote galaxies and quasars would be likely to be observed at the redshift corresponding to the coasting phase. The apparent absence of any such phenomena has deterred most cosmologists from pursuing the Lemaître universe.

Although such exotic universes are interesting and attractive, our preference must be for the simplest tenable cosmology. Practically all known astronomical phenomena can be understood in the context of big bang cosmology—if not completely, then at least to a greater degree than in any alternative framework that has yet been proposed. Thus, we shall accept the big bang models as providing a satisfactory description of the universe. The standard big bang cosmology, which we will further explore in the following chapters, will be taken to include the following three possibilities: a closed Friedmann-Lemaître universe with spherically curved space, destined to recollapse; an open Friedmann-Lemaître universe with hyperbolically curved space, destined to expand forever; and the Einstein-de Sitter universe with flat space, also destined to expand forever. As we have seen, the differences between these three models are only significant at late eras, after galaxies have already formed; for the early universe, these models are indistinguishable. This greatly simplifies our study of the evolution of the early universe, to which we now turn.

· 6 ·

THE FIRST MILLISECOND

The point of view of a sinner is that the church promises him hell in the future,
but cosmology proves that the glowing hell was in the past.

—Ya. B. Zel'dovich

What was the universe like at the earliest moment? If we trace the evolution of the universe back through time, the universe becomes progressively denser and hotter. The region of space that any hypothetical observer can view (the *observable universe*) becomes smaller and smaller. The observable universe is restricted to the distance light can travel during the time elapsed since the big bang; the actual universe is much larger.

As we noted in Chapter 3, the most distant galaxies that we can observe are receding from us at a velocity of more than one-third the speed of light, and they are at a distance in excess of 5 billion light-years. There are presently about 10 billion galaxies in the observable universe. The further back in time one goes, the greater is the apparent redshift of light, and relative to us, the velocity of the emitting source approaches that of light. In fact, galaxies could not exist until at least 1 million years had elapsed after the big bang. Nevertheless, we can move our theoretical model back, ever closer to the beginning of time. If we trace the history of the universe back to the age of about 10 years, the expansion would have progressed to the point where the matter within a single galaxy filled the entire observable universe; all the atoms of the observable universe at that time would have amounted to no more than the mass of a galaxy and would have been concentrated within a region some 10 light-years in extent. Going further back in time, a few seconds after the big bang, the observable universe would only have contained about as much matter as there is in the sun. Of course, all the matter that we see now in the universe would still have been present—it simply could not be seen by any one observer. We can imagine a vast network of hypothetical observers, however, who trace the history of different regions back through time (Figure 6.1). These hypothetical observers could approach ever closer to the big bang, until the equivalent mass of only one atomic nucleus would be contained within the observable universe. The age of the universe at this stage is but an infinitesimal fraction of a second (about 10^{-23}); the size of the observable universe is the dimension of an atomic nucleus, or about one ten-thousand-billionth of a centimeter.

THE DENSITY OF THE UNIVERSE

The ultimate instant that physics allows us to speculate about takes us back even closer to the big bang. Imagine a moment so early and a density so high that the gravitational stresses were capable

Figure 6.1   The Big Bang: A Space-Time Diagram

In a space-time diagram, space is compressed to the horizontal axis and time is plotted vertically. One space-time diagram (a) depicts the world-lines, or trajectories, of students A, B, and C. A and B attend the same class from 9 A.M. to 10 A.M., and B and C live together but attend different classes. The world-lines of A and B diverge after their class, but those B and C converge. Another space-time diagram (b) depicts the world-line of a hypothetical observer (time axis) in the standard model. A horizontal line on this diagram represents the entire universe at a particular time. The shaded regions represent the part of the universe within the hypothetical observer's horizon at any given time. As time increases after the singularity, larger regions of the universe become observable. At present, light reaches us from quasars and clusters of galaxies that are billions of light-years distant. In the future, objects that are farther away will become visible. Various eras in the evolution of the universe are also indicated. Atoms and atomic nuclei (or nucleons) are now the predominant form of matter, but hadrons, leptons, and photons predominated at earlier eras.

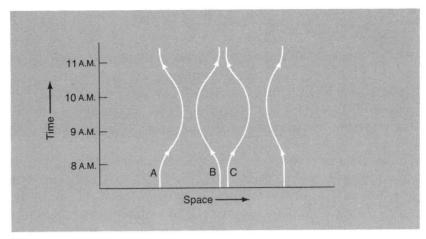

(a)

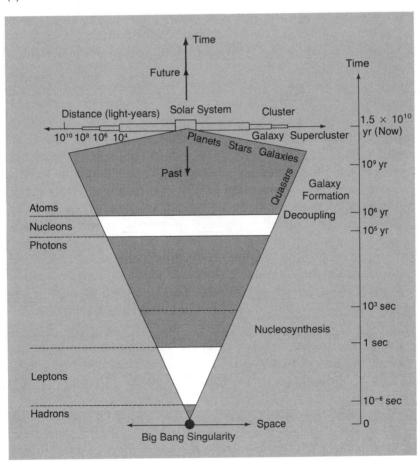

(b)

of tearing apart the vacuum. At a later era, the nuclear and electro-magnetic forces created pairs of elementary particles. If gravitational forces were sufficiently great, they also would have been capable of creating pairs of particles from the vacuum. In other words, at the moment of singularity, space-time was essentially disrupted by the gravitational forces.

To estimate the earliest instant that is amenable to our study, we must make use of the modern theory of physics of the ultimate structure of matter, *quantum mechanics*. According to *Heisenberg's uncertainty principle*, we can never precisely pinpoint the location of any elementary particle. Atomic nuclei and electrons lose their individual identity and acquire a wavelike nature on a scale known as the *Compton wavelength.* We can no longer locate elementary particles at a particular point of space; now we can locate them only in a particular region, and individual particles become indistinguishable. The dimension of this region of uncertainty is the wavelength of the particle. The larger the mass, the smaller the wavelength. Even macroscopic objects possess their wavelength of uncertainty: you, the reader, may spontaneously go through the floor, given a long enough period of time, a time that is much longer than the age of the universe. An elementary particle, however, manifests this uncertainty on a very short time scale.

Let us now consider an era so early that the entire observable universe was contained within its own Compton wavelength. This is the ultimate limit of our theory of gravity, where uncertainty reigns supreme. At this instant, known as the *Planck time*, only 10^{-43} second after the singularity, all the matter we now see in the universe, comprising some millions of galaxies, was compressed within a sphere of radius equal to one-hundredth of a centimeter, the size of the point of a needle. At this moment, the extent of the universe visible to any hypothetical observer was only 10^{-33} centimeter in diameter, far smaller even than an atomic nucleus.

If all the atoms in the present stars and galaxies were spread uniformly throughout space, there would be about one atom of hydrogen per cubic meter of space. In addition, there would be perhaps one-tenth as much helium. All the heavier atoms collectively amount to less than 1 percent of the number of hydrogen atoms. In the early universe, the density was very much higher.

One second after the bang, the density had dropped to 10 kilograms per cubic centimeter. (Ordinary rocks have a density of a few grams per cubic centimeter.) At the Planck time, the density approached 10^{90} kilograms per cubic centimeter. These physical conditions are so extreme that it seems entirely appropriate to regard the Planck time as the moment of creation of the universe.

THE TEMPERATURE OF THE BIG BANG

The big bang was extremely hot; of that we can be fairly certain. The cosmic blackbody radiation is the testament of this fiery origin. However, to understand precisely how hot the extreme environment of the early universe was, we must explore the significance of temperature, of its counterpart, energy, and indeed, of the nature of matter itself at early epochs.

A crucial role is played by the elementary particles of nuclear physics. These particles are collectively known as *hadrons* (Figure 6.2). Hadrons come in many varieties, including mesons, protons, neutrons, and heavier but short-lived particles. These particles are measured not by their mass, which can become a very transient concept, but by their total *rest mass energy*. Rest mass energy is only fully liberated when a particle is completely annihilated.

Annihilation is a common fate for a particle and its *antiparticle*— a particle of opposite electric charge but otherwise identical nature. The annihilation of a single proton and antiproton yields 1 billion electron volts of energy, which is barely enough power to run a flashlight for a billionth of a second. To make the conversion, note that 620 billion electron volts equals 1 erg, and 10 million ergs of energy expended for 1 second amounts to a power output of 1 watt. (The units of *electron volts* are chosen for the physicist's convenience, one electron volt being equal to the energy gained by a single electron accelerated through a potential of 1 volt.) The energy released from annihilation of one proton-antiproton pair may not seem like a great deal of energy, until we consider a vast number of protons and antiprotons. In fact, the efficiency of energy output per gram of material resulting from annihilation vastly exceeds anything attainable by other means of conversion. Annihi-

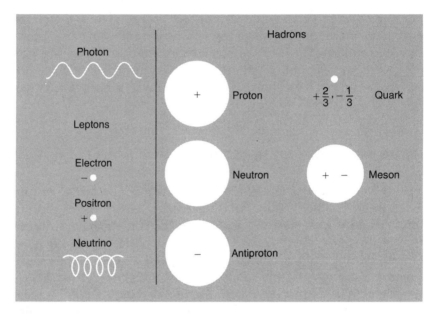

Figure 6.2 Some Elementary Particles

Hadrons are heavy elementary particles which experience the strong force, and leptons are other elementary particles which interact via the weak force.

lation is the ultimate, completely efficient energy source. No matter remains from this conversion.

Metals begin to melt at temperatures above 1000 to 2000 degrees kelvin (K), temperatures equivalent to an energy of about one-tenth of an electron volt per atom. At the center of the sun, the temperature is 10 million degrees kelvin, equivalent to about 1000 electron volts per atom. Ordinary chemical bonds between atoms have energies of about 1 electron volt, and the nuclear forces that bind nuclei together require energies of millions of electron volts to disrupt or to fuse the nuclei. Roughly a million times more energy is released in thermonuclear explosions than in chemical explosions, which explains why we measure the force of a nuclear explosion in megatons of TNT. Total annihilation releases more than 100 times more energy per gram of material than a nuclear explosion.

Not only do particles annihilate and release energy, but the reverse process may also occur: particles can be created from an intense radiation field. In one important example of this process that is responsible for ordinary nuclei of atoms, the resulting particles, mesons, ordinarily exist only for brief instants in the inaccessible interiors of atomic nuclei; these particles help to hold the nucleus together. However, in the early universe, the matter density exceeded nuclear densities, and new states of particles may have been created.

According to modern elementary particle theories, there are a finite number of elementary particles. If this were not the case, the available energy at early times and higher densities would go into creating more and more species of particles, and the temperature would level off. In effect, at earlier times and higher energies, new types of elementary particles are soon exhausted, and the temperature increases continuously back to the Planck instant, where it attains the incredible value of 10^{32} degrees kelvin. For physicists who ordinarily work with giant particle accelerators, such conditions are unattainable. The corresponding energy is 10^{19} GeV (billion electron volts): the largest planned terrestrial accelerators may smash particles together at energies of thousands of electron volts. The early universe offers a marvelous particle accelerator: we would need to build an array of superconducting magnets 1 light-year across to duplicate it. Although we have to take whatever is left over from this fiery past, fossil relics from near the beginning of time may be playing an important role in the universe we see today.

THE PHYSICS OF CREATION

The early universe was uniformly filled with radiation and neutrinos, in which a relatively small number of electrons, protons, and neutrons were interspersed. During the course of the expansion, the radiation cooled and eventually became the relic background radiation that has been measured by radio astronomers. The present temperature of the radiation is 3 degrees kelvin, which is equivalent to an energy of less than one-thousandth of an electron volt per atom.

As we extrapolate to earlier times, we find that the temperature of the universe increases in proportion to the one-fourth power of the energy density.[8] Once it rises above 1 million electron volts, something quite drastic occurs. One million electron volts is the rest mass energy of an electron plus its antiparticle mate, a *positron*. If an electron-positron pair is totally annihilated, 1 million electron volts of energy is released. Conversely, when the radiation temperature exceeds this limiting value, pairs of particles, electrons and positrons, are created from the radiation field. Annihilation then occurs by merger of an electron with a positron: two high-energy *photons* (particles of radiation) are produced, each with more than a half-million electron volts of energy.

The theory of quantum mechanics allows us to think of radiation either as being wavelike or as consisting of photons, massless particles of pure energy. The energetic photons that result from annihilation of an electron-positron pair are called *gamma rays*, and they are intensely penetrating. Each of these photons of gamma radiation has an energy of a half million electron volts. Some gamma rays can be far more energetic—for example, when a proton annihilates with its antiparticle (an *antiproton*), the energy of the resulting gamma rays is about 1 billion electron volts. By comparison, the x-ray photons used in medical diagnosis have energies of merely a few thousand electron volts. We can describe these processes of creation and annihilation by the reaction

$$\text{Particle} + \text{antiparticle} \rightleftarrows \text{gamma rays}$$

During the early moments of the universe, at temperatures above 1 million electron volts, there were roughly the same number of electrons, positrons, and photons. However, after a few seconds had elapsed, the temperature dropped sufficiently that the photons were no longer energetic enough to create any new particle-antiparticle pairs. At this point, the electron-positron pairs annihilated and could not be replenished. The result was a universe containing almost exclusively photons. As we shall see, complete annihilation did not occur, because there was a slight excess of particles over antiparticles. Consequently, a few particles were left to survive. Had the early universe contained precisely equal amounts of matter and antimatter, it would now be almost totally devoid of particles. This would be a most unfortunate circumstance, as far as we are concerned.

Now consider what happened at eras earlier than one second. The temperature was so high that more massive particles could be created. These particles included mesons and antimesons, protons and antiprotons, and far more exotic species of nuclear particles. All these particles were capable of annihilating with their antiparticle mates, and they also were capable of being created by the intense radiation field. The net result was a proliferation of different elementary particles, all of comparable abundance to the photons. Today, particle accelerators are used to obtain infinitesimal amounts of such particles; the early universe would have resembled a paradise for the elementary particle physicist, providing a prolific source of exotic particles.

As we probe back toward the initial singularity, we can make another speculation. Could not the intense gravitational field itself result in the creation of matter and radiation out of a vacuum? The very early universe might have been empty! Investigations of this possibility have found that if the universe remains isotropic, relatively little particle creation occurs. However, creation of particles and photons could occur if the initial expansion was chaotic, or anisotropic—that is, if the universe expanded at very different rates in different directions from any given point. Indeed, the form of the initial big bang probably was highly chaotic: given so many equally probable varieties of beginning, a smooth, isotropic expansion is a most unlikely possibility. The immense tidal gravitational forces that consequently resulted from the big bang can be imagined as disrupting the continuum of space-time in the process of creation. One can think of the vacuum as containing *virtual pairs* of particles and antiparticles. A sufficiently intense tidal gravitational field can disrupt these virtual pairs, releasing the particles into the real world.

The creation process was highly stabilizing. The initial anisotropy was rapidly dissipated and resulted in an isotropic universe filled with radiation. It is not difficult to understand how the creation process can produce isotropy, since the greater the anisotropy, the more prolifically are particle pairs created. The particle pairs also annihilated, and the resulting radiation flux diffused away, thereby tending to smooth out the anisotropy. We can thus speculate that such exotic processes preceded the time encompassed by the conventional Big Bang model.

IN THE BEGINNING

The initial instant of the universe coming into being lasted far less than the blink of an eye. Its duration was the minuscule fraction of 10^{-43} second. Einstein's theory of gravity can describe the history of the universe back to this instant of time, known as the Planck time (named for Max Planck, one of the founders of quantum mechanics), when the density of matter attained the incredible value of about 10^{75} tons per cubic kilometer. A cubic kilometer of lead weighs, by comparison, about 10^{11} tons. We have already noted that at the Planck time, the entire universe observable today occupied a region only one one-hundredth of a centimeter across. The distance travelled by a light ray was much smaller, because as the universe expanded, more and more matter became visible. The observable universe today is our cosmic horizon; it extends over some 30 billion light years. At the *Planck time*, any hypothetical observer's cosmic horizon only contained only about one-millionth of a gram. This mass, called the Planck mass, represents the ultimate building block of nature: at the Planck scale, we need an entirely novel concept in order to describe the state of matter. This scale is the meeting place of quantum physics and cosmology, and a new branch of physics, quantum cosmology, has emerged to meet this challenge.

Quantum cosmology represents the ultimate union of physics, in which the weakest force (gravitation), on an equal footing with the strongest force, holds nuclei together. It is not yet a well-understood regime, and speculative theories abound, including the concept of *superstrings*, which has received considerable attention. A superstring exists in a space of ten dimensions, with the extra dimensions beyond our own three dimensions of space and one of time invoked to achieve unification of fundamental forces. Just as electricity and magnetism were unified in the discovery of electromagnetic radiation in the nineteenth century, physicists dream today that the other fundamental forces of nature—namely, the weak and strong nuclear forces and gravitation—will all eventually be incorporated into a unified theory. The superstring provides a geometrical framework (stringlike in ten dimensions) that is capable, physicists hope, of encompassing all properties of elementary particles and, indeed, all of physics. Normally, the interaction force between two hypothetical point masses diverges when the two masses become arbitrarily close. Such a singular behavior is disastrous for any fundamental theory. Superstrings, being singularity-free, unlike point masses, provide the basis for a radical new theory of quantum gravity. Superstring theory is popularly described as the "theory of everything." Superstrings existed only during the first 10^{-43} second of our universe. Soon after, the six extra dimensions must have collapsed to invisibility, leaving us with our familiar four-dimensional space–time and the big bang. If during the initial, superstring phase the concept of time did not exist, neither did the universe itself.

Perhaps the perfect symmetry of the superstring was broken by some chance fluctuation, and our universe then began its headlong rush into the future, with all the laws of physics and constants of nature determined once and for an eternity to come. This at least is the dream of theoretical physicists in five continents who are feverishly joining the race to decode the mystery of the superstring. Einstein spent most of his life fruitlessly searching for the ultimate theory that would unify the fundamental forces in a mere five dimensions. We know now that we have to go to ten or, in another variant, twenty-six, dimensions to find a satisfactory theory. At present, the superstring theory corresponds poorly with the real world, and physicists pursue it for its mathematical beauty. If indeed "Truth is beauty, and beauty truth," then they may be on the right track.

INFLATION

If the result of quantum cosmology was to be the universe, why should the resulting universe resemble the one we have? To this seemingly intractable question, there has gradually developed, since 1980, an answer that promises to revolutionize our understanding of the very early universe. The story begins with Alan Guth, a young physicist in search of a permanent post who was spending a year at Stanford University, having been reluctantly initiated into cosmology by a former colleague from Cornell. Guth was appalled at the seemingly arbitrary concepts and explanations of cosmology. One day, in a flash of insight, he realized that a natural process which occurred as the universe evolved—a process somewhat analogous to the formation of ice as water cools below the freezing point—could explain several cosmological puzzles. Why is the universe so large? Why is it so uniform? So isotropic? When water freezes over a lake, energy is released so that the temperature of the lake water below the ice, as fish living there well know, remains at the freezing point. This latent heat, as physicists call it, stabilizes the water temperature and until the entire lake becomes a pack of solid ice, the temperature does not drop appreciably below the freezing point. A similar transition, Guth realized, also occurred between two phases of matter in the very early universe. The ensuing latent-heat release had a substantial effect on the expansion rate of the universe. The two phases corresponded, at very high temperature, to a state of symmetry and unification between the nuclear and electromagnetic forces. It requires energy to attain this ideal state, just as effort is needed to enter paradise: conversely, the loss of symmetry, or fall from grace, releases hidden energy. When the temperature dropped below a critical value, this symmetry disappeared, and a new, asymmetrical state of matter developed in which the nuclear force overwhelmed the electromagnetic force, just as it does today. The release of latent heat during this transition temporarily maintained the energy density of the universe, and for a brief period, light was able to traverse almost the entire universe. The light-travel distance grew exponentially large (Figure 6.3). Prior to this period of inflation, any homogenizing process at work in the universe could have acted over only a small scale—at most over the distance traversed by a light ray since the big bang—in today's units of length, this would correspond to about a centimeter. But inflation stretched this scale out to a distance of hundreds of billions of light-years, well beyond the observable edge of the universe. This was the magic trick by which the universe was able to erase all primordial imperfections, all irregularities and anisotropies, all memory of any gross defects or blemishes.

The epoch of inflation was short-lived. It set in when about 10^{-35} second had elapsed since the beginning of the expansion. At this time, the temperature had dropped sufficiently that there was no longer enough energy to maintain the perfect symmetry of the early phase. The inflation was over by about 10^{-33} second, by the time the transition to the new phase of broken symmetry had occurred. The elapsed time already was sufficient to create vast homogenized bubbles that contained ordinary matter and energy. The precise details of this transition are not fully understood. Indeed, Guth's original description of inflation was flawed: he produced bubbles that were far too small, which resulted in a highly inhomogenous universe. The essence of his idea was correct, however, and by 1984, Andrej Linde in the Soviet Union and Paul Steinhardt in the United States had independently developed successful models of inflation which allowed the formation of bubbles much larger than the present size of the observable universe. Not all problems have been solved, but the consensus among physicists is that inflation cures far more defects than it introduces. Consider the wide variety of possible initial conditions that might—and surely did, some-

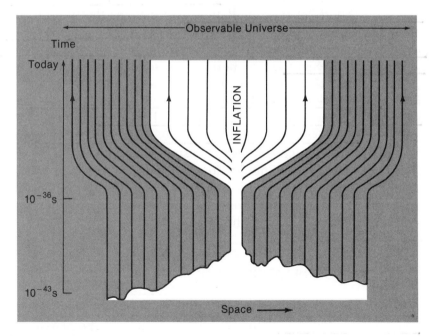

Figure 6.3 Inflation

A space-time diagram depicting inflation. Space is plotted horizontally and time vertically. The onset of inflation at about 10^{-36} second after the big bang stretches world lines; the universe we observe at present expanded from a region that once occupied a volume that was essentially zero, or more precisely was exponentially small.

where in space—emerge from the realm of quantum cosmology at 10^{-43} second: the universe could have been anisotropic, expanding much more rapidly in one direction than in another, and it might have been highly inhomogeneous, containing large density irregularities. As long as these irregularities were not great enough to induce premature collapse on horizon scales, inflation would result in a vast bubble containing our highly isotropic and homogeneous universe.

How probable was this sequence of events? There are two schools of thought. According to Steinhardt, the conditions that led to inflation should be regarded as a property of particle physics. One hopes that the ultimate theory of elementary particles will explain why the proton has a mass of precisely 1.7×10^{-24} gram. This same theory should also account for the force fields that determine how matter behaves at extreme energies and to the requisite degree of inflation. The alternative approach, from Linde, has been called "chaotic" inflation. It presumes that the physics of the beginning allowed a vast variety of particle fields, practically none of which could have induced inflation. But one infinitesimal patch of the universe, somewhere, possessed the right attribute for successful inflation. Once this patch inflated by a factor of 10^{30} or more, it had overwhelmed all other regions of the universe, and became our observable universe, and well beyond. Linde's argument is reminiscent of the anthropic cosmological principle (namely our existence selects the cosmological model), for which it provides the physical motivation. From a chaotic mix of primordial fields, there was only one force field, no matter how rare, which emerged triumphant and produced a universe within which galaxies and stars would have the time to form.

THE LEGACY OF INFLATION

Inflation not only yields the Friedmann universe, but boasts of other notable results. If the universe became too homogeneous, the galaxies could never form. The rapid-expansion phase boosts the ever present quantum fluctuations up to macroscopic scales. Quantum fluctuations are inevitable, for the simple reason that a particular quantum of energy can never be precisely localized: given the probability of locating something at some time, this inevitable uncertainty translates into energy fluctuations on the microscopic scale of the quantum itself. From these fluctuations, when inflation is over, emerge density fluctuations on the scale of galaxies. We shall see that from these occasional deviations from uniformity, at a level of only about one part in ten thousand, galaxies—indeed all large-scale structure in the universe today—originated. Without these small fluctuations, we would have no stars or planets, and with much larger fluctuations, the universe would have developed very differently. Thus, inflation sufficed to yield precisely those initial conditions necessary for our observed universe.

The inflated bubble of the big bang that we do observe must be almost precisely a flat space. This means that the gravitational potential energy within an arbitrary spherical region is just balanced by the kinetic energy of the expansion. The expansion continues forever, but only just. A slight excess of potential energy, corresponding in Einstein's language to a positive curvature of space, would mean that the universe must eventually recollapse. Inflation regulated the curvature of space: any preexisting curvature was stretched away.

To see why this consequence of inflation is philosophically appealing, suppose that the universe today were curved: as we go back in time, the deviation in density between this curved space and a flat space becomes smaller and smaller. The deceleration from the curvature, or equivalently, gravitational energy, becomes important only at a late epoch. In fact, in order for the universe to be significantly curved today, at the epoch of inflation it would have had to have been very precisely flat, to about 1 part in 10^{60}. Increase the curvature any more, and the universe might have prematurely collapsed. That would have been a disaster, because the solar system has required several billion years for its, and consequently our, evolution. Inflation explains the flatness—or equivalently, the age—of the universe.

Inflation has one other useful implication. The early phase transitions associated with symmetry breaking could have produced some undesirable objects, foremost among them the *magnetic mon-*

opole. This is a single unit of magnetic charge, whose existence is predicted by quantum field theory, the universally accepted theory of electromagnetic and weak nuclear reactions. However, because monopoles are hypothesized to be very massive, they would have been produced only at extremely high energies, the natural occasion for which was during the grand-unification phase, shortly after the Planck time. Once the symmetry broke, as the energy dropped, very few monopoles should have remained. Monopoles and antimonopoles would have annihilated, leaving behind the rare monopole that failed to find a partner. Monopoles are so heavy, however, that their relic abundance would be yet another cosmological disaster, for such ultraheavy particles would have overwhelmed ordinary matter. Observations show the contrary: searches have found that monopoles are few and far between in our galaxy. Inflation, once again coming to the rescue did away with virtually all the monopoles, leaving no embarrassing relics behind: at most one monopole might remain in all the observable universe.

STRINGS

Not all exotic fossils of the very early universe are undesirable. Monopoles can be visualized as topological knots, geometrically pointlike, that trap the vast energy density of the unification epoch. As the universe cools, the phase transition occurs, and droplets of the hot phase remain. Similar knots or defects can occur with very different topologies. The defects can be one-dimensional, linelike, objects, that are called _cosmic strings_. Two dimensional walls are also possible, but these, like monopoles, would have disastrous consequences for cosmology at later epochs. However strings are benign objects, and may even be desirable ingredients of the universe (Figure 6.4).

A cosmic string is attractive because it acts as a source of gravitational mass that seeds its surroundings. It is a condensation site, and may well provoke galaxy formation and galaxy clustering. Although inflation would remove any such strings along with monopoles, subsequent phase transitions could produce only stringlike, not pointlike, defects. Thus, we could have inflation and still have strings—albeit highly tangled ones under enormous tension. In this case nothing could happen until there is time for a string to disentangle on any particular scale. Disentangling happens only when the scale, for example corresponding to the size of a galaxy, first enters the horizon. Recall that the horizon, or the distance that

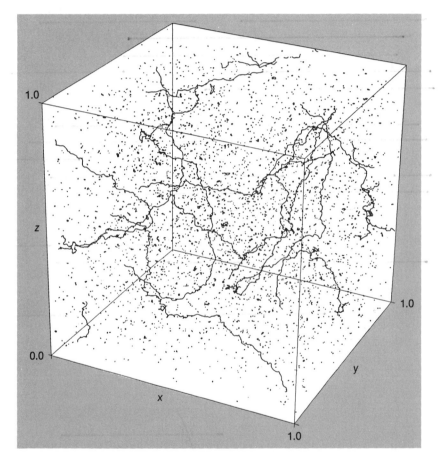

Figure 6.4 Strings

A computer simulation of a network of cosmic strings. These strings are topological defects in space; they have a very high mass, characteristic of the big bang at inflation, but are very thin. The mass per unit length of a cosmic string is about 10^{15} tons per centimeter. The strings intersect and form cosmic string loops, which are hundreds or thousands of light-years in diameter and have enough mass to accrete surrounding matter, acting as the sites of galaxy and galaxy-cluster formation. (Courtesy of D. Bennett and F. Bouchet.)

light has traveled, is growing more rapidly than the universe is expanding: larger and larger vistas of the universe therefore come into view. Within the horizon, our string reacts like a compressed spring, provoking density fluctuations as it straightens out. Strings twist and intersect, cutting themselves up into smaller pieces, or loops, that eventually decay by gravitational radiation. Still, the damage is already done: we now have fluctuations in the average density of matter in the universe. These fluctuations can extend up to very large scales indeed, to tens or hundreds of millions of light-years. For this reason, they provide a powerful source of large-scale structure; strings possess some unique characteristics.

Long strings move around nearly at the speed of light, and the density fluctuations they induce are much like supersonic wakes, trails left by high-flying jet airplanes, except that the wakes are sheets; the result is a universe with sheetlike condensations intersecting one another. The gravitational radiation produced by dying loops is the only remnant, the smile on the Cheshire Cat, of earlier generations of strings that may have seeded galaxies and galaxy structures. This radiation results in a cosmic background of long-wavelength gravitational waves, transient disturbances in the gravitational field that move around at light speed. The crests of the string-induced waves may be separated by up to tens of light-years, which has led to a rather dramatic prediction: pulsar "jitter." One of the best clocks we have is a distant pulsar, or spinning neutron star, which has been monitored over a period comparable to the passage of several of these waves. The first millisecond pulsar, which was discovered in 1982 and which rotates 600 times every second, is perceptibly slowing down. The slow-down rate, resulting from electromagnetic-radiation losses by the spinning, highly magnetized neutron star, is very steady, however, and the pulsar is keeping time to microsecond precision over a timespan of several years. This accuracy rivals that of the bank of atomic clocks at the Bureau de l'Heure in Paris, which sets the world's time standard. The passage of gravity waves causes a jitter in the pulsar that modulates its signal. In 1989, we are on the verge of detecting any residue of gravity waves that might, for example, have been produced from dying cosmic strings.

Other factors, such as perturbations of the earth's motion, could account for pulsar jitter. Simultaneously seeing the modulation in two or more millisecond pulsars would be a unique signature of a very long-wavelength gravitational wave. A grid of millisecond pulsars would provide a superb detector of any cosmological background of gravitational waves. Of course, these might not be associated unambiguously with strings in the early universe. However, one other characteristic of strings is absolutely unmistakable. A straight segment of a string does not curve nearby space, in the einsteinian sense that it acts like a localized mass. Whereas space near a string is flat, it has the topology of the surface of a cone, not a plane (spread a cone out, and you will recover a plane with a missing segment). Light rays that travel on either side of a string emerge on distorted paths, and an image, say, of a background star or galaxy, may appear doubled. The cosmic microwave background radiation should be discontinuous across a string. The effects are very small: the doubling is only a few arc-seconds, and the discontinuity in background temperature is one-thousandth of 1 percent, but these are unambiguous predictions. Only strings would produce linelike distortions: galaxy pairings and double images should appear along a line, and there should also be linelike temperature anisotropies in the cosmic background radiation. Although such signatures remain to be found, they should become experimentally verifiable within the next few years. Strings may yet make the transition from the twinkle in a cosmologist's eye to astrophysical reality.

THE PARTICLE ZOO

The universe need not be what it seems to be. More precisely, the distant galaxies that we see as shining beacons in space may be islands in a sea of more exotic matter than the ordinary baryons of which stars are made. The physics of the very early universe produced a plethora of weakly interacting particle species, many of which are candidates for the role of stable relics that may dominate the mass density of the universe today. The only necessary requirement is that of survival, for many of the strange new particles are very short-lived. In fact, every theory that postulates a new type of interaction and generates a family of new particles does have a survivor—a particle at the bottom of the ladder into which its more massive siblings have decayed.

One hypothetical stable, weakly interacting massive particle is the photino. According to a theory called *supersymmetry* (affectionately called SUSY by its disciples), all elementary particles should be paired, with each partner fitting one of two categories, *fermions* or *bosons*, recognized by quantum mechanics. The principle difference between fermions and bosons is spin: the quantum of light, the photon, is a boson of spin 1, whereas the electron, the proton, and the neutron are fermions of spin $\frac{1}{2}$, in units of Planck's constant. SUSY requires that for every boson, there should be a corresponding fermion, and vice versa. The *photino*, our candidate for the massive particle, is the hypothesized spin $\frac{1}{2}$ superpartner of the photon. Unlike the photon, the photino is expected to be very massive, anywhere between one and one hundred times the mass of a proton. Unlike the proton, however, which interacts strongly by virtue of the nuclear interaction, the photino interacts very weakly with ordinary matter. We would be hard-pressed to discover photinos even if the universe were full of them. Ordinary matter can be strongly compressed relative to the diffuse state of the intergalactic medium. The heat generated in the compression is converted into radiation, and compression continues until luminous stars form. With weakly interacting particles, however, the ability to convert heat into radiation is greatly reduced, so that photinos if they exist remain diffusely distributed, in the depths of intergalactic space, and are very rarely trapped inside a star. The photino's—or some similar massive particle's—lurking place is the halo of our galaxy and the spaces between the galaxies. There is abundance astronomical evidence for the existence of dark matter beyond the luminous confines of galaxies, and photinos are one of the candidate particles to make up dark matter.

How could we ever test such implications of the latest particle-physics theories about the structure of the universe? There are ways. First, the reality of a hypothetical particle such as the photino must be demonstrated by particle accelerator experiments at CERN, SLAC, Serpukhov, and elsewhere. The budgets for these experiments are becoming enormous, and billions of dollars will be required for the next step forward in accelerator power. When experiments are undertaken with the generation of machines presently being planned, they will likely confirm or definitively reject the existence of particles such as the photino. Second, direct astronomical measurements can assess the distribution of specific particles such as the photino. The reason such invisible particles can occasionally become visible is that very rarely, a photino collides with another photino and annihilates. Like the meeting of matter and antimatter, the collision of a pair of photinos produces a burst of energy comparable to the explosion that occurs when an ordinary atom is converted into pure energy. We know precisely how frequently this annihilation process occurs, because once, long ago in the very early universe, it occurred very frequently. In fact, only after the temperature and density dropped and annihilation finally ceased were we finally left with the photinos that remain today. In fact, we can completely specify the probability of a photino's undergoing annihilation once we know how many of these weakly interacting particles are left over, distributed too sparsely to meet partners and annihilate. Although annihilation once occurred predominantly at very high density, today, when the density is very low, it still occurs, but very rarely indeed. When photinos become concentrated in the dark halo of a galaxy, the probability of annihilation once more increases. It is still low: the photinos are safe. But the occasional, rare annihilation leads to potentially observable byproducts. If photinos exist, and this is a very big if, they should be detectable by astronomers.

These byproducts may include energetic antiprotons. We observe high-energy particles that move near light speed as cosmic rays that impact the earth. Very few of these are antiprotons, fewer than one in ten thousand. An annihilation can produce a proton–antiproton pair, created out of pure energy, if the available energy is large enough. The resulting antiprotons are trapped, along with all other cosmic rays in the interstellar medium. Now, it is exceedingly difficult to produce cosmic ray antiprotons. Indeed, the only generally accepted mechanism produces very high-energy antiprotons through interaction with ambient heavy atoms in the interstellar gas by high-energy cosmic rays. We might hope to find a unique contribution at lower energies from the photino annihilations. Another byproduct of the annihilations is the production of gamma rays: very energetic photons that can penetrate the interstellar medium and that are visible throughout the entire galaxy. Detection of diffuse cosmic gamma rays provides another means of searching for exotic relics from the big bang.

One tantalizing possibility for detecting dark matter has arisen in the core of our own sun. As the sun orbits around the galaxy, it passes through the dark matter of the halo. Occasional dark-matter particles are then trapped inside the sun, and they build up gradually into an appreciable concentration. In Chapter 8, we will describe how this may lead to observable consequences for neutrino detection.

GRAVITONS

The gravitational analog of a photon is a *graviton*, or quantum of gravitational radiation. One would expect a sufficiently chaotic early universe to contain copious numbers of gravitons, which are produced by rapidly changing gravitational fields. However, there should now be a significant background of short-wavelength cosmological gravitons if our hypothesis that the temperature rises without limit as the singularity is approached indeed describes the physics of the first millisecond. In this scenario the high temperature and radiation density enable the gravitons to become intimately coupled with the radiation and approach equilibrium. Just as with the photons (but at a much earlier era because of the exceedingly weak interaction of gravitons), the gravitons attain a characteristic energy distribution. We are uncertain about the nature of quantum gravity and graviton interactions, and this need not be a blackbody distribution. As the density drops, the gravitons come out of equilibrium and decouple from the matter. They subsequently expand freely, decaying in energy until the present era, when the typical energy of a primordial graviton would correspond to a temperature of somewhat less than 1 degree kelvin or to a wavelength of about 1 millimeter.

It is also possible, according to a more radical viewpoint, that more and more particle states become accessible at very high energy. In this case, the temperature and radiation density of the gravitons would remain finite, and the gravitons would never achieve thermal equilibrium with the matter: essentially no short-wavelength cosmological graviton background would be produced. Thus, we have, in principle, a remarkable means of probing physical conditions in the very early universe.

Unfortunately, the cosmological gravitons are probably too low in energy to be detectable. There are two types of gravitational-radiation detectors. Interferometers are designed to monitor slight perturbations to orthogonal beams of light, reflected many times and hence equivalent to very long arms of a gravity-wave antenna. Mechanical detectors consist of carefully suspended cooled aluminum bars, which are monitored for infinitesimal oscillations in the gravitational field as a pulse of gravitons passes through the bar. The goal of existing detectors, hitherto unaccomplished because of lack of sufficient sensitivity, is to detect gravitational-radiation pulses produced as stars collapse into black holes. The predicted pulses from such events are vastly more intense than any feeble relic graviton background from the very early universe. The only hope for performing a cosmological measurement with a graviton detector would be if prolific fluxes of gravitons were produced in chaotic events involving gravitational collapse during the later stages of the evolution of the universe. Because of the extreme symmetry of these later stages, which we infer from studies of the isotropy of the cosmic blackbody radiation, this possibility seems unlikely, but we cannot totally dismiss it. One can conclude from this isotropy that the energy density today in horizon-size gravitational waves is at most a tiny fraction, about 10^{-4}, of the closure density of the universe. Such waves would correspond to a distribution of gravitons of the Planck epoch that would be produced by a very mild degree of chaos, roughly that consistent with the density fluctuations that subsequently gave rise to galaxies.

PRIMORDIAL QUARKS

The events of the first millisecond of the universe need not remain entirely in the domain of speculation. Physicists have actively sought observable evidence for at least two consequences predicted by the big bang theory. The most promising of these is the search for *primordial quarks*. A quark is a fundamental particle whose existence has been postulated by nuclear physicists to help explain the proliferation of different types of particles discovered in recent years with the giant particle accelerators. The distinguishing feature of quarks is a fractional charge (either one-third or two-thirds of the charge of an electron or proton), and quarks may be the basic building blocks from which all the hadrons are constituted. Although quarks have not been isolated, strong evidence for their existence has been forthcoming from high energy accelerator experiments. The existence of quarks remains an important hypothesis of the theory of elementary particles.

The early universe provided a unique environment where quarks and antiquarks were abundant. When the temperature exceeded about 1 billion electron volts (or 10^{13} K), quark–antiquark pairs were created and annihilated in vast numbers. After the temperature fell, as the universe expanded and cooled, new quark–antiquark pairs were no longer created, and the quarks and antiquarks almost completely annihilated. Whether any trace of them remains depends on the detailed (and uncertain) nature of the interaction between quarks and other particles. If the strong attractive force between quarks vanishes at sufficiently large separations, an exceedingly small number of quarks will remain; their density will have dropped down so far that they are subsequently unable to annihilate. The residual abundance of quarks (and antiquarks) amounts at most to only one quark for every billion hydrogen atoms. But this ratio remains almost constant as the universe evolves. An individual quark can be destroyed only by a collision with an antiquark, and the time necessary for all such collisions to occur is estimated to be longer than the present age of the universe. The predicted quark abundance in the universe may be as large as that of gold.

Experiments have shown that the actual quark abundance in terrestrial rocks and seawater is far less than the predicted upper limit. The fractional charge of quarks greatly aids the attempts to distinguish them from ordinary matter, and very sensitive results have been obtained. Oysters process large volumes of seawater and are excellent repositories for any rare cosmological relic with anomalous charge or mass. There appears to be at most one quark for every 10^{20} atoms of hydrogen in seawater. One experiment has reported detection of quarks in a rare metal; it seems conceivable that very heavy materials may retain free quarks more easily than lighter materials. Although this result has not yet been confirmed, the possible survival of free quarks remains an issue capable of experimental resolution.

Where this leaves us is not presently very clear. If quarks exist and if their separation from antiquarks enables them to avoid annihilation, the fact that significant numbers have not been detected may imply that the early universe did not experience temperatures as high as 1 billion electron volts or 10^{13} K. This would have profound implications for cosmology, and we might have to seek a theory that avoided a very hot big bang. Alternatively, we might be compelled to constrain our ideas about the nature of quarks, for the hypothesis of a hot big bang seems as reasonable as the assumptions about quarks. Indeed, our current theories of elementary particles have explained so much that physicists would be very reluctant indeed to release quarks from their confinement inside nuclei. It seems valid to conclude that the effect of the strong nuclear forces must have greatly, but not totally, diminished the chances of survival of primordial quarks as free particles.

MINI–BLACK HOLES

A second consequence of the earliest phase of the big bang may be the production of *mini–black holes.* To understand how they arise, we must consider a powerful argument often applied to the early universe.

Because there is essentially no direct and unambiguous experimental consequence of our assumptions about the first seconds in the big bang, we may question the model of a simple, uniform, and isotropic big bang. Surely, the metaphysical conjecture continues, a highly irregular and chaotic beginning seems the most likely of the infinite set of possible models of the early universe. The one constraint, of course, is that such models must eventually decay to a uniform state of expansion in order to provide an adequate description of the presently observed universe. This postulate of *primordial chaos* can be framed quantitatively. One such model, originally formulated by Charles Misner, is known as the *mixmaster model*. According to this model, the early universe underwent alternating cycles of simultaneous expansion in two directions and of contraction in a perpendicular direction. (Figure 6.5).

One consequence of a highly chaotic and inhomogeneous model of the universe is that when the amplitude of an irregularity exceeds a critical value, collapse of all the matter in the local region

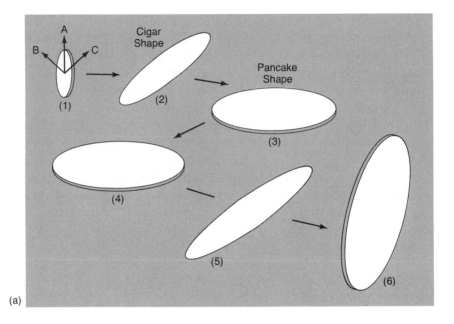

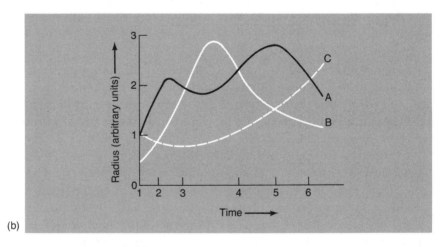

Figure 6.5 The Mixmaster Model

The chaotic mixmaster model describes the most complex behavior of any cosmological model. The universe is pictured as a ball of fluid that expands with time, evolving through a series of randomly oriented cigar-shaped and pancake-shaped configurations (a). As the universe expands in one direction, it contracts in a perpendicular direction. Cycles of expansion and contraction alternate in each direction. The variation in radii of three orthogonal axes, denoted by A, B, and C, shows the transition from pancake shapes to cigar shapes as the volume increases with time (b); six time sequences are shown.

of space will ensue. This collapse will not be halted until a black hole forms—that is, until the gravitational field of the collapsing region becomes so intense that light is trapped within it (Figure 6.6). If the sun were ever to collapse and form a black hole, it would have to shrink almost a millionfold from its present radius of 1 million kilometers. However, the masses of any cosmological black holes formed less than 10^{-3} second after the big bang would be much less than the mass of the sun. The mass of a primordial black hole is determined by the size of the observable universe at the time of formation. If the collapsing region were any larger than this, something very drastic would happen: no particles or light rays could have escaped from it or traveled to it. This means that the collapsing region would be effectively disconnected from the rest of the universe. Eventually, it would have formed a "separate universe." Thus, we consider only black holes of mass comparable to the mass within the horizon. The masses of the first black holes, formed at the Planck instant, are about one millionth of a gram. Larger black holes formed later.

Stephen Hawking has suggested that the smallest black holes may evaporate. The evaporation works in much the same way as particle annihilation in the earliest moments of the universe. The immense forces near the surface of a black hole can disrupt the virtual particle pairs of space-time and produce a shower of complex nuclear particles. These black holes would die in an intense, catastrophic burst of high-energy gamma radiation.

The smaller the black hole, the more rapidly it will tend to self-destruct. Only the largest primordial black holes remain—those with mass in excess of 10 billion tons, roughly the mass of a small mountain. The size of such a hole is only a hundred-billionth of a millimeter, making it difficult to detect directly. Such black holes could be very common, however, and satellite experiments have even been proposed to trap such objects and use them as an exotic energy source. Perhaps a more realistic experiment would be to try to detect the evaporation of slightly smaller primordial black holes, those that are just now on the threshold of final disruption. Satellites have detected many cosmic gamma-ray bursts for which no satisfactory explanation has been provided. It is conceivable that dying primordial black holes could be the sources of these bursts, according to some theorists.

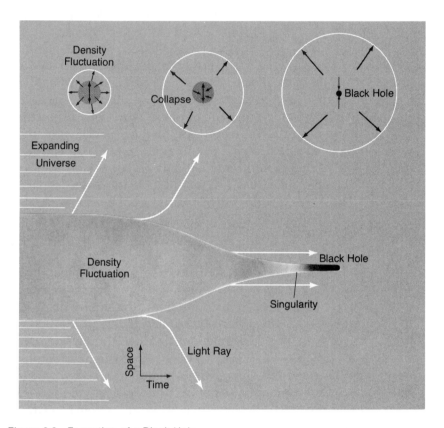

Figure 6.6 Formation of a Black Hole

A large density fluctuation in the very early universe could have collapsed to form a black hole (*top*). We can also depict this in a space-time diagram (*bottom*) in which two of the three space dimensions have been suppressed. The horizontal dimension is time. When the matter collapses to a critical radius, light can no longer escape but remains trapped (*arrows denote light rays*), hovering at that radius forming the boundary of the black hole.

MATTER AND ANTIMATTER

One of the early accomplishments of quantum theory was the prediction of the existence of *antiparticles*. Thus, the positron was discovered to be the counterpart to the electron, and the antiproton was discovered to be the counterpart to the proton. The earth is predominantly constituted of matter rather than antimatter. Antiparticles that are produced in cosmic-ray showers or particle accelerators do not survive very long in the terrestrial environment. As soon as they slow down, their inevitable fate is rapid annihilation with a corresponding particle.

One of the great puzzles of cosmology is whether matter or antimatter particles are preponderant in the universe. Is our terrestrial laboratory typical of the rest of the universe? Within our galaxy, we can be confident that there are no stars of antimatter; otherwise, the pervasive interstellar medium would instigate annihilation and ensuing gamma-ray emission at a rate far in excess of that observed. But it may be quite another matter in the vast depths of intergalactic space. Indeed, there could be entire antigalaxies consisting of antistars made up of antimatter. These systems would be indistinguishable in appearance from ordinary galaxies. If such objects exist, however, they must, we now argue, be extremely rare.

One difficulty with the idea of antigalaxies lies in maintaining their separation from galaxies. Empty space may now separate them, but in the early universe, these regions must have been in relatively close contact. Annihilation seems difficult to avoid, particularly because we now know that many regions of intergalactic space are occupied by a tenuous gas. Interaction with the gas would make annihilation inevitable in antimatter regions, with the consequent emission of observable gamma radiation (Figure 6.7). Thus, we infer that the universe is likely to be fundamentally asymmetric, with a large excess of matter over antimatter.

The required excess is now very large, in accord with limits imposed by gamma-ray observations, but in the very early universe, the amounts of matter and antimatter must have been almost equal. The reason is that at very early eras, the radiation field copiously created pairs of particles and antiparticles. Embedded in this hot environment were occasional protons and electrons. However,

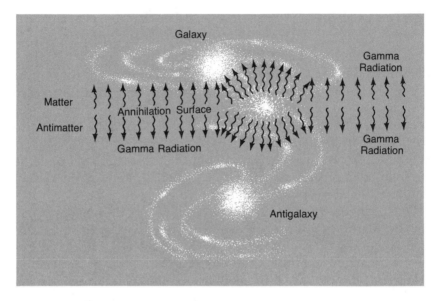

Figure 6.7 Galaxies and Antigalaxies

When matter and antimatter are in proximity, annihilation must occur. This annihilation results in intense emission of gamma radiation, which can be detected by space telescopes. The lack of much observed gamma radiation indicates that only a small fraction of the observable universe can be made of antimatter. It is conceivable, but not likely, that some unknown process inhibits annihilation by separating galaxies from the hypothetical antigalaxies.

there was only 1 excess proton for every 100 million photons and particle pairs. The symmetrical content of the universe (the particle pairs and photons) overwhelmed the asymmetrical part (the surplus protons and electrons) by a factor of 100 million. The photons were sufficiently energetic that their associated mass density was the predominant form of matter (recall that $E = mc^2$). Yet, as the radiation cooled and the pairs annihilated, the excess matter eventually became the predominant constituent as the energy of each photon decreased. The energy of a typical photon of the cosmic blackbody background radiation is now only one-thousandth of an electron volt. As a consequence, the atoms now provide the bulk of the mass density.

Once, seconds after the singularity, the universe was highly symmetric, It is very important that the universe was not completely symmetric, however—our existence depends on it! Yet the lesson we learn from particle physics is that the universe began in a symmetric state. Elementary particles come into existence when the symmetry is broken, as the universe cools down. Imagine steam condensing into droplets of water: the droplets are the elementary particles that have suddenly come into existence when thermodynamic conditions permit. But why matter rather than antimatter? A better analogy is with a ferromagnet. When cooled below a critical temperature, magnetism develops spontaneously, with the direction of the field (analogous to the sign of matter) chosen randomly in different domains. Recent developments in elementary particle physics have indicated a possible clue to the origin of matter. According to a new and speculative theory that attempts to unify the strong nuclear interaction with both the weak and electromagnetic interactions, supermassive particles, formed soon after the Planck time, decay asymmetrically, producing slightly more particles than antiparticles.

How reasonable is such a primordial asymmetry? The theory is entirely plausible, because any unifying theory can only treat hadrons and leptons on an equal footing at very high energy. This symmetry between heavy and light particles is only natural at sufficiently high temperature; it requires energy to convert one species into another. The temperature declines, the symmetry breaks, and we are left with occasional rare hadrons that can make only a one-way voyage: decay without ensuing creation. The decays release some symmetric constituents, with equal parts of matter and antimatter, but there is an imbalance: a slight preponderance of matter over antimatter. In true equilibrium, this would be of no consequence, because for every decay a corresponding particle would be created. But the fall from grace is inevitable as the temperature drops and decay predominates over creation. The debris annihilated, but the residue of excess matter (and possibly some regions of antimatter) left behind now constitutes all the matter in the observable universe.

There is one consequence which has led to an observable effect. The initial state of symmetry means that at high energy, hadrons converted into leptons, and vice versa. Today, in the low-energy universe, this process still occurs, but exceedingly slowly. This has led to the following conjecture: protons should decay into leptons over a time-scale that can be computed for specific models of the grand unification interactions to be in excess of 10^{30} years. Despite the immensity of this time-scale, experiments have been run to search for proton decay. Take a sufficiently large number of protons—say 10^{32} as the number in a hundred tons of water—and decays could be a weekly or even a daily occurrence. The experiments have been established deep underground in mines to avoid cosmic-ray–induced backgrounds, and search for rare flashes of light triggered by the energetic by-products of the decay. Water detectors in a gold mine in India and in a salt mine in Cleveland have failed to detect any proton decays, but require the proton lifetime to exceed 10^{32} years. This is a comfortingly long time-scale for cosmologists not to be unduly concerned, but it is perfectly compatible with grand unification theories that allow an explanation of the matter-antimatter asymmetry as a consequence of asymmetric decays during the first 10^{-36} second or so of the universe.

Whether these decays resulted in a preponderance of matter or of antimatter is a somewhat arbitrary result. We call our surroundings matter, but it is conceivable that there are vast domains of antimatter and other domains of matter. All we can say is that observationally there is very little evidence for more than the merest trace of antimatter around us. The ultimate proof that matter and antimatter once existed in comparable amounts is that the annihilated material transformed itself into the background radiation and eventually cooled to its present temperature of 3 K. Our existence testifies to a slight (1 part in 100 million) excess of matter. In the chapters that follow, we shall trace the evolution of this excess matter from the big bang that survived annihilation and now constitutes the universe we observe.

· 7 ·

THERMONUCLEAR DETONATION
OF THE UNIVERSE

If I had been present at the creation, I would have given
some hints for the better arrangement of the Universe.

—Alfonso the Wise

After the first millisecond, the pairs of hadrons, or heavy
elementary particles that feel only the strong nuclear interactions,
had practically all annihilated. The *hadronic era*, dominated by
strong interactions (the forces that help to bind the nucleus together), was over; subsequently, the *weak nuclear interactions*
came into play. The weak interactions determined the decay of
free neutrons that remained from the hadronic era into electrons
and protons (as well as certain other processes involving radioactive decays). A crucial trace amount of neutrons was left from the
previous era, because there were no corresponding antiparticles
for them to annihilate with. Weak interactions also involve ghostly
particles called *neutrinos* and *antineutrinos*, which are generally
considered to possess no mass, yet are characterized by their spin
and energy. These particles, along with electrons and positrons,
are known as *leptons* (light particles), and we can speak of the
leptonic era as commencing after the hadronic era drew to a close.
During the leptonic era, the universe consisted of a mixture of
photons, nucleons, neutrinos and antineutrinos, and, for a brief
initial period, electron-positron pairs.

NEUTRONS

One of the most important weak interactions is the merger of a proton and an electron to yield a neutron and an antineutrino.[9] Neutrons formed prolifically by this reaction and attained an abundance comparable to that of protons. However, a large number of electrons are required for neutron formation to proceed, and this favorable situation rather abruptly changed when the universe was about one second old. At this stage, the temperature dropped below 1 million electron volts, or 10 billion degrees kelvin. It requires an energy of about 1 million electron volts to create a new electron-positron pair. Creation ceased, and the electron-positron pairs annihilated. The neutron-creating reactions stopped, but a considerable number of neutrons remained. There was in fact just one neutron for every six protons. This ratio of one neutron to six protons depends largely on the mass difference between the neutron and the proton, and it relies on the competition between two rates: the weak nuclear interactions and the expansion. Applying Einstein's famous $E = mc^2$ relation, we can think of this mass difference as an energy. The neutron is more massive than the proton by about 1.3 million electron volts. When the temperature of the universe, of the energy of any particle, drops below this typical energy, reactions involving the heavier neutrons are suppressed and the neutron-proton ratio is frozen out: neutrons are neither produced by weak interactions nor destroyed, except by decay of free neutrons.

The neutrons were the vital component in the next sequence of events. Free neutrons are unstable particles. A free neutron will spontaneously disintegrate in about eleven minutes. The role of neutrons is also essential in thermonuclear fission and fusion. The combining of neutrons with nuclei results in more massive and stable nuclei. Because of the strong nuclear bonding forces between neutrons and protons, there is a very small reduction in mass of the new nucleus relative to that of its constituent parts. The energy that is bound up in holding the nucleus together reveals itself as a mass deficiency and will be released when the nucleus is synthesized. In this way, an enormous amount of energy can be released. (Similarly, in a fusion bomb, hydrogen is converted to helium, and the tiny difference in mass between one helium nucleus and four protons provides the source of energy.) We describe this process, whereby heavier elements are synthesized from lighter elements, often accompanied by release of energy, as nucleosynthesis.[9]

NUCLEOSYNTHESIS

The early universe behaved much like a fusion bomb, although then no atomic-trigger analogue of the atom bomb was needed to produce a high enough temperature for fusion to occur. At first, the neutrons remained free; the radiation field was so intense that it instantaneously destroyed the first nucleosynthetic products that might have formed. Only after about a minute, when the temperature had fallen to a billion degrees kelvin, did the reactions commence. First, a neutron captured a proton to make a nucleus of deuterium (heavy hydrogen). Deuterium readily absorbs neutrons. Heavy water (HDO), in which one of the hydrogen atoms in H_2O is replaced by a deuterium atom, is commonly used in nuclear reactors because of its high absorptivity for neutrons. The deuterium next captured another neutron to make *tritium*; finally, the tritium reacted with a proton to make helium. Almost all the neutrons ended up in helium nuclei. Since each helium nucleus contains two neutrons and two protons, the result was one helium nucleus for every ten hydrogen nuclei (Figure 7.1). The big bang had created helium. The predicted ratio of helium to hydrogen can be changed, but only by severely modifying the big bang theory. We shall refer to this uniform and isotropic big bang as the standard model.

There are radical alternatives to the standard model. One alternative postulates enormous amounts of turbulence present during the *nucleosynthetic era.* Another alternative postulates that the gravitational force was much larger in the past. In either case, the effect would have been to accelerate the rate of expansion of the early universe and not to allow enough time for the neutrons to react effectively before they decayed. Thus, essentially no helium would have been produced. A more modest acceleration of the expansion would actually have increased the production efficiency

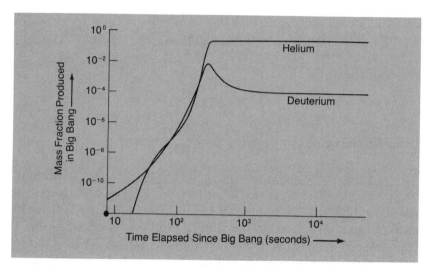

Figure 7.1 Nucleosynthesis in a Closed Universe

The gradual synthesis of deuterium and helium in the early universe is illustrated in this graph. Nuclear reactions effectively started after 1 minute had elapsed. Deuterium attained a peak abundance of about 1 percent by mass, relative to hydrogen, then was reduced to a very low value as helium was synthesized. The helium abundance built up to roughly 25 percent by mass.

at which the neutrons initially formed, thereby raising the predicted helium abundance.

HELIUM ABUNDANCE

Astronomical observations of the present helium abundance permit a surprisingly strong constraint on conditions in the early universe at an age of only one minute. Helium has been found throughout our galaxy and in many nearby galaxies. The regions where helium is seen include the atmospheres of old stars and nebulae of ionized gas surrounding young stars. It is also a constituent of the energetic particles called cosmic rays that pervade interstellar space. Helium has even been detected in the luminous and very distant objects known as *quasars*. Other elements are found to be strongly variable in abundance in different sources, as we might expect from the widely differing evolutionary histories of these regions. Yet, in all cases, the evidence strongly suggests that there is everywhere one helium nucleus for every ten hydrogen nuclei, neither significantly more nor less. (Actually a slight variation is found. The more chemically evolved, or more-metal-rich, gas clouds have a slightly higher helium abundance, because of the contribution from stellar nucleosynthesis. Ideally, one would like the primordial or pregalactic helium abundance, and this is found in the most primitive, or most-metal-poor, galaxies.) When confronted with such strong evidence for a universal abundance of helium, the hypothesis of a primordial origin in the big bang becomes compelling.

The helium abundance attained in the big bang is relatively insensitive to small changes in the standard model. There is little difference between the amount of helium produced in an open model (one that will expand forever) and that produced in a closed model (one destined eventually to recollapse). At first sight, this result is rather surprising. The nucleosynthetic era is primarily determined by the value of the temperature, which has to be about 1 billion degrees kelvin in either cosmological model when nucleosynthesis occurred. Consequently, the principal difference between these two models is that, in an open universe, the density during the nucleosynthetic era is much lower than in a closed universe. However, because neutrons are consumed with great efficiency in combining with protons to form helium nuclei, lowering the density by one or two orders of magnitude merely lowers the reaction rate but does not significantly affect the final abundance of helium. The predicted helium abundance is reduced by only 1 or 2 per cent.

DEUTERIUM

Although the production of helium is similar for open and closed big bang models, it turns out that an important byproduct of the reactions, deuterium, is extremely sensitive to density (Figure 7.2). In both the open and closed models, large amounts of deuterium were produced but were then destroyed by collisions with protons. Because the density was lower during the nucleosynthetic era for an open model than for a closed model, the amount of collisional destruction of deuterium is reduced in the open model. Conse-

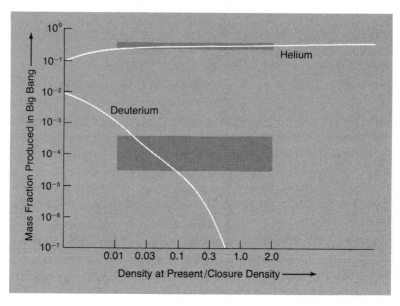

Figure 7.2   Dependence of Nucleosynthesis on the Mean Density of Matter in the Universe

The amount of helium produced in the big bang is relatively insensitive to the present density of the universe, or to whether the universe is open or closed. However, the deuterium abundance is extremely sensitive to this parameter. The boxes indicate the range of abundances that is consistent with observations. An open (low-density) model produces ample deuterium to explain the observed abundance, but a standard closed model (density greater than about 5×10^{-30} grams per cubic centimeter) produces essentially no primordial deuterium.

quently, the resulting deuterium abundance can be greatly increased in an open model, relative to the amount predicted in a closed model.

Deuterium is not a very abundant isotope. About 1 deuterium atom is found for every 30,000 hydrogen atoms (Figure 7.3). This low abundance makes deuterium sensitive to a small change in the much larger abundance of helium. Unlike most other heavy atoms, however, it cannot be made in ordinary stars. At the high temperatures in the center of the sun, the relatively fragile deuterium isotope would be entirely destroyed. Thus, all the deuterium we observe in the galaxy was probably synthesized in the first minutes of the big bang. Because essentially no deuterium is produced in

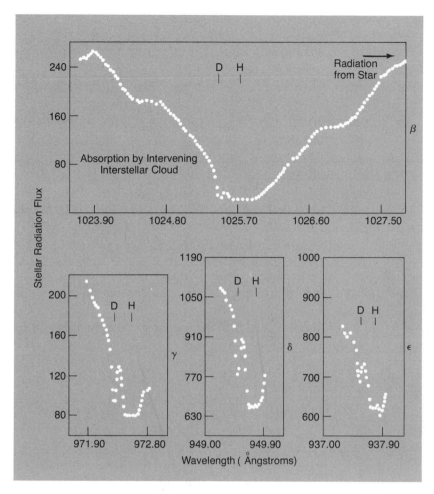

Figure 7.3 Observations of Deuterium

Interstellar absorption-line profiles in the ultraviolet spectrum of a bright star, obtained with the *Copernicus* Orbiting Astronomical Observatory space satellite, show the presence of narrow deuterium absorption features. An intervening cold interstellar gas cloud absorbs light from the star, and the broad lines are due to the far more abundant hydrogen atoms. The lines shown correspond to absorption from the ground state of the hydrogen (and deuterium) atoms to the second (Lyman beta), third (Lyman gamma), fourth (Lyman delta), and fifth (Lyman epsilon) excited states. The Lyman alpha transition to the first excited state due to hydrogen results in such a broad feature that the deuterium is masked. Each of these transitions removes radiation from the incident starlight at a specific wavelength, which corresponds to the energy of the excited electron. The result is a series of dark lines superimposed on an otherwise bright spectrum. The illustrations show tracings of the spectrum, represented as the amount of light measured at any particular wavelength. A substantial amount of deuterium must be present in the interstellar gas, amounting to 1 deuterium atom for approximately every 3 $\times$ 10^4 hydrogen atoms.

the standard closed model, we might conclude that the universe is open.

We see one possible loophole in this argument when we consider a nonstandard variation on the big bang. During the transition from a quark-dominated to a hadron-dominated universe, which occurred when the temperature was about 200 million electron volts, large bubbles of the new hadron phase may have formed before the transition was completed. The neutrons in these regions of ordinary hadronic matter tend to diffuse more rapidly than do the charged particles. Hence, there are pockets of neutron-poor and also neutron-rich material. Variations in the neutron abundance from place to place can modify the ensuing deuterium and helium synthesis, and a dense universe can mimic a low-density universe. Another possibility is that a local speedup in the expansion rate during the nucleosynthetic era over an area comparable in size to the observable universe at that time would inhibit destruction of deuterium by collisions with protons, because there would not be enough time for many collisions to occur. Large amounts of deuterium would accordingly survive. Even if this happened only in a small fraction of the universe, the deuterium produced, after mixing with the unperturbed regions, would suffice to explain what we observe. Any excess helium that might result would be suitably diluted by the bulk of the matter in the universe to yield its standard abundance.

These possibilities are not appealing. Indeed, they generally result in the production of excessive lithium, another trace element produced in very small amounts (only 1 part in a billion), which also is generally destroyed rather than produced in ordinary stars. The standard big bang succeeds remarkably well in simultaneously accounting for the abundance of helium, deuterium, the isotope of helium of mass 3, and the lithium isotope of mass 7. In fact it has been argued that one can use this success to constrain unknown parameters in the standard model, such as the number of neutrino species. The presence of additional neutrino types helps to speed up the expansion and to make more helium: for each additional species, about 1 per cent more helium is produced. With astronomers' estimates of the primordial helium abundance ranging from 22 to 24 per cent, relative to amounts of hydrogen, little room remains to add neutrino species to the three species known to exist.

Experiments underway at SLAC and Fermilab can directly measure the total number of neutrino species, because these contribute to decay channels for the Z boson, a newly discovered unstable massive particle weighing about 90 proton masses which helps mediate weak interactions. Current limits already come close to the value allowed by the standard big bang model.

It is important to realize that big bang nucleosynthesis only constrains the baryonic content (that is, for our purpose, the protons) of the universe. The universe need not be open if weakly interacting nonbaryonic particles provide the requisite contribution to the density. However the baryonic density is inferred to be about one-tenth of the critical value for closure in order for the simple model to account for the origin of the light elements.

Elements heavier than helium are not produced in any significant abundance in the big bang, except for certain isotopes of lithium and boron. Nature has arranged matters so that there are no stable elements of atomic masses 5 or 8. Thus, the process of neutron capture, which is a step-by-step process, must break down at this hurdle. A helium nucleus cannot capture a proton or another helium nucleus and form a new stable nucleus. The only other way to overcome this hurdle is to use a different nuclear process. The principal alternative involves the capture of helium nuclei by two other helium nuclei. This process requires a physical condition (a high enough density for a sufficiently long time) that is not attained in the big bang theory. By the time helium had been produced, the density was too low and the universe was thinning out and cooling too rapidly to provide the time for any further nucleosynthesis to occur. Astronomers believe that elements heavier than helium have been created in the cores of evolved stars or in supernova explosions at relatively recent eras. We shall pursue this topic in Chapter 15.

· 8 ·

THE PRIMEVAL FIREBALL
EMERGES

This is not the end.
It is not even the beginning of the end.
But it is, perhaps, the end of the beginning.
—*Winston Churchill*

A few minutes after the big bang, the nuclear fireworks ceased. The expansion of the universe continued rather uneventfully afterward for better than 300,000 years. We call this period the *radiation era,* and it marks the emergence of the primeval fireball.

How do we know that this hot, dense soup of primordial matter and radiation evolved in a quiescent and regular manner? As we saw in Chapter 7, the uniformity and isotropy of the cosmic expansion is theoretically constrained by our direct observations of the ubiquitous and more or less invariant helium abundance throughout the present universe and by observations of a much smaller but still significant amount of deuterium. Thus, the expansion could not have been too dissimilar from the standard model.

But we would like to have direct evidence from this early era of the universe. Of course, we cannot directly observe the primeval fireball, and in fact direct observation would have been impossible even by a hypothetical human observer, for the universe did not become transparent until after 300,000 years. Direct observation of the early universe could not be feasible until the density and temperature had fallen to the point at which matter could form and radiation could propagate freely. Before 300,000 years had elapsed, observing the early universe would have been like trying to peer into a dense fog. We do, however, have a means of partially penetrating this fog, but before we describe it, we first turn to an important and pervasive constitutent of the early universe that has never been directly detected.

THE CASE OF THE ELUSIVE NEUTRINOS

Once the electron-positron pairs had annihilated, when the universe was about one second old, radiation became the dominant constituent of the universe. At first, gamma rays were abundant. These photons with intense penetrating power are ordinarily produced only in nuclear explosions and by radioactive decay of unstable nuclei. Neutrinos and antineutrinos were also present. These elementary particles have zero rest mass but are characterized by energy and spin. Neutrinos have an extremely weak interaction with matter and are consequently very difficult to detect. The core of the sun is believed to be a prolific source of energetic neutrinos, according to the theory that nuclear reactions provide the source of solar energy.

Neutrinos are the only direct link we have to the fiery core of the sun. Because they interact so weakly with matter, we can effectively peer into the innards of the sun by studying neutrinos. Detection of solar neutrinos therefore would provide the ultimate verification that the sun is a gigantic thermonuclear fusion reactor. However, attempts to measure a sufficient number of solar neutrinos have thus far been unsuccessful.

To detect neutrinos produced by the nuclear reactions that power the sun, scientists have utilized an enormous neutrino "telescope," which consists of a tank of cleaning fluid surrounded by a jacket of water deep underground inside a gold mine. A nuclear reaction occurs very rarely, when a neutrino passes through the cleaning fluid; this reaction enables the experimenters to infer the neutrino's passage through the fluid. The method of detection is based on a reaction between an isotope of chlorine and a neutrino, which produces a nucleus of a rare radioactive argon isotope and an electron. Despite the large flux of neutrinos that we believe are emitted by the sun, very few are absorbed on their passage through the detector. Although the reaction is infrequent, the few times per month that it is expected to occur can be verified by means of a very sensitive technique for detecting tiny traces of argon. The mine was chosen as a location so that the neutrinos from the sun would be the only possible source of argon. Cosmic rays, for example, were a contaminant that had to be excluded. The experiments to detect neutrinos appear recently to have met with some success; however, relatively few neutrinos were detected as compared with the predicted value. Only about one-third of the flux of neutrinos predicted by the standard model of the solar core was observed.

Some bizarre theories have been developed to account for the discrepancy. The readjustment time for any temperature change in the solar core is about 10 million years, whereas neutrinos propagate to us in about 6 minutes. Perhaps the solar core is now slightly (by a few percent) colder than our model predicts; no change in the solar luminosity at atmospheric temperature would occur for a long time. Unfortunately for the theories, we believe that the sun is highly stable against the occurrence of any such changes.

Another theory of the neutrino shortfall invokes dark matter. Suppose that the halo dark matter, described in Chapter 6, consists of exotic relic particles from the very early universe and that these particles, which occasionally become trapped inside the sun, do not annihilate, for example. Very massive neutrinos would be a suitable candidate. Neutrinos annihilate only with antineutrinos. In this case, over the age of the sun, the concentration building up would amount to about one hundred billionth of the mass of the sun. This is a tiny fraction, but it leads to an intriguing result: these dark particles, heavier than protons, collect in the innermost core of the sun, where one consequence of their accumulation is that heat diffuses a little more readily, because these weakly interacting particles travel much further than protons before undergoing any collisions. Thus, the center of the sun is slightly cooler than it otherwise would be, which affects the rate of the nuclear reactions that power the sun. There are two pathways for hydrogen to burn into helium and release nuclear energy. Because one of these, involving carbon, nitrogen, and oxygen as catalysts, requires a slightly higher temperature, it is suppressed and the number of neutrinos expected to be emitted from the sun is reduced.

In the search for the elusive solar neutrinos, the novel idea is that the trapping of dark matter in the solar core could be responsible for suppressing the flux detectable in the chlorine experiment. The next step will be to use a new detection material, gallium. Because gallium is sensitive to lower energy neutrinos than chlorine, a gallium experiment can test whether the sun is slightly cooler in the center than current models allow. The neutrinos must be produced in order for the sun to maintain its observed luminosity. To mount a gallium experiment requires a considerable amount of gallium, perhaps 50 tons, which is equivalent to the entire world production for about a year. But the gallium is returned when the experiment is completed, and at least two solar neutrino gallium experiments are being planned. If the gallium experiment were unsuccessful in detecting solar neutrinos, we would have to resort to an extreme solution indeed—for example, we might have to allow the electron neutrinos produced in the sun to change, or oscillate, into other types of neutrinos such as muons or taus (in all, there are at least three species of neutrino). With such assumptions we could evade the detection technique, which is sensitive only to electron neutrinos.

We expect that most plausible candidates for the dark-matter particles will eventually annihilate, resulting in a negligible heat input into the solar core. Although only about 1 part in 100,000 billion of the sun now consists of dark matter, this is sufficient to have one observable consequence: the high-energy neutrinos emitted during the annihilations escape unimpeded from inside the sun and may be observable on the earth with a suitable detector. The same deep underground detectors used to search for proton decays (see below) are sensitive to these high-energy neutrinos; so far, none has been detected from the sun, but the search is continuing.

Although there are 100 million neutrinos for every atom in the universe, the neutrinos left over from the big bang have decayed in energy as the universe has expanded. Their energy is presently only one-thousandth of an electron volt, more than a billion times less energetic than the predicted solar neutrinos. Unfortunately, the only means of detecting neutrinos requires that they possess energy high enough to trigger certain nuclear reactions. Consequently, the presence of a sea of cosmological background neutrinos remains an unverified prediction of the big bang theory. No direct observations of this sea of neutrinos that pervades the universe are contemplated in the foreseeable future. (An effect would have resulted on the evolution of the universe, however, in a nonstandard Big Bang model in which the neutrinos had vastly outnumbered the antineutrinos in the early moments. In this case, the production of neutrons would have been inhibited, and subsequent nucleosynthesis would have been correspondingly affected—little or no helium could have been produced. The converse situation, an excess of antineutrinos over neutrinos, would inhibit proton production and would at first also result in little helium.) The most direct way we could measure neutrinos from the early universe would be by their gravitational effect if they were massive enough, as they are in some models of dark matter, to have an appreciable influence on cosmological evolution, or if they are unstable, and decay into observable photons. In either case, it seems possible that their presence might yet be inferred.

CHARACTERISTICS OF THE BACKGROUND RADIATION

As the universe has expanded, the background radiation has passed through the entire spectrum, from gamma rays to x-rays to ultraviolet to optical to infrared and finally to photons of radio wavelength. At any given moment, the temperature had a certain value, and we specify the instantaneous radiation spectrum of the entire universe in terms of that temperature.

Radiation and matter were in intimate contact, and the only surviving characteristic of the radiation is its temperature. Imagine taking a rare, expensive automobile and completely pulverizing it. The resulting heap of scrap could not be distinguished from the wreckage of an ordinary automobile. Similarly, we can characterize the radiation field of the early universe only by specifying its temperature; all other information is lost in the primordial soup.

We have previously referred to the background radiation as

blackbody radiation. The distribution, or spectrum, of radiation peaks at a particular wavelength, which we call the color of the radiation. The color depends only on temperature (of the walls, in the case of a black enclosure); that is, the color depends on the average energy of the atoms that have been in close contact with the radiation. Blackbody radiation is always produced in *thermal equilibrium,* when there is a complete exchange of energy between the radiation and its surroundings. Many of our efforts on the earth, such as insulating our living environment or even staying alive, are aimed at staving off thermal equilibrium. The early universe was a unique environment, where high density and high temperature guaranteed thermal equilibrium.

The blackbody radiation has several basic properties. Its characteristic wavelength depends only on temperature. As the temperature decreases, its spectrum maintains a similar shape but peaks at a longer wavelength. This distribution of the intensity of the radiation with wavelength is known as the *Planck distribution.* The average energy of a photon of blackbody radiation is proportional to the temperature of the radiation, or inversely proportional to the wavelength of the photon. The typical separation between photons of blackbody radiation is roughly equal to their average wavelength.

One consequence of these properties of the background radiation is that the number of blackbody photons in a given volume must vary as the inverse cube of their average wavelength, or cube of their characteristic blackbody temperature. As the volume of the universe expanded, the wavelength of the radiation was stretched out by a proportional amount. This phenomenon is nothing more than the Doppler shift, which we discussed in Chapter 3. Thus, photons lose energy, lengthening in average wavelength as the universe expands, and the temperature of the radiation decreases.

Another consequence of these properties of the background radiation is that the total energy of the radiation in a given volume is proportional to the average energy of a single photon multiplied by the number of photons in the volume. Because the energy of a photon is proportional to the temperature, and because the number of photons is proportional to the cube of the temperature, the *energy density* of the radiation is proportional to the temperature to the fourth power. We have already seen in Chapter 5 that the density of the matter (in the form of atoms) is proportional to the inverse of the comoving volume being considered. It follows that the density of the matter is proportional to the number of blackbody photons in the volume, or to the cube of the temperature.

DENSITY OF MATTER AND RADIATION

Comparison of the density of matter with the mass-equivalent density of radiation can tell us much about the conditions in the early universe. Radiation is a form of energy and of mass, and we can convert energy to mass values if we divide energy by the square of the velocity of light, according to Einstein's relation, $E = mc^2$. Presently, the radiation temperature is only 3 degrees above absolute zero, and the density of the radiation is small compared with the density of the atoms. The mass-equivalent density of the radiation amounts to approximately one ten-thousandth of the total mass density that we can observe in the form of galaxies and stars. However, at progressively earlier epochs, the radiation density played an increasingly greater role. The ratio of the density of radiation to the density of matter is proportional to the radiation temperature. Thus, when the temperature was roughly 10,000 times higher than at present (when the universe was one ten-thousandth of its present size), the mass-equivalent radiation density was equal to the mass density. At still earlier eras, the radiation density outweighed the mass density completely. This means that the effective gravity in the early universe was a consequence of its radiation density. Tracing back in time, matter played an increasingly negligible role until about the first second, when the temperature was so high that substantial numbers of pairs of particles and antiparticles were created. During this stage, the masses of the leptons were roughly comparable to the equivalent energy of the photons. There were roughly equal numbers of electrons, positrons, and photons during the first second of the expansion.

   Although as we trace back to still earlier eras, the energy density of radiation increases more rapidly than the mass density, the number of photons, like the number of particles, depends only on the size of the comoving volume being studied after one second, when pair annihilation has terminated. By the number of particles, we mean the *net* number of particles minus the number of antiparticles. Although particle pairs (particles and antiparticles) are created (and destroyed), the net particle number changes only as the volume changes. This leads us to conclude that the ratio of the number of photons to the number of nuclei or electrons does not change as we follow the universe back in time, until we reach the era of pair creation. The process of pair creation creates (and destroys) photons by creating (and annihilating) pairs of particles and antiparticles.

There are at present between 100 million and 1 billion blackbody photons for every atom observed in the universe. We know this directly, because we have measured the present temperature of the radiation and the average density of atoms in the universe. However, in the earliest moments of the expansion, there were a comparable number of particle pairs and photons, so that the universe contained almost equal amounts of matter and antimatter. There was then an excess of matter over antimatter by only 1 part in 1 billion. This tiny excess is responsible for the resulting photon-to-particle ratio. After a millisecond or so, the protons and antiprotons annihilated, and the electron-positron pairs in turn annihilated after about one second. The particles that remain must be due to this infinitesimal asymmetry in the big bang. The residue of the annihilation process took the form of 1 billion blackbody photons for every surviving atomic nucleus. As we saw in Chapter 6, the origin of this ratio of 1 billion blackbody photons per particle is one of the unsolved mysteries of big bang cosmology.

Besides this background radiation, another form of energy may have survived from the first second. However, this form of energy is very difficult to measure. Once again, we encounter the sea of neutrinos that is predicted to have survived from the earliest seconds. The neutrinos are a relic of the era of the weak interactions. These were dominant when the temperature exceeded about 10 billion degrees kelvin. Slightly later, the electron-positron pairs annihilated, dumping their energy into the blackbody radiation. Consequently, the energy density of this neutrino background would be similar to (but slightly less than) the blackbody radiation, and we can characterize it by the concept of a neutrino temperature, which should be about 2 degrees above absolute zero at present. As is the case with the photons, these missing neutrinos must be the pale remnant of highly energetic particles that played a dominant role in the evolution of the early universe. Could we but measure these feeble neutrinos, we would expect their abundance to confirm our picture of the radiation era revealed by the blackbody radiation.

RADIATION TEMPERATURE

To understand fully the concept of *radiation temperature* in the context of the expanding universe, we must return again to the first millisecond. Many types of particles and antiparticles were present; they annihilated to produce radiation and were also being created by the radiation. Because these reactions were simultaneously destroying and creating particles and photons, a balance between matter and radiation was maintained: we had thermal equilibrium. When protons and antiprotons were in thermal equilibrium, the temperature must have been above 10,000 billion degrees (or 1 billion electron volts of energy per particle); when only positrons and electrons were in thermal equilibrium, the temperature had dropped to 10 billion degrees. As the universe expanded, the radiation cooled, and the photons filled a progressively larger volume, thereby increasing in wavelength. When the temperature dropped below about 5 billion degrees, the electron-positron pairs almost entirely vanished. Radiation remained, initially at a temperature characteristic of the electron rest mass (about 5 billion degrees). This radiation was pure blackbody radiation because all the photons were created when matter and radiation were in equilibrium.

SCATTERING

Although it became progressively more difficult to maintain this equilibrium after the electron-positron pairs annihilated, little could affect the spectrum of radiation except the cooling that resulted from expansion. The blackbody spectrum was difficult to modify, because the residual density of matter and radiation was sufficiently high to ensure that the nuclei and photons remained well coupled. This intimacy was maintained because the photons were frequently scattered by the electrons (Figure 8.1). The photons could not travel freely and remained coupled with the electrons, which were in turn coupled by electromagnetic attraction to the protons. Energy was freely shared, and this *scattering* process was so effective that the particles of matter remained at the blackbody radiation temperature.

To understand how scattering occurs, we must consider an impinging photon as a pulse of electromagnetic energy. A free electron is momentarily accelerated by the sudden impulse of an electric field and gains momentum. This in turn removes momentum from the photon. Although the *energy* of the photon does not change appreciably, it does lose momentum. This results in a change of the propagation direction of the wave, or a scattering of the radiation. For the average impinging photon, bound electrons (electrons in orbit around nuclei) are rather poor scatterers, because they respond primarily to the atomic binding forces that keep them in orbit.

As the universe expanded, the blackbody radiation temperature gradually dropped, but the radiation always remained pure blackbody radiation. It would have been very difficult for the nature of

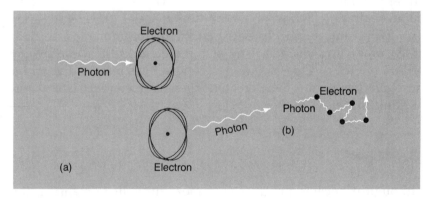

Figure 8.1 Electron Scattering

An incident photon interacts with a free electron as a momentary electric impulse (*a*). The electron accelerates slightly and gains some momentum. Consequently, the photon loses momentum and changes direction. Essentially no transfer of energy occurs in this process, because the electron is so much heavier than the photon. Each photon undergoes many scatterings (*b*).

the blackbody radiation to change, unless some photons could have been absorbed or new photons could have been created. Once the temperature had dropped below 10 million degrees (the universe at this stage was about six months old), the density had fallen sufficiently that significant absorption or emission processes could not occur. The fact that the blackbody photons so greatly outnumbered the protons and electrons made it unlikely that any plausible subsequent source of radiation would significantly perturb the blackbody spectrum. Consequently, when observers study the cosmic blackbody radiation, they are looking at radiation that effectively originated when the universe was about 6 months old.

We could always try to imagine some other prolific source of radiation occurring during a later era. One possibility might be the heat and radiation produced during the decay of primordial turbulent eddies. Another possibility might be the ejecta from exploding mini–black holes. These options seem speculative to most cosmologists, however, since it is extremely improbable that such phenomena could simulate the blackbody characteristics of the cosmic radiation. The protons and electrons would have been capable of only scattering photons; they could not have created new photons or destroyed old photons. The blackbody characteristics of the radiation show up in the energy, or wavelength, distribution of the photons; a blackbody spectrum has a unique shape, once the radiation temperature is specified. The near-perfect nature of the cosmic blackbody spectrum tells us that the big bang must have been relatively quiescent ever since the universe was 6 months old. Once blackbody, it remained blackbody, with the radiation spectrum shifted to a lower characteristic temperature as the universe expands.

PRIMORDIAL CHAOS?

Let us now examine one of the more speculative scenarios that could lead to modification of the standard model. Consider a very chaotic and turbulent early universe. Suppose this turbulence is driven by gravitational collapse in localized regions. The primordial motions would eventually dissipate and produce heat, in much the same way as the airflow around a supersonic missile generates heat. Thus, more radiation would be generated. The photons produced would be mostly of higher energy, or frequency, than the average blackbody photon. Consequently, a shift to the blue, or reduction in wavelength, of the blackbody-radiation spectrum would occur. Once the radiation spectrum became distorted from its original distribution, it would subsequently remain bluer relative to the radiation temperature (Figure 8.2). At these early eras, the universe would still be highly opaque because of the many scatterings undergone by the photons. Thus the distorted blackbody radiation would continue to be scattered. No further distortions of the spectrum would occur. Once the scattering ceased, a

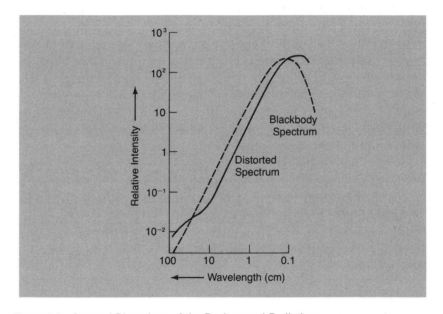

Figure 8.2 Spectral Distortions of the Background Radiation

According to the nonstandard model, excessive turbulence in the early universe would generate energy that becomes part of the general background radiation. If turbulence occurred after the universe was 1 year old, the radiation is only scattered; the density would have fallen too low for it to be absorbed and attain thermal equilibrium with matter. Consequently, the blackbody spectrum would have become distorted. Actual measurements have failed to reveal any appreciable distortions from a blackbody spectrum, which implies that the role of primordial turbulence in dissipating energy must have been minimal.

"blue glow" would be detectable and we would expect to discern a short wavelength (blue) distortion of the observed blackbody radiation.

To understand why this should be an observable phenomenon, let us examine the fog analogy a little more closely. Imagine two light sources, one red and one blue, surrounded by fog. An observer would not be able to see the light sources but would see the color of the scattered light. The red light would be seen from a greater distance than the blue light, because red light is of longer wavelength and scatters less than blue light. (This phenomenon explains why the sky is blue and the sun appears to be red when it is near the horizon; the atmosphere scatters more of the blue light and less of the red light.) By examining the scattered blue light, we could draw conclusions about the dust content of the atmosphere and about the nature of the light source, which we could not discern directly. Similarly, we can analyze the cosmic background radiation to determine the conditions in the universe during the radiation era. We should expect to see an excess of "blue" scattered photons. However, observations of the background radiation yield no definitive evidence for any such distortion; more precisely, they do not provide evidence for a phenomenon at radio wavelengths that is analogous to seeing "blue," namely, a preferential enhancement of the intensity of short wavelength photons relative to photons of longer wavelength. The best estimate is that at most 10 percent of the energy in the blackbody spectrum could have been distorted in this fashion. From this evidence we infer that any turbulence or dissipation in the radiation era (from age 6 months to about age 1 million years) must have been relatively modest.

THE END OF THE RADIATION ERA

As the expansion proceeded, the significance of the radiation gradually diminished. There were still about 1 billion photons per proton, and the ratio remains constant today. However, the photons gradually lost energy as the temperature dropped, whereas the rest masses of the protons remained unchanged. Thus, the protons make a progressively greater contribution to the total density of the universe. Mass density dominated energy density after about 10,000 years.

As the universe expanded during the radiation era, the density of protons and electrons decreased, and the radiation was scattered less and less frequently. When the temperature dropped to about 4000 degrees, the electrons and protons combined into hydrogen atoms. Cosmologists customarily refer to this era as the *decoupling era;* it marks a significant stage in the evolution of the universe.[10] Subsequently, radiation moves independently of matter because scattering no longer occurs. The matter must be mostly in atomic form for scattering to be suppressed. The temperature was simply too high for atoms to form any earlier; any atoms that formed would have been instantly split, or ionized, into protons and electrons by the hottest blackbody photons. As the temperature dropped, however, the number of these hot photons rapidly declined, and the hottest photons became incapable of ionizing hydrogen atoms. Hydrogen atom formation, or *recombination,* occurred rather suddenly, beginning when the universe was about 300,000 years old, and the process was almost completed before the age of 1 million years (Figure 8.3). The process was so efficient that only one electron and proton remained apart for roughly every 100,000 atoms.

The most dramatic aspect of decoupling is that, once the electrons became bound, the universe became entirely transparent— the fog had lifted. From this point, the blackbody photons continued to cool but never again deviated in their motion. No longer were enough free electrons present in space to scatter the blackbody photons significantly. (However, if the intergalactic gas atoms subsequently, at a much later era, became reionized into electrons and protons by a sudden injection of heat or ionizing radiation before the density had fallen too far, further scattering conceivably could have occurred at a relatively recent era.) The radiation temperature continues to drop as the universe expands; at present, the blackbody temperature has dropped to 3 degrees above absolute zero. We can be fairly certain of this, because radio astronomers have measured both the spectrum and the temperature of the radiation. For the radiation temperature to have fallen from 3000 to only 3 degrees Kelvin, we infer that the wavelength of the typical photon, and the universe itself, has expanded by a factor of approximately 1000 since the decoupling era.

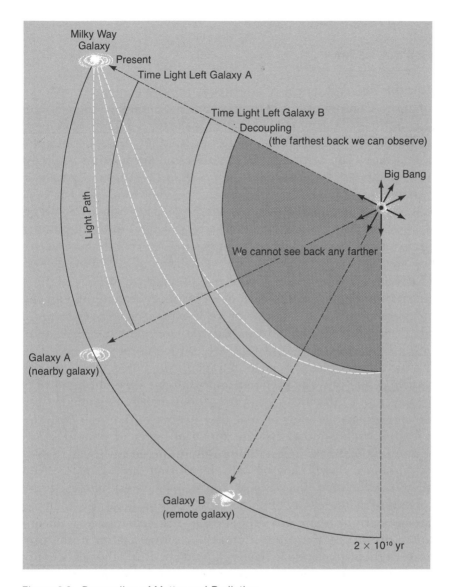

Figure 8.3   Decoupling of Matter and Radiation

At the decoupling era, when the universe was about 300,000 years old, the radiation temperature had dropped sufficiently that the protons and electrons formed hydrogen atoms. Once the free electrons had formed atoms, scattering ceased, and the radiation was subsequently able to travel freely and decouple from the matter. When we study nearby galaxy A, we see it as it was some time (perhaps millions of years) ago, whereas a more remote galaxy B would appear to us as it was billions or even tens of billions of years ago. The universe was already billions of years old when the galaxies that are observed today formed. However, the blackbody radiation comes to us directly from the decoupling era. The background radiation has a redshift far greater than that of any known galaxy or quasar, probably as large as 1000.

COOLING OF MATTER

After the decoupling era, matter cools rapidly compared with the radiation. Cooling of the hydrogen atoms occurs because the random motions of the atoms fail to keep pace with the rate of general expansion of the atoms away from one another. Imagine an automobile race. The cars start out together and are moving at almost the same speed, but gradually they drift farther and farther apart. It takes longer and longer for any car to overtake another as time goes on: we can equally well say that the relative velocity between an adjacent pair of cars declines with time. A similar argument applies to the random motions of atoms in the expanding universe: they also decline with time as the separation of the atoms grows.

Random motions of atoms give rise to what we ordinarily think of as temperature. For the case of an ordinary expanding gas, the rate of decrease of the temperature is equal to twice the rate of expansion. For the case of blackbody radiation, the temperature of the radiation decreases only linearly with the rate of expansion. The difference arises because of a result predicted by the theory of special relativity: as particles approach the speed of light, they become more difficult to compress. Less energy is gained. Conversely, expansion leads to a smaller decrease in energy for *relativistic particles* (moving at the speed of light), or photons, than for nonrelativistic particles. Photons therefore decay less in energy than do more slowly moving particles during the expansion. (Recall, however, that, because photons are massless particles, the equivalent mass-density of photons declines more rapidly than the mass-density of nonrelativistic particles, in accordance with our earlier result.) By the present era, the residual matter in the universe might be expected to have retained very little thermal energy (in particle motions) and to be extremely cold, only a fraction of a degree above absolute zero. In actuality, because of the intervention of other sources of heat and energy associated with galaxy formation and related activity in the past, such low temperatures are probably not attained.

Study of the angular distribution of the blackbody radiation on the sky yields direct information on the distribution of matter when the radiation was last scattered. As we measure the radiation, we are looking back into the remote past, when the universe was less than a million years old. At that time, the separation of any two distant galaxies (or, of the atoms now contained within any two galaxies) was only one-thousandth of its present value. The wavelength of the radiation present at decoupling has since increased by 1000. As we saw earlier in this chapter, an increase in wavelength in the optical part of the spectrum is equivalent to a reddening of the light. This concept can be generalized; the redshift is the relative increment in wavelength of the radiation spectrum. Decoupling is said to have occurred at a redshift of about 1000 (a more accurate value is 1200, according to numerical computations: the precise value depends weakly on the adopted parameters of the cosmological model), because the radiation that suffered a last scattering at decoupling is stretched in wavelength by this factor by the present era (actually, it is just the increment or shift in wavelength that equals the redshift multiplied by the initial wavelength).

Redshift is commonly used as a label for an earlier era, and it is also a measure of distance—the greater the redshift, the more remote the object and the more difficult it is to detect. The most remote objects that we have found, the quasars, have redshifts as high as about 4.5. (We shall discuss quasars further in Chapter 12.) These are exceedingly luminous objects, however, and the most remote galaxies hitherto detected have redshifts of about two. In such galaxies, the wavelength of any spectral feature would be three times that of the same feature in a nearby galaxy of low redshift. The background radiation yields observations of one component of the universe at a redshift far higher than observations by any other means.

Studies of the cosmic blackbody radiation have hitherto found it to be completely uniform and smooth. Irregularity or graininess has not been detected to upper limits of less than 1 part in 10,000. However, there seems little doubt that this search for irregularity will ultimately meet with success. The blackbody radiation provides us with a remarkable window to the early universe; it takes us farther back in time toward the big bang than any known object. The primordial fluctuations, from which the galaxies formed, must have left their imprint on the blackbody radiation. Detection of these fluctuations would provide impressive confirmation of the entire theory of galaxy formation, which we will discuss next.

· 9 ·

ORIGIN OF THE GALAXIES

It seems to me, that if the matter of our sun and planets, and all the matter of the universe, were evenly scattered throughout all the heavens, and every particle had an innate gravity towards all the rest, and the whole space throughout which this matter was scattered, was finite, the matter on the outside of this space would by its gravity tend towards all the matter on the inside, and by consequence fall down into the middle of the whole space, and there compose one great spherical mass. But if the matter were evenly disposed throughout an infinite space, it could never convene into one mass; but some of it would convene into one mass and some into another, so as to make an infinite number of great masses, scattered great distances from one to another throughout all that infinite space. And thus might the sun and fixed stars be formed.

—Isaac Newton

The most imposing feature of the universe must be its structure. Until now we have almost completely ignored this aspect, describing the universe instead as a uniform, homogeneous distribution of particles. On sufficiently large scales, we can justify this assumption of homogeneity. Observations of the most distant galaxies in different directions from Earth reveal similar abundances of galaxies for comparable volumes of space. Measurement of the expansion rate of galaxies also yields a similar value (the Hubble constant) in different directions in space. These data imply large-scale isotropy of the universe, as viewed from our vantage point on earth. Of course, this conclusion is only as valid as the accuracy of the astronomical data. Determinations of the Hubble constant by different astronomers differ by as much as a factor of 2; thus, observations of galaxies probably do not demonstrate isotropy much more accurately than this crude level.

Observation of the cosmic background radiation has produced a considerably more compelling result: the background radiation is uniform to better than 0.1 per cent throughout space. Observations have been performed over the range of angular scales, from arc minutes to entire quadrants of the sky, with similar results. Indeed, between angular scales of 1 arc minute and 90 degrees, the best limits on any fluctuations are at the level of 0.01 per cent. We saw in Chapter 8 that the distribution and degree of uniformity of the observed radiation is produced by the spatial distribution of matter at an early era. Therefore, we conclude that matter at the decoupling era was of comparable uniformity. This assumption, when combined with the copernican argument, that our position as an observer is in no way a preferred location, has profound implications for our theory of the structure of the universe. These assumptions provide the basic justification for applying big bang cosmology to the observed universe.

THE CONSERVATIVE APPROACH

How can we reconcile the idealized conditions of big bang theory with the complex particulars of the universe we observe? One approach has been to study the stability of the universe with respect to expansion and its effect on the growth of small irregularities. Imagine that infinitesimal fluctuations in density were present in the early universe. How would these *density fluctuations* evolve during the expansion? The answer is that the expansion of the universe must have exerted a stabilizing influence on such irregularities. The expanding universe has the effect of greatly impeding what otherwise might have been catastrophic forces. A static universe that contained areas of excess density would be highly unstable and subject to local collapse. However, an expanding universe is only very slowly perturbed by the formation of local regions of enhanced density. This is fortunate for our sake because recollapse of the denser regions in the early universe would have prohibited the formation of galaxies.

At early times, the newly formed dense regions would have been much denser than any present galaxies. The expansion greatly retarded any amplification of these small fluctuations. Nevertheless, the process of growth of fluctuations went on for a very long time, and the initial degree of inhomogeneity need not have been large for huge fluctuations to develop that could eventually recollapse. In a region with only a slight excess of matter, compared with its surroundings, the local gravitational field will be increased by a

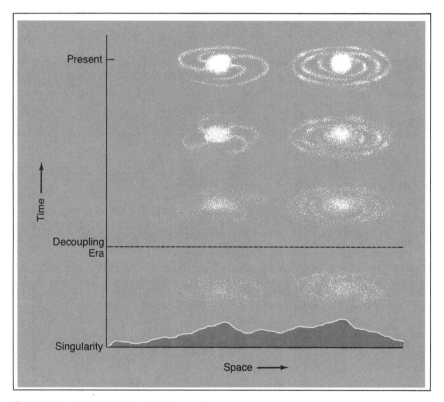

Figure 9.1 Growth of Density Fluctuations

In this space-time diagram, time is denoted by the vertical axis, and the three spatial dimensions are compressed into the horizontal axis. Galaxies developed from small density fluctuations in the early universe; these can be traced back to infinitesimal density fluctuations in the initial singularity. As the universe expanded, the fluctuations grew larger as their self-gravity attracted the neighboring matter. Eventually, after the decoupling era, the density fluctuations collapsed to form galaxies.

small amount. This field exerts a force on adjacent regions. If density fluctuations with amplitudes between 1 per cent and 0.01 per cent were present prior to the decoupling era, the natural accretion resulting from this process of gravitational instability would have ensured eventual collapse—perhaps to form galaxies (Figure 9.1).

These fluctuations are rooted in our quantum past, in the first 10^{-35} second of the big bang. We have already seen how the universe inflated from the microscopically small to the macroscopically vast, which accounts for its remarkable isotropy and homogeneity. At the same time, tiny density fluctuations were present and indeed were inevitable, according to quantum theory, which tells us that we can never be sure precisely how many particles or quanta of energy are in a specified region. If this region is macroscopically large, the quantum uncertainty is completely irrelevant: statistics gives us as precise an answer as we could desire. But if the region is sufficiently small, then the quantum fluctuations will be large. Our theory of the big bang commences with the Planck instant, at 10^{-43} second, the threshold of classical cosmology, when quantum fluctuations were huge. Prior to this instant, the universe was described by the still poorly understood theory of quantum cosmology. Inflation set in as the universe cooled down and made a transition in the type of matter it contained, shortly after the Planck time. The density fluctuations were amplified, from the quantum scales up to the vast bubble formed by the inflating universe. Inflation soon ceased, and the gentler big bang expansion resumed, but the universe was permanently scarred: it was homogenized and isotropized yet also permeated by density fluctuations that would be the seeds of future galaxies.

THE REVOLUTIONARY APPROACH

The conservative approach to large-scale structure emphasizes growth from infinitesimal seed fluctuations that aggregate matter under the inexorable influence of gravity. An alternative, which I shall call the revolutionary approach, takes a very different tack. The structure that we see today could have formed as a consequence of gigantic explosions sweeping up and compressing matter to form galaxies and stars. In fact, on a smaller scale within our own galaxy, we have considerable evidence that the interstellar medium, the gas between the stars, has a very violent existence punctuated by occasional great explosions that sweep it up into dense shells of gas. These shells fragment into clouds, which themselves eventually form stars. The deaths of stars weighing ten or more times as much as the sun trigger the explosions. We shall see later that such massive stars have a fiery fate, releasing a very considerable amount of energy when they finally exhaust their nuclear fuel and collapse into compact remnants.

Although such stars burn out within millions of years, successive generations of these stars form, and the life cycle of the interstellar gas spans billions of years, as stars are born and die. Perhaps the intergalactic medium, over tens of billions of years, suffered a similar history. It is possible that similar processes were repeated on a vastly larger scale when the first galaxies formed. The guiding concept of one current scheme is *explosive amplification,* the idea that a few, very rare and massive objects formed, perhaps a million years or so after the big bang. Why were the first objects likely to have collapsed in this epoch rather than an earlier or later time? Before decoupling, the outward pressure of the radiation in a region of locally enhanced density was very great. After decoupling, however, the radiation moved freely through matter that was almost wholly in atoms; matter accordingly had a very low pressure. A sudden lowering of pressure could have led to implosion or collapse of large regions. This would have resulted in the formation of shock waves, which would compress the gas.

The resulting clump of compressed gas would contain more than 1 million solar masses of material. Its fate can be catastrophic. Black hole formation is one possibility. Collapse to a black hole is inevitable at a much earlier epoch for large inhomogeneities, on the scale of the horizon, because light itself becomes trapped by the gravitational field of the inhomogeneity. Indeed, a black hole can be defined as an object whose gravitational field is so intense that light cannot escape. We know that the universe is not full of large black holes; say, of galactic mass or larger. The formation of any such primordial black holes must be restricted to very early eras (within the first year of the big bang), when the mass within the horizon was much less than that of a galaxy.

Subsequent accretion onto such a supermassive black hole

would release considerable energy. Perhaps such an object acts as the nucleus of a forming galaxy. The accreting gas fragments into stellar-mass chunks as a galaxy forms. Before long, many massive stars would have fragmented out of the collapsing cloud. These would explode after a few million years had elapsed, and the resulting blast waves would sweep out dense shells of gas. In a forming galaxy, so many explosions would be likely that they would reinforce each other as the swept-out shells overlapped. A wind of very hot gas would be driven out from the forming galaxy by the collective effects of many exploding stars. This wind would sweep up a dense shell of intergalactic gas, a shell containing perhaps 10,000 times more mass than in the original galaxy.

The shell breaks up into galactic-mass clouds, each of which collapses to form new galaxies, which in turn drive winds and form new shells. Galaxy formation becomes self-propagating. Very few initial seeds are needed, and the final distribution and properties of the galaxies depend on the details of the explosive-amplification mechanism. This scheme begs the question of the origin of the seeds, but these may have originated in bizarre types of irregularities in the very early universe, such as cosmic strings.

Indeed, it is entirely possible that the very early universe could have been highly chaotic. Although the notion is very seductive that inflation removed all extreme primordial irregularities, leaving us with a smooth and regular universe, we have to remember that it is merely a hypothesis. We have practically no direct information on the first second of the big bang, let alone the first 10^{-35} second, when inflation was presumed to have occurred. Some cosmologists say that inflation was an unlikely event: they argue that out of the set of all possible initial universes, only an infinitesimal subset could have been isotropized and homogenized by inflation. Some other assumption, which they conveniently call "initial conditions," is needed to specify the very early universe. In this case, we might as well begin with the concept of primordial chaos and search for physical processes that could have helped remove or regulate it. Perhaps, for example, the universe expanded much more rapidly in some directions than in others. It could even have expanded in one direction and simultaneously contracted in another direction. A group of particles initially on the surface of a sphere might have deformed into a pancakelike or cigarlike surface and then changed into a different configuration. By the most revolutionary speculation, an effect analogous to friction could have smoothed out the anisotropy in the early universe, resulting in a standard big bang expansion.

This account attributes the friction to enormous numbers of neutrinos. To understand how these massless particles, which travel at the speed of light, could play such a drastic role in the early universe, it is helpful to recall that neutrinos ordinarily have a very weak interaction with matter. In the first second of the universe, however, the density of matter was so high that neutrinos were entirely absorbed by atomic nuclei. During this early phase, the neutrinos could have exerted considerable pressure on the rest of the universe. Only while they are being absorbed by matter can neutrinos exert a considerable force on the matter (just as light can exert pressure on dust particles). If the expansion of the universe had been initially highly anisotropic, the neutrinos would have tended to smooth out much of the initial anisotropy, making the expansion more uniform and isotropic. As the universe expanded, neutrino absorption would have become less effective. Once the electrons and positrons had annihilated, the neutrinos no longer would have interacted strongly with the surrounding matter.

Chaos can comprise an infinite variety of amplitudes and scales. Early large-scale disturbances may have acted like stones dropped into a pond, causing a series of waves to be propagated through the water. Similarly, we can imagine waves (analogous to sound waves) propagating throughout the radiation field in the early universe. The waves are driven by pressure disturbances, which could be due to the formation of black holes triggering sudden releases of energy, that is to say, to explosions; the waves also could be simply the initial conditions of the very early universe, unable to propagate effectively until enough time had elapsed since the big bang for a disturbance to travel across one wavelength. Only if the disturbance were on a large enough scale would the self-gravitational force of a local region dominate the outward pressure of the radiation. In this case, collapse into black holes most likely would result.

A still more extreme variant of chaotic cosmology holds that the cosmic background radiation may have been produced by dissipation of primordial chaos. This scheme hopes to account for the number of photons per atomic particle that we observe, although to use it, we would have to abandon other achievements, notably helium production, of the standard model.

In the remainder of this book we will take the conservative approach, which offers us, until it is proved wrong, the least drastic revision of the cosmological principle. It also provides us with the arrow of time. A fundamental law of physics is that entropy, or the degree of disorder in a physical system, can never decrease. As the universe becomes more and more complex with the development and evolution of structure, time and entropy proceed irrevocably onward. It therefore seems preferable to postulate relative homogeneity and order rather than chaos in the early universe.

A COSMIC FILTER

Our picture of the early universe prior to decoupling is evidently more complex than naive application of the big bang theory leads us to expect. We adopt the conservative approach, in which the early universe was approximately homogeneous. However, some degree of mild inhomogeneity was crucial if structure was eventually to evolve. The fluctuations represented only small deviations in density from that of the expanding background, but they encompassed large amounts of matter, enabling the large-scale structures that we now observe to develop.[11] At any point in space, both large-scale and small-scale irregularities coexisted. Indeed, there was a hierarchy of fluctuations within larger fluctuations within still larger fluctuations. A convenient measure of a fluctuation is the mass it encompasses. Thus, we can refer to star-sized, galaxy-sized, or even cluster-sized inhomogeneities.

The fluctuations were not static. Any slight density enhancement exerted a small gravitational attraction on the surrounding matter, which caused it to flow toward the local inhomogeneity. The density fluctuation grew in amplitude, provided that pressure forces resulting from the thermal motions of the protons did not impede the accretion. There was always a continuous battle between the outward forces of pressure and the inward pull of gravity (Figure 9.2). The outcome of this struggle determines the lifetime of the sun and other stars, as we shall later see. The same struggle was

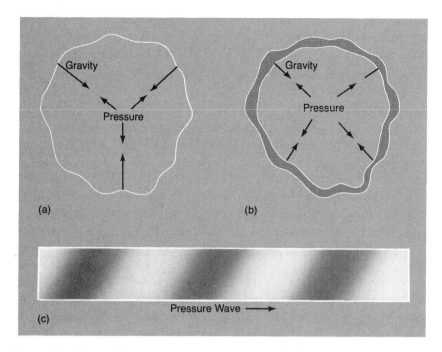

Figure 9.2 Gravity Versus Pressure in a Density Fluctuation
In a large fluctuation, gravity is the dominant force (*a*). The fluctuation grows and will eventually collapse. However, a small fluctuation will be dominated by pressure forces, which tend to expand it (*b*). The result of the competition between gravity and pressure is an oscillation of the fluctuation, which propagates much like a sound or pressure wave (*c*), as long as pressure remains the dominant force.

also crucial in the earliest stages of the universe for the growth of tiny density fluctuations. For sufficiently large-scale fluctuations, gravity was always the dominant force, and the density fluctuations continued to grow. As the expansion continued, however, the pressure forces became increasingly important. Over any particular scale of interest, such as that of a galaxy or a galaxy cluster, pressure forces prior to the decoupling era became larger than gravitational forces. This resulted in the cessation of the growth of the fluctuation. Once pressure forces dominated gravitational forces, the fluctuations can be visualized as being rather like pressure (or sound) waves. They become transient disturbances that pass by any given point, rather like waves on the surface of a pond.

The pressure forces in the radiation era were really due to radiation. Disturbances could actually have propagated in the radiation field at almost the speed of light. Recall that the size of the observable universe at any era was restricted to the distance light could travel within the age of the universe. Consequently, this boundary defines the region where pressure forces just balance gravitational forces.

Imagine a disturbance from an inhomogeneity during the radiation era. It will produce a pressure wave. Now, an ordinary sound wave travels a finite distance before dissipating itself. We know that even shouting will not enable us to communicate over too great a distance. In much the same way, these pressure waves are also subject to dissipation. The photons, or quanta of radiation, must be squeezed along with the protons and electrons in a pressure wave. The vast bulk of the pressure is due to the photons rather than to the random motions of the protons, which are greatly outnumbered by the photons. Hence, the photons must be scattered by electrons for effective trapping and subsequent squeezing to occur in the pressure wave. However, the time for a photon to be scattered and trapped can be comparable with the time of the compression phase of the wave. In fact, this time clearly depends on the size (or wavelength) of the disturbance. Long wavelengths can efficiently trap photons, but short wavelengths are ineffective at compressing the radiation. Photons therefore can leak out during the wave cycle. Consequently, the wave will lose energy, weaken, and dissipate itself into heat. In other words, the shorter the wavelength, the more rapidly will disturbances disappear (Figure 9.3). We say that they have been smoothed out, or *damped*.

There is actually a critical scale below which the waves are strongly damped. Longer wavelengths can more readily contain the photons. If no leakage of the photons occurs, the waves are not damped. The critical scale increases as the universe expands, and longer and longer wavelengths suffer damping. Eventually, at decoupling, the critical scale for damping to occur amounts to roughly 1000 billion solar masses. The visible regions of a typical galaxy contain about 100 billion solar masses. By the time of decoupling, it seems that radiative damping would have destroyed fluctuations of mass even larger than a typical galaxy. After decoupling, this process abruptly stopped. Once scattering of the photons ceased,

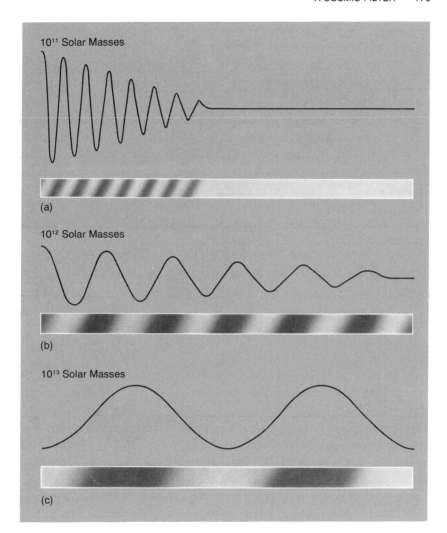

Figure 9.3 Radiative Damping of Adiabatic Density Fluctuations

Disturbances in the early universe behave rather like sound waves in air, causing alternate compressions and rarefactions of the matter. If a region of the universe is compressed, both matter and radiation are squeezed together. This compressed region will try to reexpand. During the cycles of expansion and compression, photons can leak out of such small-scale adiabatic fluctuations. Because the radiation exerts a strong pressure on the matter, fluctuations of short wavelength are damped (a). Fluctuations of longer wavelength (b) can go through more cycles of compression and rarefaction before being damped. Very long wavelength fluctuations (c) barely have time to oscillate, and they survive unscathed through the radiation era. After decoupling, the radiation can move freely and no longer exerts any force on the matter.

the radiation was no longer trapped, and damping of pressure waves no longer occurred.

The preceding argument applies to primordial pressure waves in the radiation-dominated gas. We can also conceive of a physically distinct type of inhomogeneity. In this case, only the density of protons and electrons is allowed to vary; the radiation remains undisturbed. We refer to these inhomogeneities as *isothermal* (constant-temperature) *fluctuations*, in contrast to *adiabatic* (energy-conserving) *fluctuations*, which heat up when matter is compressed (Figure 9.4). Only the adiabatic variety is subject to radiation-pressure diffusion and damping.

Nothing significant happened to isothermal fluctuations throughout the entire radiation era. As in the preceding discussion, the self-gravity of the fluctuation tended to pull the matter inward. However, let us focus on an individual electron. Once it is subjected to any force, it encounters an enormous flux of radiation. Only when the electron is at rest does the radiation flux vanish. Consequently, the radiation exerts an immense drag, or friction, as it is scattered by the electrons, which themselves are under the gravitational attraction of the fluctuation. The electrons in turn cannot separate from the protons because of the attractive force between positive and negative charges. The net result is that the matter fluctuations are frozen. No motion can occur, and fluctuations can neither amplify nor dissipate. From the perspective of an observer comoving with the fluctuation, it would appear rather like trying to move through a very viscous fluid such as treacle or molasses.

After decoupling, scattering ceased. The matter could now move freely through the radiation. Consequently, the radiation field abruptly lost its viscosity. The radiation drag vanished. Local enhancements in the matter distribution could then begin to amplify, as they moved together under their self-gravitation. We describe this process as the initiation of *gravitational instability*.[12]

This process of fluctuation growth actually begins only when the mass-density of the universe is predominantly ordinary matter rather than radiation. It is notoriously difficult to bottle up quanta of radiation along with matter particles whose random motions are practically the speed of light. After about 10,000 years have elapsed, however, the universe becomes matter-dominated, and

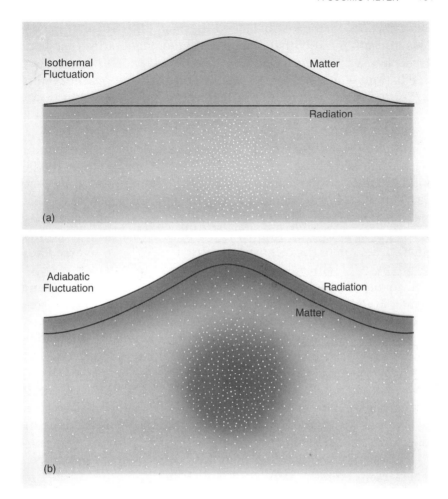

Figure 9.4   Two Types of Density Fluctuations
Isothermal density fluctuations (*a*) maintain a uniform temperature, or radiation density. The background radiation field is completely unperturbed, and only matter is compressed in the fluctuation. Adiabatic fluctuations (*b*) compress both matter and radiation, and the radiation density within the fluctuation is slightly higher than the background radiation density.

gravitational forces can now effectively augment the contrast of surviving density fluctuations. Consider an inhomogeneity of matter that is large enough for the gravitational force to dominate the pressure force. Growth occurs by accretion of surrounding matter. The critical mass of an inhomogeneity at decoupling, when these two opposing forces just balanced, is about 100,000 solar masses. It seems rather likely that most of the inhomogeneities that began to recollapse at this era would have had masses of this order; larger fluctuations would have been liable to break up into smaller lumps.

Physical processes in the radiation era have, according to our model, apparently filtered out certain preferred scales of mass (Figure 9.5). In particular, two preferred scales of mass have emerged from the murky depths of the radiation era. These correspond to primordial isothermal and adiabatic fluctuations of about 100,000 and 1000 billion solar masses, respectively. It may be a coincidence that globular clusters contain about 100,000 solar masses and that the most massive galaxies and small clusters of galaxies have masses of around 1000 billion solar masses. However, to cosmologists in search of the origin of the structure of the universe, these results provide clues that we would be rash to overlook.

A novel variation on this scenario describes the predominance of nonbaryonic dark matter. As we saw earlier, the ubiquitous presence of such matter is predicted by inflationary cosmology. The dark matter responds to gravity, and initial density fluctuations grow in contrast, just as they do with ordinary matter. The final outcome depends in a general way on the nature of the dark matter. Two distinct possibilities may be envisaged. If the dark matter consists of weakly interacting heavy particles, generally weighing as much as a proton or more, then at the epoch in the universe when structure first starts developing, the temperature was about 10 electron volts, much less than the rest mass of these particles, which consequently have negligible random motions. They are said to be cold, and we speak of cold dark matter as a generic category that includes photinos and other exotic candidates from the particle zoo. Just as with isothermal fluctuations, cold dark matter density fluctuations are neither erased nor augmented during the radiation era. Radiation damping smoothes out inhomogeneities in the baryon distribution but leaves the weakly interacting cold dark matter unaffected. After matter domination, the fluctuations grow in contrast, eventually collapsing to form clouds of dark matter in which baryons are entrained. Surviving scales range from about 1 million solar masses, which is the smallest scale on which pressure forces allow the baryons also to clump, up to clusters or superclusters comprising 10^{15} or 10^{16} solar masses. Cold dark matter actually predicts a very specific distribution of fluctuations.

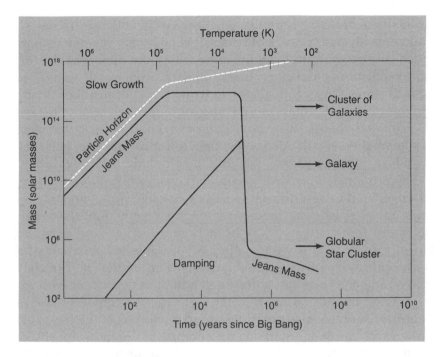

Figure 9.5 Preferred Mass Scales

Any density enhancement that reached a minimum value of 10^{15} solar masses when the universe was roughly 3×10^4 years old (when matter and radiation densities were equal) had enough self-gravity to overwhelm the expansion pressure of radiation. Such a density enhancement then entered a lifetime of uninterrupted growth, culminating in the formation of a large cluster of galaxies. In an intermediate mass range (from 10^{12} to 10^{15} solar masses), density fluctuations persisted until decoupling, when radiation subsequently ceased to interact effectively with matter. These surviving density enhancements may have become individual galaxies, following a period of fragmentation. Below this mass range, radiation pressure damped adiabatic density fluctuations. However, isothermal density fluctuations could have survived the radiation era and would be subsequently gravitationally unstable if their mass exceeded 10^6 solar masses (roughly the mass of a globular star cluster). The particle horizon denotes the boundary of the observable universe, where the recession velocity approaches that of light. The Jeans mass defines the mass at which gravity balances pressure.

Below a scale of about 50 million light years, as measured today, fluctuations would have had similar amplitude. All these scales initiated growth at the same era, namely, at the onset of matter domination, whereas larger scales commence to grow in amplitude only at a later epoch. In fact, the larger the scale, the later growth commences, so that the epoch of matter domination imprints a natural scale on the fluctuation distribution. The history of the smaller-scale fluctuations is complicated, because if collapse occurs simultaneously over a range of scales, it is no longer so obvious how much substructure survives. On large scales, the picture is clearer; galaxy clustering develops hierarchically, from small groups to large clusters. On galactic scales, however, it is not clear how many of the million solar-mass clouds survive to form globular clusters. Such objects are relatively rare, and astronomers believe that most of the smaller clouds merge together to form luminous galaxies. Certainly, merging plays a key role in cold dark matter evolution.

The principal alternative to cold dark matter is *hot dark matter*, a term coined to describe dark-matter particles whose random motions are very large at the onset of matter domination: these particles can freely stream away in random directions and suppress any developing density fluctuation. An example of hot dark matter is the neutrino, with mass about one-ten-thousandth that of an electron or about 50 electron volts. Such a mass is consistent with experimental bounds, at least for the rarer species (muon or tau) of neutrinos, and because neutrinos are so numerous in the big bang, they result in enough dark mass to dominate the mass density of the universe. Adiabatic fluctuations in the baryon (ordinary-matter) content of the universe are another example of hot dark matter. Only after a considerable time has elapsed since matter domination occurred have the neutrino random motions slowed sufficiently to permit any galactic scale structure to develop. In fact, the free-streaming erased all primordial density fluctuations on scales below that of a rich cluster of galaxies, or 10^{15} solar masses. Before galaxies can form, these scales must first collapse and fragment into smaller objects. Hot dark matter results in a top-down scenario for large-scale structure evolution, as for adiabatic baryon fluctuations, whereas cold dark matter and isothermal fluctuations result in a bottom-up evolution scenario.

As we shall see in the next chapter, observations favor a bottom-up scenario for galaxy formation. This means not that we must necessarily abandon hot dark matter, but that we at least need to add a new ingredient in order to seed it with the capability of forming galaxies at an early epoch. Cosmic strings are one possible ingredient, and we will address their possible role in large-scale-structure evolution.

· 10 ·

EVOLUTION OF
THE GALAXIES

The evolution of the world can be compared to a display of fireworks
that has just ended: some few red wisps, ashes, and smoke. Standing
on a cooled cinder, we see the slow fading of the suns, and we try
to recall the vanished brilliance of the origin of the worlds.

—Georges Lemaître

After decoupling, the surviving density inhomogeneities began to grow. A density fluctuation exerts a slight gravitational pull on its surroundings. The local matter tends to fall toward the fluctuation at a very slow rate. The entire region around the fluctuation lags slightly behind the expansion of the rest of the universe because of this attractive force. The amount of this lag gradually increases, as more and more matter falls toward the fluctuation. The fluctuation in turn grows larger and exerts an increasingly greater gravitational pull. Eventually, the gravitational self-attraction becomes so large that the inhomogeneity stops expanding; it reaches a maximum extent and subsequently recollapses.[13] From this point, we will effectively ignore the rest of the universe, which of course continues to expand, and examine the subsequent evolution of this collapsing gas cloud (Figure 10.1).

In Chapter 9, we saw that such gas clouds either are very massive, comparable in size to a large galaxy or to a small galaxy cluster, or else share a range of scales of mass, from low mass (comparable to

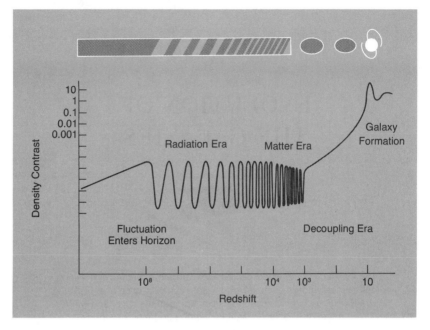

Figure 10.1   Evolution of an Adiabatic Fluctuation

The density contrast describes the enhancement of a fluctuation relative to the background. At large redshifts, the fluctuation increases in amplitude. However, the Jeans mass, or mass at which gravity and pressure forces are just in balance, is growing with time (Figure 9.5). At a redshift of 10^8 or so, a galaxy-sized fluctuation first drops below the Jeans mass, which means that pressure forces begin to dominate. The fluctuation subsequently oscillates like a pressure wave or sound wave. Suppose the wavelength is large enough so that damping does not occur. During the radiation era, the amplitude of the sound wave stays constant; during the matter era, it begins to decrease slowly. At the decoupling era, the pressure forces are drastically reduced, and the Jeans mass changes from being much larger to being much smaller in scale than a galaxy-sized fluctuation. Subsequently, the fluctuation grows; when the density contrast becomes large enough, it collapses to form a galaxy.

a globular cluster) to the mass of a luminous galaxy. We shall first consider the collapse of relatively massive clouds. What happens when such a *protogalactic gas cloud* collapses?

COLLAPSE OF A PROTOGALACTIC CLOUD

A protogalactic cloud could not possibly have collapsed smoothly and symmetrically. For one thing, gravity must have dominated pressure forces. Thus, the speed at which matter collapsed was much greater than the speed at which a sound wave can travel. (The random motion of sound waves is essentially what constitutes pressure.) We infer that the collapse was highly supersonic. But supersonic gas flows are very liable to create turbulence, and any small irregularities would grow rapidly. As small irregularities are amplified, the collapse becomes increasingly chaotic. (Supersonic aircraft must be designed with immense care in order to keep excessive turbulence from developing in their airflow pattern.)

A turbulent gas flow can be thought of as the ultimate manifestation of chaos. To illustrate, let us imagine stirring a cup of coffee. The regular motions of the spoon drive smaller and smaller eddies that will eventually mix the molecules of cream and sugar with the coffee. Similarly, a turbulent gas flow consists of a hierarchy of transient eddies of all scales that ultimately dissipate in heat (Figure 10.2). The gravitational collapse of the large cloud is the driving force, and this energy becomes subdivided into the motions of the eddies. A graphic description of turbulence is contained in the following verse by L. F. Richardson:

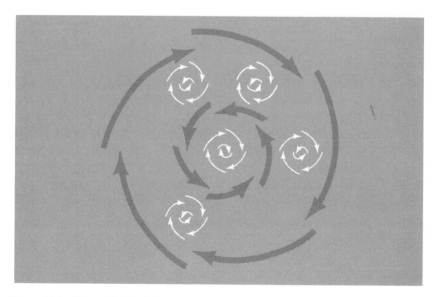

Figure 10.2 Primordial Turbulence
A hierarchy of transient eddies of many different scales represents turbulence in the early universe. Eddies die away (and new ones reform) after moving through a distance comparable with their size. In reality, the eddies are moving in random directions.

> Big whirls have little whirls,
> That feed on their velocity;
> And little whirls have lesser whirls,
> And so on to viscosity.

Viscosity is the process, similar to friction in solids, that occurs as eddies slide by other eddies and eventually dissipate their energy into heat.

   Let us focus on what may have been happening at a given point in the collapsing gas cloud. We can always think of any net motion as consisting of three perpendicular components of velocity. It seems probable that at a given point, the collapse would have proceeded faster in some direction than in either of the two perpendicular directions. Why indeed should nature conspire to have all three components of motion identical? Such a result would correspond to a perfectly spherical collapse, which seems implausible. It seems more likely that a pancakelike blob of gas formed as the gas rushed rapidly in the preferred direction (Figure 10.3), for even if the degree of anisotropy was small to begin with, it would rapidly grow as the collapse proceeds. Many such transient pancakes may have developed at different places in the gas cloud. Pancakes like this would not last; they would have developed a high density,

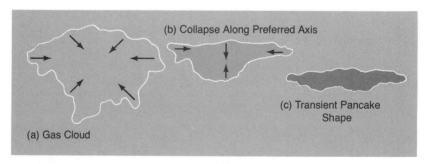

Figure 10.3   Collapse to a Pancake
The collapse of a gas cloud (*a*) will not be spherically symmetric; because any initial asymmetry tends to become amplified during the collapse, the collapse occurs preferentially in one direction (*b*). A flattened, pancakelike gas distribution results (*c*). This highly transient stage of the collapse will be followed by fragmentation and either further collapse or reexpansion, depending on how great a role the pressure forces play.

heated up, and fragmented into smaller gas blobs. This seems to be a very random sort of process, but as we shall see, this process of *gaseous fragmentation* provides specific conditions for galaxy formation.

FRAGMENTATION INTO PROTOGALAXIES

Here we must pause and review the general characteristics of observed galaxies. If we restrict our attention to the more prominent galaxies, those that make a substantial contribution to the average luminosity density of the universe, then it is apparent that galaxies span a very limited mass range. The Milky Way galaxy has a mass in visible stars amounting to about 100 billion solar masses. Other galaxies that are easily observable appear to have masses that lie within a factor of 100 of this mass. It would seem that our galaxy is neither exceptionally large nor very small. The smallest galaxies often have low surface brightness and are difficult to detect. A very substantial number of dwarf galaxies may well be present in the universe, although in aggregate they contribute very little luminosity and probably very little mass to the mean density.

Many galaxies also appear to be fairly uniform in size. The Milky Way has a radius of about 30,000 light-years. Most other spiral galaxies have similar dimensions. Elliptical galaxies exhibit a wider range of masses, and irregular galaxies are often slightly smaller. Any satisfactory theory of galactic evolution must provide an explanation for these properties of the galaxies.

The process of gaseous fragmentation of a massive cloud does lead to the formation of galaxy-sized fragments. The breakup of the original gas cloud into smaller fragments is not in itself, however, a sufficient condition for the fragments to survive. The fragments could subdivide further, or they could collide with one another and be destroyed as the cloud collapses. Such turbulence would involve only transient structures, and we are searching for structures that can survive the collapse.

The key to survival is the ability of a fragment to radiate away the energy contained in the random motions of the atoms. This capacity enables it to contract and become a more cohesive, tightly bound structure. The amount of this energy in atomic motions determines the temperature of the gas. A gas cloud cools by losing some of the kinetic energy of its constituent atoms. Individual atoms collide with one another, and the electrons, bound in the atoms, acquire energy, or become excited. This energy is almost immediately lost by radiation, for the electron in an excited atom rapidly drops to the lowest orbit it can attain. When it does this, a quantum of radiation is emitted. The net effect of atomic collisions is to convert the kinetic energy of the atoms into radiation.

Under ordinary conditions, the radiation is free to escape from the gas cloud. We can therefore visualize a gas cloud as cooling because of the collisions between its atoms. The greater the density of its atoms, the more collisions occur and the more radiation is emitted. Consequently, at higher densities, a gas cloud can cool more rapidly to a lower temperature.

If a gas cloud can cool, it can also contract. If it can contract, it will be less liable to disruption by colliding with other fragments. Even more significantly, if a fragment can cool rapidly enough, it will be capable of further fragmentation, because the role of gravity relative to pressure becomes progressively more dominant. It will continue to subdivide into many smaller subfragments, which eventually form stars.

We know from observations that collisions between systems of stars often leave the bulk of the stars unaffected. Stellar systems can interpenetrate at high velocity without doing any gross damage. There is so much space between the stars in a galaxy that it is as if two ghosts were to collide. Collisions therefore play a greatly diminished role in the later stages of fragmentation, after stars have formed.

At first, the collapsing gas cloud is very tenuous and diffuse. Efficient cooling, which requires a rather high density, cannot occur. However, high densities are eventually achieved, once regions of the gas cloud collapse into thin sheets or pancakelike substructures, and efficient cooling will take place in the denser regions. A more detailed study of this process indicates that two conditions must actually be satisfied for efficient cooling to occur. The fragments must have masses less than about 1000 billion solar masses; they must also possess characteristic radii of less than about 150,000 light-years (Figure 10.4). Fragments not satisfying these

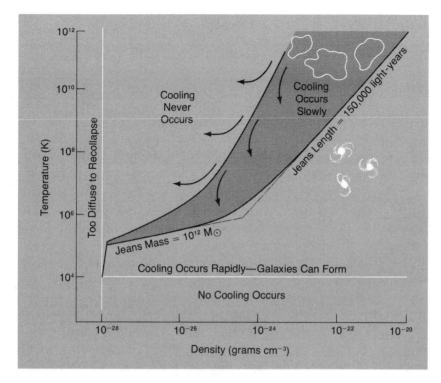

Figure 10.4 Why Are Galaxies the Preferred Scale?
The fate of a gas cloud is decided by its average density and temperature. If the cloud is initially very diffuse, it will never collapse (the limiting density at present is about 10^{-28} grams per cubic centimeter). If the cloud is initially too hot, it may begin to collapse but will be unable to cool. Thus, its pressure will remain great, and collapse will be halted. If sufficiently dense, even hot clouds can cool slowly. The clouds that are fairly cool and not too diffuse (*shaded region*) can cool most efficiently by radiation, so they can rapidly collapse to form galaxies. However, if the gas is too cold, it becomes neutral, and no significant cooling can occur (unless heavy elements are present). Two preferred scales emerge from these processes. Gas clouds that collapse (*downward pointing arrows*) will first be able to collapse rapidly at either a particular mass (about 10^{12} solar masses) or at a particular radius (about 150,000 light-years). Fragments will tend to form from a large cloud with these characteristic dimensions. These fragments are likely to become the first galaxies to form.

constraints will be either too diffuse or too hot to cool effectively. The likely fate of such fragments is to run into other fragments and be destroyed before they can break up into stars. In other words, the survival of structures of galactic dimensions is favored over more massive or larger-scale structures.[14]

FORMATION OF GALAXIES AND GALAXY CLUSTERS

The theory of gaseous fragmentation suggests that galaxy-sized structures are the preferred outcome of the collapse of a gas cloud. Fragmentation is a theory of top-down evolution, however, and we have valid reasons for favoring a bottom-up scenario. Many aspects of the evolution of galaxies cannot yet be determined with any certainty. One key unresolved issue is the question of whether or not galaxies formed before galaxy clusters. A brief perusal of the observational evidence appears to favor the hypothesis that galaxies formed first because galaxies are much denser systems than clusters. Let us suppose that both galaxies and galaxy clusters formed purely by the gravitational aggregation of matter in the expanding universe. Suppose also that relatively little energy dissipated into heat and radiation during their formation. If these assumptions are valid, then isolated galaxies must have formed at an earlier era (when the universe was denser) than the clusters. We can be fairly confident that most galaxy clusters formed at a redshift less than or of order 3, because, from a dynamical point of view they appear to be very young systems. Many galaxy clusters are irregular; they evidently have not had sufficient time to become dynamically relaxed systems.

However, isolated galaxies appeared to have formed between redshifts of 3 and 30. This estimate is based on the fact that only at an earlier epoch corresponding to this redshift range was the universe about as dense as the protogalactic clouds from which galaxies formed.[15] (However, some galaxies could have developed as a result of the mergers of small, denser systems that individually formed at much earlier epochs, namely a redshift of 100 or 1000. Some also could have formed more recently and, indeed, could still be forming today as a consequence of mergers. We discuss the possibilities later.) If little energy was radiated during the collapse, then we can directly infer the protogalactic density from the observed density. Energy conservation plays a crucial role in this argument, and the reasoning is similar to what we would use to predict how high a ball will bounce—we know we need a low-friction surface if energy is to be conserved. However, the skeptic would argue that galaxies that now appear to be isolated may originally have formed in clusters or groups. As a cluster of galaxies evolves, some galaxies (often the more massive galaxies) become more highly concentrated toward the cluster center, and outlying members are ejected. This process is called *relaxation.* Isolated galaxies may be castoffs from clusters. With respect to our own galactic neighborhood, the Local Group of galaxies may be a far-flung member of the Virgo cluster (or Virgo supercluster; see Chapter 11). Recent evidence suggests that vast voids, empty of luminous galaxies, exist between the great clusters of galaxies. One means of explaining these apparent holes in the galaxy distribution would be to assert that galaxies formed preferentially within great clusters.

We could argue that cluster-sized clouds initially condensed from inhomogeneities in the density of the early universe and fragmented into galaxies. This approach would explain the masses and dimensions of galaxies and certain other properties of clusters. However, one perplexing property of the distribution of galaxies remains that cannot easily be interpreted from a fragmentation hypothesis: the abundance of the galaxy clusters themselves.

There are very few extremely large galaxy clusters, and there is a great preponderance of small clusters and groups of galaxies. The separations between pairs of galaxies have been quantitatively computed in an analysis of the *Shane-Wirtanen Catalogue,* which lists the distribution in the sky of almost 1 million galaxies brighter than seventeenth magnitude. The catalogue is believed to be complete over a large volume of space. It is found that the probability for a galaxy to have a neighbor within a certain distance r decreases systematically as r increases. The resulting probability can be expressed in terms of the separation r. This *galaxy correlation function* appears to decrease roughly as the inverse square of r (Figure 10.5). This result holds over a range of 1000 in r, from separations of tens of thousands of light-years to tens of millions of light-years—that is, from the dimensions of galaxies to the dimensions of large clusters of galaxies.

The main theory hitherto advanced to explain the correlations of the distribution of galaxies uses the concept of *hierarchical clustering.* We can visualize hierarchical clustering as being nearly the inverse process to gaseous fragmentation. This is not to say that gaseous fragmentation played an unimportant role: in fact, it was almost certainly responsible for forming stars within a collapsing

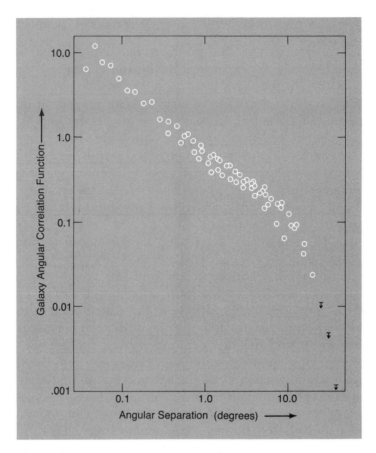

Figure 10.5   The Galaxy Correlation Function

Consider any randomly chosen galaxy. What is the probability that another galaxy will be located at some specified angular distance? This probability, the galaxy angular correlation function, should give us a measure of how frequently galaxies occur in pairs or small groups as opposed to being distributed randomly. The probability can be computed for an entire catalogue that contains almost 10^6 galaxies and samples the universe to a depth of about 10^9 light-years. What is plotted on this figure is actually the correlation function: choose any galaxy, move some specified angular distance away, and count the number of galaxies at that location. This is done for all angles and for every galaxy in the sample. Only when the correlations are weak can we identify the correlation function with the probability of finding a galaxy at some specified angular separation. The angular correlation function (denoted by data points) is found to decrease roughly as the inverse of the angular distance. When this angular dependence is deprojected into three dimensions, the galaxy correlation function is found to decrease as the square of the galaxy separation. Over very large scales, the correlation function decreases more rapidly. This means that the galaxy distribution becomes increasingly random. An angular scale of 10 degrees corresponds to a linear distance of 10^8 light-years. On smaller scales, galaxy positions are highly correlated; that is, there is a high probability of finding a close neighbor to any given galaxy. From the relative smoothness of the correlation functions, we infer that no particular scale seems preferred.

protogalaxy. However, from the perspective of primordial density fluctuations, it is simplest to consider the evolution of isothermal or of cold dark matter fluctuations. In such a picture, galaxies first form and then gradually aggregate as a result of their mutual gravitational attraction. It is possible that the initial fluctuations destined to become galaxies possessed a wide range of masses from globular-cluster dimensions of a million solar masses to giant-elliptical dimensions of 1000 billion solar masses. A group develops first, a cluster develops subsequently, and a rich cluster or even supercluster (a collection of several clusters) develops eventually. A hierarchy of clustering is produced: of groups within weak clusters within rich clusters. The resulting distribution of galaxies can account for the observed correlations, provided that the initial distribution of newly formed galaxies is approximately randomly distributed. If galaxies began to form at a redshift of approximately 10, correlations would subsequently develop that correspond to what we now observe. This presumes that the universe is not of very low density, compared with the critical value for closure; if it is, gravity is less important today and galaxies must have formed at a somewhat higher redshift.

The rival theories of clustering and fragmentation have been studied by means of computer simulations. The gravitational interactions of many thousands of mass points have been followed as the universe expands. The outcome is shown in Figure 10.6. The pancake-fragmentation, or hot dark matter, theory results in an excessive number of great galaxy clusters, whereas the hierarchical clustering, or cold dark matter, theory produces a universe more closely resembling what we observe. Indeed, a comparison of the cold dark matter theory with observations (Figure 10.7) shows that it is difficult to distinguish the simulated universe from the real universe.

It has been inferred from similar numerical computer simulations of galaxy clustering that the required initial fluctuation distribution systematically decreases at larger and larger scales. Specifically, the initial density contrast must vary about as $M^{-1/3}$ on galactic scales, and $M^{-2/3}$ on much larger scales, where M refers to the total mass encompassed within a given length scale. Cold dark matter, rather remarkably, gives rise to precisely the desired form of the correlations that are measured in the galaxy distribution. The universe was initially very smooth at galaxy-cluster scales but

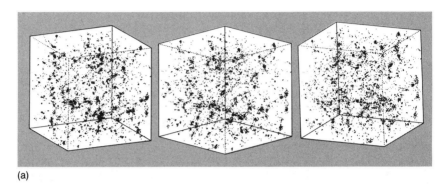

(a)

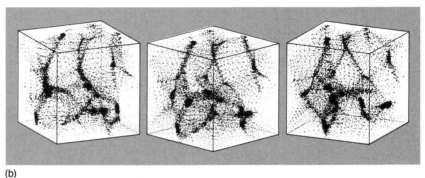

(b)

Figure 10.6 Galaxy Clustering

Computer simulations of galaxy clustering shown in three different projections: (a) hierarchical clustering, with the mass points, each representing a galaxy, initially laid down at random and (b) a top-down scenario, similar to what we would expect in the pancake-fragmentation theory, in which the mass points are initially clustered only on large scales, all small-scale structures having been suppressed. With these two sets of initial conditions, the universe is allowed to expand; the final outcome, corresponding to the universe observed today, is shown in (a) and (b).

progressively more inhomogeneous at smaller and smaller scales. At the decoupling era, at a redshift of 1,000, the required fluctuations at the scale of galaxy clusters are about 0.1 per cent. In principle, such fluctuations should be measurable if we search for temperature fluctuations in the cosmic microwave background radiation. This effectively probes the decoupling era; since then, the microwave photons have streamed freely to us. If we simply extrapolate the inferred fluctuation spectrum on cluster scales to smaller mass scales, the inference is that clumps of mass of the order of 100 million solar masses would have condensed somewhat

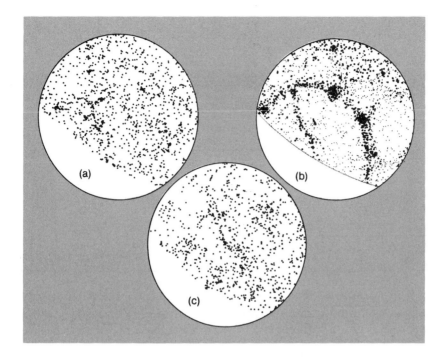

Figure 10.7 Galaxy Distributions
A comparison of the observed (c) and simulated (a) and (b) galaxy distributions. The north galactic pole is at the center of these equal-area projections. One computer simulation shows the outcome of pancake (hot dark matter) initial conditions, in the top-down scenario (b), the second simulation (a) shows the results of hierarchical, or bottom-up, clustering, as in the cold dark matter theory.

after decoupling. Some cosmologists have argued that these clumps were the precursors of galaxies. They could have resulted from primordial isothermal fluctuations that collapsed and formed star clusters or compact galaxies at a redshift of 100 or even 1000. Such fluctuations, as we have previously seen, were expected to be of the order of 100,000 solar masses and larger. These first objects to collapse subsequently merged into galaxies.

The cold dark matter theory makes a more specific prediction. The shape of the primordial fluctuation distribution is completely specified, and we infer that the first nonlinear objects collapsed at a redshift of about 30. Hierarchical clustering occurred as larger and larger aggregations pulled away from the cosmic expansion and collapsed. Clumps came together and merged. Clouds of gas or dark matter merge very easily because much of the relative kinetic energy in an encounter is absorbed by internal motions, and the colliding clouds inevitably spiral together. Computer simulations have revealed that almost all traces of substructure are erased by such gravitational mergers. We find that a whole hierarchy of systems developed, many merging into larger and larger galaxies, until a wide range of galaxy masses is attained.

How can we elucidate the kind of initial density fluctuations that must have been present in the very early universe to satisfy this theory? At early eras, before galaxies formed, the randomly distributed density fluctuations that pervaded the universe resembled a form of noise. Ordinarily, we use the concept of noise to characterize sound, but noise is also a convenient way to characterize any distribution of fluctuations. There are many possible varieties of noise—speech, for example, is highly correlated, nonrandom noise. Whether the noise spectrum that describes the fluctuations in the early universe is completely random and uncorrelated *white noise* has not been definitely resolved. According to computer studies of galaxy clustering, there may be a slight tendency to favor a model in which the noise was systematically correlated toward the larger scales, in which case the initial density fluctuation spectrum varied more or less as is predicted by the cold dark matter hypothesis. White noise, which resembles radio static and appears to have the same intensity at all frequencies, does not provide as good a match to the observed galaxy correlations. If the noise is increasingly correlated on larger scales, it would increase in intensity toward low frequencies, or longer wavelengths. The universe appears to have initially contained correlated rather than random noise.

As we saw in Chapter 9, the source of the cosmic noise, in terms of gravitational effects in the early universe, is still a matter for speculation. Inflationary cosmology does predict that quantum fluctuations amplified into a spectrum of small fluctuations resembling a form of correlated noise. Something of this nature must be postulated to account for the origin and spatial distribution of the galaxies. Whether this spectrum extends to mass scales as small as 1 million solar masses (corresponding to primordial fluctuations in cold dark matter or to primordial isothermal fluctuations) or only ranges from 1000 billion solar masses to larger scales (corresponding to primordial adiabatic fluctuations in baryons or in hot dark matter) is an issue that is still vigorously debated, although the former hypothesis is favored.

Numerical simulations support cold dark matter, although there are still unresolved issues, notably concerning large-scale structure, that will be described later. Hot dark matter is out of favor for the following reason. The top-down evolution means that galaxy formation has to be a recent phenomenon, because galaxies can form only after clusters have collapsed. Now, large-scale clustering develops rather rapidly, and only if galaxy formation occurred at a redshift less than 1 would the galaxy distribution be compatible with the observed clustering. In fact, galaxies are found at redshifts as large as 2, and quasars, objects that are closely related to galaxies, are seen at redshifts up to 4 or larger. It follows that a hot dark matter, or gaseous-fragmentation, scenario for large-scale structure and galaxy formation is unacceptable. Moreover, correlations of the galaxy distribution definitely favor a hierarchical clustering model, in which small galaxies formed first.

Variations on the gaseous-fragmentation scenario are still feasible, however. According to one of the revolutionary approaches described in Chapter 9, gigantic explosions, on the scale of an entire galaxy, may have swept up vast shells of intergalactic gas. These shells fragmented to form a new generation of galaxies, which may in turn have released energy explosively early in their lives. This mechanism of explosive amplification resulted in galaxy formation by fragmenting gaseous shells, with only rare seeds required to initiate the process, and would also have led to the formation of great voids with no luminous matter. We will later consider evidence for these voids.

The gaseous fragmentation model has not yet been developed to the point where we can say much about any predictions of the generation of correlations among galaxies. Whether galaxies formed before or at the same time as clusters and whether hierarchical clustering or gaseous fragmentation was the dominant process remain controversial and unresolved issues. A synthesis of these viewpoints can be attempted, however, and we shall return to this issue in Chapter 11 after further examining some characteristics of galaxies that our theory must explain.

THE LUMINOSITY FUNCTION

Correlations between the separations of the galaxies are not the only regularity we find when we analyze their distributions. A further intriguing symmetry can be expressed in terms of the galaxy *luminosity function,* or distribution of galaxies with luminosity. There are more faint galaxies than bright galaxies, and there are exceedingly few very bright galaxies. This distribution in luminosity has been evaluated for different clusters of galaxies. Clusters are the easiest systems to study, because all the galaxies in any given cluster are at the same distance. The luminosity function appears to be universal; no very significant differences are found when different clusters of galaxies are studied. Furthermore, there are indications that a similar luminosity function may even apply to galaxies outside of rich clusters (Figure 3.6).

The origin of the luminosity function of galaxies is another mystery of galactic evolution. One intriguing possibility is that either gaseous fragmentation or multiple mergers of *protogalaxies* that formed hierarchically may have led to the development of the mass distribution (and thereby the luminosity distribution) of newly formed galaxies. Newly formed protogalactic fragments were largely gaseous; at first, they would not have acquired substantial random motions, and their relative motions would have been subsonic and gentle. Thus, emerging fragments may have tended to collide and coalesce. After a number of protogalaxies (particularly the smaller systems) had merged into larger systems, the collision rate would have diminished. Protogalaxies would have developed very large random motions in the gravitational field of the cluster. These motions would have been supersonic, and collisions would have been violent. Any subsequent collisions certainly would not have led to coalescence. Any remaining diffuse gas in the galaxies might be heated and ejected, but the stars would freely interpenetrate and would survive, more or less unscathed. It is quite conceivable that initial mergers of protogalaxies led to the development of the luminosity function of the galaxies. However, a detailed theory of this process remains to be developed.

FORMATION OF THE STARS

Once a protogalactic fragment formed that was capable of cooling during collapse, it must have rapidly subfragmented into smaller, denser substructures (Figure 10.8). We can readily imagine that gaseous fragmentation was a chaotic, turbulent process. The turbulence ensured that the gas distribution remained highly irregular. Turbulent eddies continually collided and dissipated their

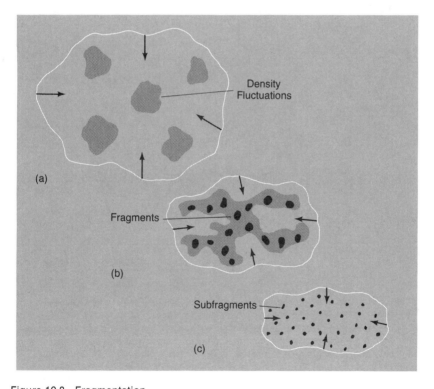

Density
Fluctuations

(a)

Fragments

(b)

Subfragments

(c)

Figure 10.8 Fragmentation
A protogalactic cloud with density fluctuations cools (a), collapses (b), and fragments into smaller and smaller subfragments (c).

energy of motion into heat and radiation. As long as the gas continued to radiate effectively, it would not have heated significantly. If the temperature did not rise, pressure gradients would have remained small, and the collapse would have continued unimpeded. As fragmentation proceeded, more and more subfragments formed, and they became smaller and denser. Eventually, subfragments were produced that were extremely dense and opaque. Radiation was trapped within them, and further cooling was inhibited. Cool-

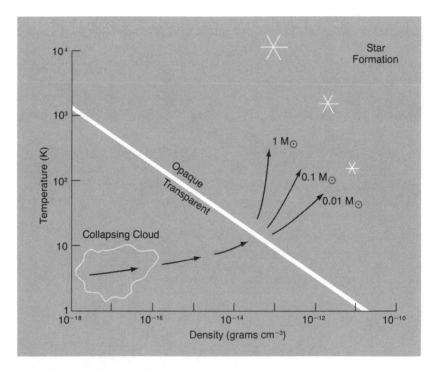

Figure 10.9   Star Formation

The process of fragmentation ceases only when sufficiently dense, opaque fragments form. As the density increases during the collapse (*heavy arrows*), smaller and smaller fragments form. Eventually, at a sufficiently great density, the fragments become opaque (*diagonal line*). Radiation is consequently trapped. Because the loss of radiation and pressure allowed fragmentation to continue and the collapsing cloud to subdivide, subfragmentation now ceases. The heavy arrows show the smallest fragment sizes that can form and become protostars. If heavy elements are present in the form of dust grains, which would then determine the opacity, the smallest protostars that form are about 1 percent of the mass of the sun. Much larger protostars can also form.

ing would occur, but rather more slowly, as the radiation gradually leaked out. The gas would then heat up as it continued to collapse, raising the temperature in the interior of the fragments and developing a pressure gradient. The difference in pressure between the hot interior and the exterior of the cloud helped to resist the gravitational collapse, and the pressure force gradually decelerated the collapse. As the collapse decelerated, the opaque subfragments gradually contracted.

No further subfragmentation could occur, and these final fragments are presumed to be stars. We have some confidence in this conclusion, because when a self-gravitating fragment becomes opaque, we can compute its mass. Other processes, involving for example, magnetic fields and energetic stellar winds, must certainly intervene to produce the detailed mass distribution of stars. However, it is a remarkable and entirely unexpected coincidence that the final mass calculated for such fragments is similar to the mass of a typical star (Figure 10.9).

ELLIPTICAL AND SPIRAL GALAXIES

Some galaxies, notably elliptical galaxies, are fairly round and are often concentrated in rich clusters. Others, such as spiral galaxies, are very flat and tend to be more uniformly distributed in space. What are the implications of this morphological distinction (see Figure 10.10) for a theory of galactic evolution?

Imagine a massive collapsing cloud that is destined to become a great cluster of galaxies. Or, taking the hierarchical-clustering viewpoint, we can simply imagine a region where the frequency of occurrence of protogalaxies is much higher than the average density of these systems. A region destined to become a great cluster must contain a larger number and a higher density of galactic fragments than a smaller cloud contains. We have already seen that they could have survived mutual collisions, once they had subdivided into stellar fragments. Collisions of this sort would have occurred more or less simultaneously with the fragments' detachment from the large cloud.

Simple arguments about star formation suggest that typical fragments were likely to be fairly flattened systems. Densities in re-

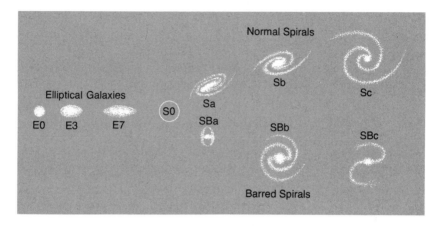

Figure 10.10 The Morphology of Galaxies

The classification scheme developed by Edwin Hubble in the early 1930s organizes galaxies according to shape. They range from amorphous, relatively uniform elliptical systems containing many red stars and little gas (*left*) to highly flattened spiral disks with prominent nuclei, many blue stars, and lanes of gas and dust (*right*). The elliptical galaxies range from spherical systems (E0) to flattened ellipsoids (E7). The spiral galaxies form two sequences, depending on whether the nuclei are round (*upper right*) or barlike (*bottom right*). At one end of the spiral sequence are galaxies dominated by large bright nuclei (Sa, SBa); at the other end are galaxies with smaller nuclei and prominent spiral arms (Sc, SBc). The category S0 denotes a disklike galaxy with no spiral arms or young stars; these galaxies have many of the properties of both intermediate ellipticals and gas-poor spirals.

gions of greatest compression would continue to rise, until subfragmentation into stars could occur. Such high densities were first attained when regions of the cloud collapsed into pancakelike structures. Thus, the collapsing cloud remained gaseous until galactic pancakes developed. Once pancakes formed, rapid fragmentation into stars probably followed. Since angular momentum tends to be conserved, the motions acquired during collapse in the direction perpendicular to the plane of rotation are preferentially dissipated. This process resulted in the formation of flattened, disklike systems of stars.

As mentioned previously, initial collisions between such fragments may have led to coalescence. Protogalaxies could not have developed large relative velocities until they had fallen through the cluster center, and early collisions would tend to be rather gentle and sticky affairs. The interacting protogalaxies merge together, and the resulting merged system would rapidly become spheroidal, like an elliptical galaxy (Figure 10.11). After a few close encounters and ensuing mergers had occurred, a protogalaxy would have acquired sufficient speed to subsequently survive any collisions, passing more or less unscathed through other galaxies.

Even if mergers are not the dominant process in elliptical-galaxy-formation, another effect will tend to create round galaxies in dense clusters where collisions occur. There is a small acceleration of the stars in the protogalaxy by the *tidal forces* generated during such an encounter. This effect will tend to strip stars from the loosely bound outer regions, or *halos*, of galaxies. There is also a small transfer of energy from the relative motion of the two systems into the internal motions of the stars. If stars in an initially flat galaxy are induced to move more rapidly, the system must thicken. We can visualize this process as a heating of thin systems. Disks are initially cold because the random motions of stars, which are responsible for the thickness of a stellar disk, are relatively low compared with the rotation velocity of the stellar system, which accounts for the radial extent of the disk.

Away from rich clusters of galaxies, we would expect collisions of protogalaxies to occur much less frequently. These galaxies should, for the most part, remain flat if they are flat initially. We indeed observe that elliptical galaxies are usually found in clusters; outside clusters, relatively few ellipticals are observed. However, we would expect the old components of galaxies to be spheroidal. According to the hierarchical-clustering theory, if galaxies formed by mergers of many smaller substructures, it is likely that round systems rather than highly flattened systems were produced. Star formation was initiated before any significant degree of flattening could occur, and the resulting galaxies would tend to be spheroidal rather than disk shaped. It is only later that the disk develops.

In fact, a flattened subsystem or disk develops naturally. As the stars evolved, they shed considerable amounts of gas. Within galaxy clusters, interactions among galaxies ensured that this gas would be swept out into the intergalactic medium. Outside clusters, however, the gaseous debris would spiral in toward the galactic centers, tending to collect into a disk, where it subsequently would fragment into a second generation of stars. In this way, disk galaxies are thought to form outside the great clusters, and predominantly

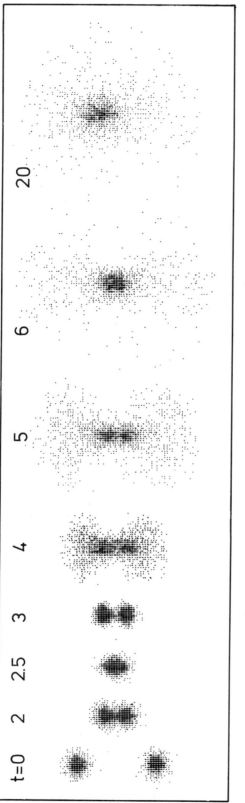

Figure 10.11 Galaxy Collisions

The merger of two galaxies may initially lead to the formation of a very irregular galaxy. The irregularity will rapidly disappear, and the galaxy will settle down to become a much fatter elliptical galaxy. This process takes hundreds of millions of years—a short time compared with the ages of the oldest stars and many galaxies. Thus, relatively few galaxies are expected to be caught in this act of digestion. The figure shows a computer simulation of such a merger between two galaxies, each containing 1000 stars. The projected density is shown at seven successive instants. Figure 10.15*b* may represent such a system. (Courtesy of T. van Albada.)

Figure 10.12 Spiral Galaxies

(a) A bright giant face-on spiral of Hubble type Sc, object M101 in *Messier's Catalogue of Nebulous Objects* (1784). (b) The nearby edge-on SBc spiral NGC 891. (c) A composite photograph of the Milky Way, also seen edge-on. (d) The Sombrero galaxy, a giant edge-on Sa spiral (NGC 4594). (e) NGC 4565, type Sb.

elliptical galaxies are thought to form within the clusters. More-over, since both the disk material and the intergalactic gas are sec-ondary ejecta of stellar origin, we would expect them both to be relatively abundant in heavy elements, which have been synthe-sized and recycled by successive generations of the stars. In fact, the disk populations of spiral galaxies have abundances of heavy elements that, on the average, approach that of the sun. Recent evidence suggests that the hot intergalactic gas observed in rich galaxy clusters also possesses near-solar abundances of heavy ele-ments. We shall take up this topic again in the following chapter.

Let us return to the morphological distinction between spirals and ellipticals. Elliptical galaxies acquired their name because their light distribution (or more precisely, the contours of constant brightness) appears to have a smooth elliptical shape. No very flat-tened elliptical galaxies are found. The most extreme cases (known as E7 galaxies) are flattened in the ratio of 3 to 1. Spiral galaxies, however, are dominated by a highly flattened disk with a distinctly different light distribution from that of an elliptical galaxy (Figure 10.12). Spirals are dominated visually by their spiral arms, which consist of gas and young stars, but these form an insignificant part of the mass distribution of the disk. The disk is uniform in surface brightness toward the center and falls off in brightness rapidly at a rather well-defined edge. In contrast, an elliptical galaxy grad-ually decreases in intensity from the center until it fades away and becomes undetectable against the sky background (Figure 10.13).

Why are there no very flattened elliptical galaxies? The answer is almost certainly contained in the issue of stability. A galaxy that is too flat will not maintain a symmetrical shape for many rotation periods. Instead, it will develop a central barlike condensation. Many spiral galaxies have central bars that may be manifestations of this phenomenon. Most other spiral galaxies, including the Milky Way and the Andromeda galaxies, are probably stabilized by the presence of massive spheroids of old stars that dominate the gravity of the system in its inner regions. The Sombrero galaxy (Figure 10.13) is a beautiful example of a spiral galaxy with a dom-inant spheroid.

Nature evidently will not tolerate the presence of highly flat-tened ellipticals. If such systems form, they probably become un-stable. A central condensation of stars develops, which may even-

Figure 10.13   An Elliptical Galaxy, M87

M87 is the brightest elliptical galaxy in the Virgo cluster; it is one of the brightest galaxies known. The mass of M87 is more than 10 times larger than our own galaxy. The fuzzy dots scattered over the outer parts of the galaxy are actually globular star clusters, each containing a million or more stars. A peculiar jet is present in the nucleus; it can be seen on a short exposure photograph (*bottom*). M87 is also an intense x-ray emitter and a source of intense radio emission, which emanates from both the jet and a more extended halo.

tually settle into a disk-shaped system. If these systems are able to retain their gas and continue to form young stars, they may develop a spiral structure and be recognizable as spiral galaxies. Consequently, spiral galaxies can develop later than (as well as during) the initial collapse phase of a protogalactic cloud. This develop-

Figure 10.14 A Galaxy of Type S0

Galaxy NGC 2685 (the "Spindle Galaxy") resembles a typical S0 galaxy in its inner regions, with a flattened, smooth, amorphous light distribution. However, luminous helical filaments, also containing dark dust lanes, surround the inner spindle. The outer filamentary structure is more characteristic of a spiral galaxy, and it has been suggested that the Spindle Galaxy is in the process of merging with a smaller spiral system.

ment could most easily happen outside rich clusters, where collisions with other galaxies or with hot intracluster material would not strip spiral galaxies of gas. Inside clusters, we might expect to find disklike galaxies that have been stripped of gas and are devoid of new stars or of prominent spiral structure. Such galaxies are indeed observed, and in many of their properties they are intermediate between ellipticals and spirals. These galaxies are designated S0. Although S0 galaxies are highly flattened systems (Figure 10.14), they share some characteristics, such as color and spatial distribution, with ellipticals. However, whether most S0 galaxies have formed from stripped spirals or instead owe their morphological characteristics to the conditions in the protogalactic gas cloud from which they condensed is a matter of continuing debate. The issue is one of nature or nurture, and is unresolved.

THE ANGULAR MOMENTUM OF GALAXIES

All galaxies are rotating, but ellipticals are rotating less rapidly than spirals. The origin of galactic angular momentum can be attributed to the action of tidal torques between neighboring protogalaxies. As long as the protogalaxies are not spherically symmetric, differential accelerations are induced by neighbors, and these result in the acquisition of angular momentum. No net angular momentum is produced: one object spins clockwise, and its neighbor spins counterclockwise. The spins induced this way are very small, amounting to a rotational velocity that is perhaps only a tenth of the critical value needed for centrifugal balance against gravity.

As the protogalaxy collapses, however, it spins more and more rapidly. Exactly how rapidly depends on whether a halo of inert dark matter is present. The dark halo acts as a support against which the gas can torque as it loses energy and falls toward the inner regions. After collapsing by about a factor of 10 in radius, the gas is supported centrifugally in the plane of rotation: perpendicular to this plane, the gas distribution flattens because there is no support. In this way, a rotating gas disk develops that now fragments into stars to form a spiral galaxy.

In the absence of a dark halo, centrifugal forces do not become dominant; instead, stars form when the protogalaxy has contracted

(c) ARP 241

(b) ARP 315

(a) ARP 271

(d) ARP 243

(e) ARP 244

(f) ARP 147

Figure 10.15   Various Interacting Galaxies

Several pairs of interacting galaxies from Arp's *Catalog of Interacting Galaxies* are shown. (*a*) Arp 271 is a pair of spiral galaxies (NGC 5426 and NGC 5427). These are interacting tidally, pulling material from one another. (*b*) Arp 315 (NGC 2832) is an elliptical pair in intimate contact, perhaps on the threshold of the final stages of galactic cannibalism. Some pairs have suffered very close encounters; examples are (*c*) Arp 241, (*d*) Arp 243 (NGC 2623), and (*e*) Arp 244 (NGC 4038 and NGC 4039), also known as the Antennae, characterized by ejection of gigantic plumes of material. Ring galaxies such as Arp 147(*f*) are a related phenomenon. A head-on collision between two galaxies may have resulted in the formation of a great ring of stars in the outer regions of one galaxy.

to form a flattened spheroid. The flattening of an elliptical is due not to systematic rotation but to anisotropic random motions of the stars: the motions are somewhat smaller in the direction perpendicular to the symmetry plane. A merger of two disk galaxies also produces a spheroidal stellar system, much like an elliptical galaxy. Whether the usual elliptical-galaxy formation mechanism is through loss of a dark halo, followed by contraction of the gas cloud, or through mergers is not known, but both processes are likely to occur during the formation of galaxy groups and clusters.

GALAXY GOBBLING

Some galaxies are giant systems that dominate an entire cluster. The largest galaxies of this type extend to a radius of up to 1 million light-years. They are perhaps 100 times more luminous (and more massive) than our own Milky Way galaxy. Clusters of galaxies usually contain one such system at most, although not all clusters contain a giant galaxy. It seems that the giant galaxies have grown at the expense of other galaxies in their cluster.

Initially, the central galaxy may have been only slightly more extended than its neighbors. Collisions between galaxies in a cluster then tend to remove the outermost stars, which moved around freely in the intergalactic medium. Galaxies experience a dynamical friction when passing through the diffuse material (ejected stars and gas) in intergalactic space. For example, if a star is gravitationally deflected toward a passing galaxy, the star gains energy at the expense of the galaxy, which gradually loses its kinetic energy of motion. After many such encounters, the galaxy tends to spiral in ever-decreasing orbits toward the center of the cluster. This process may take billions of years, and it seems to have occurred to an appreciable extent only in the oldest and most centrally concentrated clusters of galaxies. These clusters show the greatest tendency toward relaxation (the heavier galaxies having had time to slow down), and these clusters often contain a giant galaxy near the center. The giant galaxy probably forms when the original central galaxy in the cluster swallows a host of smaller systems. The final stages of the digestion of a galaxy occur relatively rapidly, in one or two orbital periods. The galaxies that appear to have grown in this cannibalistic manner generally have very regular and smooth extended stellar halos. They show no signs of indigestion.

A few unusual galaxies appear to have suffered a recent collision or merger (Figure 10.15). The rarity of such systems is consistent with the speed of stellar relaxation that follows a merger. An interesting example of a galaxy collision is the head-on impact of one galaxy with another. When this occurs, the central regions of one galaxy may be ejected, forming an expanding shell or ring of stars. Rings are extremely unstable systems, and they do not survive very long. Nevertheless, a few *ring galaxies* are known, and the collision hypothesis appears to account for their structure. A more widespread relic of mergers that occurred in the remote past is the prevalence of shells in the extreme outer regions of many isolated elliptical galaxies; these will be described in the next chapter.

Understanding the early evolution of galaxies lies within our grasp. The hypothesis of hierarchical clustering of primordial aggregates under gravitational forces cannot by itself account for the complexity of observed structures. Galaxy interactions with ensuing dissipation and collapse must also play a crucial role. Gaseous fragmentation may have occurred as protogalaxies formed stars; the complex physics of the associated dissipation accounts for the structure of the luminous cores of galaxies. In studying galactic evolution, we often find at least two alternative theories for any observed fact. Is the distinction between elliptical and spiral galaxies the result of collisions and mergers of initially flat galaxies in clusters or the result of accretion of gas into intrinsically round galaxies outside clusters? In the next chapter, we shall cite observations of the great clusters. These enable us to take a tentative stand on this issue. It is nevertheless true that we would expect galaxy formation to depend strongly on the environment if galaxy interactions play a role. Thus, the morphology of galaxies should differ when they are observed in clusters and in the field, and further observations of this sort may help us to discriminate between alternative models.

The great clusters contain only a small percentage of all galaxies. However, because of their high central concentration, they provide an attractive laboratory for study of galactic evolution, as we shall next learn.

· 11 ·

GIANT CLUSTERS
OF GALAXIES

Stellar systems . . . are concentrations of stars, dust, and
gases within a tenuous but continuous distribution of matter
distributed throughout the whole universe.
—*Fritz Zwicky*

Almost every galaxy has a neighbor to share the extreme
solitude of space. Our Milky Way galaxy has two close companions,
the Large and Small Magellanic Clouds. They are rather small galaxies, with their respective masses 10^{10} and 2×10^9 solar masses,
compared with the 2×10^{11} solar masses of the Milky Way galaxy.
The Andromeda galaxy is the nearest galaxy of comparable size to
our own. Andromeda and the Milky Way, together with a number
of smaller companion galaxies, form a group of galaxies known as
the *Local Group.* Many other groups of distant galaxies are visible
from photographs of the deep sky.

Groups of galaxies are generally considered to be gravitationally
bound systems. Just as the planets orbit around the sun, galaxies
in a group orbit around one another. They remain in the same
region of space by virtue of their mutual gravitational attraction.
Groups of galaxies occur very frequently in space, and a typical
group contains between 10 and 100 galaxies (Figure 11.1). Occasionally, large concentrations of galaxies occur. These are the great
clusters, which can contain over 1000 galaxies. The nearest rich

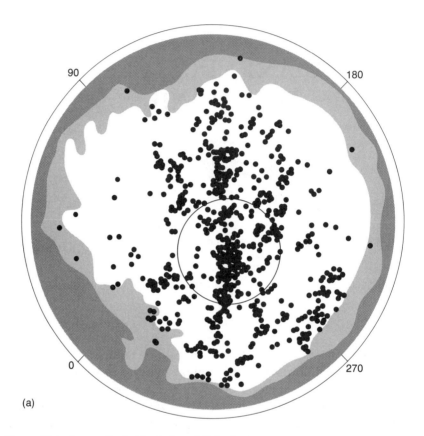

(a)

Figure 11.1   Large–Scale Clustering of Galaxies

(a) A map plotting the distribution of all galaxies brighter than thirteenth magnitude in the Northern Galactic Hemisphere shows The North Galactic Pole at the center. The shaded area marks the boundary of observation because of gas and dust in the Milky Way, which extends around the circumference of the circle. The inner circle is centered in the Virgo cluster of galaxies. The flattened distribution around Virgo is the local supercluster, which extends for about 50 million light-years. Space is sampled to a distance of about 250 million light-years. (b) A map of the North Galactic Cap illustrates the distribution of bright galaxies down to fourteenth magnitude and uses a stereographic projection about the North Celestial Pole (NCP). Space is sampled to a distance of about 400 million light-years. Approximately 1000 galaxies are shown, each point representing one galaxy.

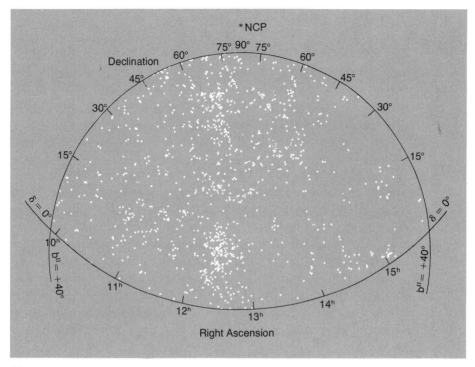

(b)

Figure 11.1 *Continued*

cluster to us is the Virgo cluster of galaxies, some 60 million light-years away. The still richer Coma cluster of galaxies is about 400 million light-years away. In a giant cluster such as Coma, the galaxies are moving relative to one another at velocities of thousands of kilometers per second. These velocities are much greater than the velocities of stars within individual galaxies; for example, the sun orbits the center of our galaxy at a speed of 250 kilometers per second.

In this chapter, we shall examine some properties of galaxy clusters and galaxies to see how they fit our emerging theory of galaxy formation. As we shall see, many questions of galaxy formation remain unresolved, and further observations will be required before we can definitively choose between alternative models. Meanwhile, our tentative hypotheses can show us where to look for conclusive evidence.

GALAXY CLUSTERING

Galaxy clusters, the great aggregations of galaxies, contain thousands of luminous galaxies, and uncounted hordes of dwarf galaxies, the lesser systems. Gravity alone appears responsible for their origin. Scientists have tested this hypothesis by performing computer simulations. Take a few thousand point masses, each one representing a galaxy and all of them initially expanding away from one another with the Hubble flow. Local gravity inexorably exerts itself, and this expansion of space that determines the universal expansion is gradually overcome. At first, a few extra points are accreted into the excess density fluctuation, so that the density enhancement becomes more pronounced. The points are still expanding from one another but are undergoing a systematic deceleration relative to the Hubble flow. Soon, enough mass has accumulated to make most of the points fall into the center of the local density peak. Nearby, other density peaks have developed in a similar way, and these various clumps eventually fall together. The form of this final agglomeration of mass points resembles a galaxy cluster.

Such a cluster is a stable entity. Its galaxies orbit in random directions, but lack enough energy ever to escape from the cluster environment. Clusters vary considerably in appearance. Some have dense, spherically symmetric cores, with hundreds of galaxies crammed into a few million cubic light-years. These galaxies are almost exclusively elliptical galaxies and S0 galaxies (with features in between elliptical and spiral). Others are looser, more irregular aggregates of predominantly spiral galaxies. Some clusters have a central giant elliptical galaxy with a large diffuse stellar envelope, known as a cD galaxy, whereas others may have a central pair of more-normal giant ellipticals. Some cD galaxies have multiple nuclei, direct evidence that they have grown by cannibalizing other, smaller galaxies, not all of which have yet been fully digested. It can take as long as a billion years or more for the final consummation of the act of cannibalism.

Some clusters are so inhomogeneous that they must be systems too young to have undergone much dynamical relaxation. Dynamical interactions are important as the initial collapse of the cluster occurs; they eventually result in a relatively homogeneous and centrally condensed cluster. Continuing infall of neighboring clumps of galaxies destroys this simple picture, however, because few clusters are likely to have formed in a truly isolated region. Environment seems to be the ultimate arbiter in cluster as well as in galaxy morphology. Many clusters appear to extend indefinitely until the next nearest cluster, with galaxies spread out between them in bridges and sheets. Larger regions containing several clusters are called superclusters.

Superclusters are the largest known aggregates in the universe, extending for tens of millions of light-years. The best studied supercluster is centered in the Virgo cluster of galaxies, and our own Milky Way galaxy is on its outer periphery. Although we are not sure how the dark matter is distributed, the Hubble-flow deviations in the Local Supercluster are known: their measurement requires determination of galaxy distances and redshifts. Our Local Supercluster contains enough mass to decelerate the motion of our Local Group of galaxies by about 10 percent of the normal Hubble expansion. Had it undergone a substantial gravitational contraction, it now would be effectively detached from the Hubble expansion and the random component of the galaxy motions would be much greater than is observed. Evidently the Local Supercluster is just a density enhancement in the galaxy distribution and is still predominantly expanding. Whether our galaxy is destined to eventually fall into Virgo depends on how much matter there is in the Local Supercluster region. Our future fate will be determined by the presence of dark matter whose presence we cannot verify by direct observation. We can say that if the dark-matter-distribution traces that of the luminous galaxies, then the galaxies will recede forever.

As Figure 11.2 demonstrates, a closed universe, in which the galaxy recession will eventually reverse, requires that the dark matter in our Local Supercluster be more smoothly distributed than the luminous matter. This would allow a higher background density, sufficient perhaps to close the universe. Whether this could be the case cannot be resolved by observations of the Local Supercluster; instead, we must explore the structure of the universe on still larger scales.

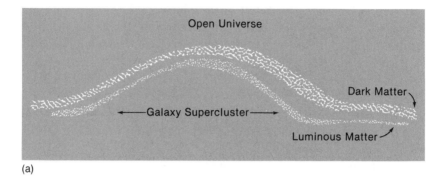

(a)

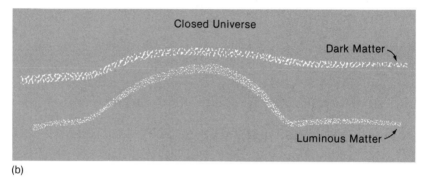

(b)

Figure 11.2 The Distribution of Dark Matter

The luminous-matter distribution is known from observation and is schematically the same in this figure of the density in an open (a) and a closed (b) universe. However, the dark matter must be much more uniform if the universe is closed; otherwise it would have been observed to affect the dynamics of galaxy clusters.

HUBBLE BUBBLES

Beyond the Local Supercluster, there are other, even larger superclusters of galaxies. The most surprising discovery, however, is that the universe contains many giant holes, resembling quasi-spherical bubbles, or more precisely voids, in the galaxy distribution. These appear in redshift surveys, which use Hubble's law to convert redshift into a distance measure and thereby yield a three-dimensional map of the galaxy distribution. The maps are then distorted in the radial direction, for the simple reason that any peculiar velocities of galaxies are treated as being part of the usual Hubble expansion, or flow; distances are overestimated or underestimated, depending on whether the random component of a galaxy's motion is directed away or toward the observer. This stretching in the radial direction produces what has been called the "fingers of God" effect, the appearance of filaments of galaxies as pointing toward us (Figure 11.3). If astronomers had a reliable measure of distance, we might hope to distinguish peculiar velocity from the Hubble component of velocity that is associated with the expansion of the universe. Despite the distorted perspective from which we view them, most galaxies appear to be distributed in filaments and in thin sheets that surround large voids, or bubbles, up to 300 million light-years in diameter. Perhaps 90 percent of the volume of space appears to be empty.

But is it really empty? The precise statement is that luminous galaxies are absent. Already, several strong-emission-line dwarf galaxies have been discovered in voids. It is possible that nonluminous forms of matter are also present. It has even been suggested that the voids contain many dwarf galaxies of such low surface brightness that they would not show up in the usual surveys. An alternative form might be matter that has not condensed into galaxies or stars but is still in its primitive, gaseous form. Intergalactic gas in the nearby voids should be observable. If the gas is distributed in clouds, sophisticated space observations are needed to detect them, and these should be forthcoming with the Hubble space telescope at the end of the 1980s. Because such primordial gas clouds would not contain any metals, there is no absorption in the visible part of the spectrum. These clouds are still detectable because the atomic hydrogen can absorb light from background sources, but this absorption occurs in the far ultraviolet part of the spectrum, at the *Lyman alpha line* of hydrogen. This is the fundamental emission line of atomic hydrogen that results from the radiative transition by an electron from the ground state to the first excited state. Lyman alpha is in the far ultraviolet at a wavelength of 1216 Ångstroms, which is observable only above the earth's atmosphere.

In the spectra of highly redshifted quasars, the Lyman alpha absorption is shifted to the visible region of the spectrum, and distant parts of the universe are found to contain many such clouds. Whether the voids contain such clouds remains to be seen. It is also possible that the voids might truly be empty. The collapse of

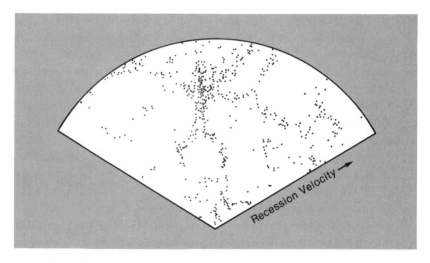

Figure 11.3 A Slice of the Universe

In this wedge-shaped map, recession velocity is the radial coordinate and ranges from 0 (at the origin) to 15,000 kms⁻¹. This 6-degree slice of a redshift survey contains about 1000 galaxies brighter than magnitude 15.5. Elongations pointing toward the origin are due to intrinsic velocities in rich clusters which contribute to the Hubble velocities, thereby either increasing or decreasing the distance estimate relative to the true distance.

large-scale regions of the universe to form the thin sheets and filaments where the luminous galaxies are concentrated would have evacuated large regions of matter. The theoretical basis for these diverse models of large-scale structure will be described in a later section.

INTERGALACTIC GAS

Gas is present in galaxy clusters, just as it is among the stars of a spiral galaxy. A variety of possible mechanisms heat this *intergalactic gas*. Perhaps the nearby passage of other rapidly moving galaxies induces supersonic shock waves; more probably, the heating occurred during the early history of the cluster. Once heated, the intergalactic gas becomes so diffuse that it cannot easily cool. At temperatures of hundreds of millions of degrees Kelvin, it is extremely hot and consequently emits x-radiation. Experiments have been conducted from space satellites to study the x-ray emission from rich clusters (Figure 11.4). We must go into space to study

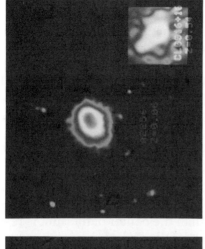

(a)

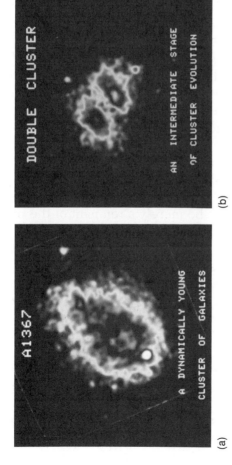

(b)

(c)

Figure 11.4 Distant Clusters of Galaxies

These x-ray photographs of remote clusters of galaxies, taken with a small x-ray telescope in space, reveal considerable diffuse x-ray emission throughout the clusters. A range of x-ray photon energies, from 0.25 keV to 2 keV, is represented in the images of these clusters. (a) An irregular cluster, inferred to have collapsed recently. (b) A double cluster. (c) A symmetrical, centrally concentrated cluster, inferred to be relatively mature.

x-radiation and ultraviolet radiation because the earth's atmosphere filters out and absorbs this short-wavelength radiation.

A considerable amount of hot gas is found in rich clusters. It is generally the cD-galaxy-dominated clusters that have substantial amounts of diffuse gas and are strong x-ray emitters. There appears to be about as much diffuse matter in these clusters as is present in visible stars. The x-ray spectrum of the intergalactic gas has characteristic emission lines produced by highly ionized iron nuclei (Figure 11.5). The nuclei are stripped of all but one or two electrons. (Compare this structure with a terrestrial iron nucleus, surrounded by twenty-six electrons.) The high temperature of this gas, about 10 times hotter than the center of the sun, is responsible for the unusual stripped state of the atom. In a cooler gas, the atomic electrons are not bombarded as frequently by collisions with neighboring particles. Consequently they will be less easily removed, and the atoms will not be as highly ionized.

The abundance of iron in intergalactic gas in clusters of galaxies is not much lower than the abundance of iron (relative to that of hydrogen) in the sun, and iron is not the only heavy element to be detected through its x-ray emission. It happens to produce the strongest lines in the spectral region accessible to x-ray satellite experiments. However, in one or two nearby galaxy clusters, other elements—notably oxygen and silicon—have also been detected via their x-ray line emission. This result came as a considerable surprise to many astronomers, because they had speculated that the intergalactic gas might be merely a remnant of the primordial matter from which the galaxies condensed. After all, why should the process of galaxy formation be so efficient as to exhaust entirely the primordial gas clouds? We know, for example, that star formation in our own galaxy is not a very efficient process. Only about 10 percent of the mass of a dense molecular cloud will be converted into stars; the rest is returned to the interstellar medium.

The discovery of iron in the intergalactic gas has shown that such gas cannot be completely primordial. It must somehow have been enriched by the ejection of iron, one of the end products of stellar evolution, from galaxies. The origin of this enriched intergalactic gas must have been from stars in the cluster galaxies themselves. The galaxies in large clusters are not presently shedding mass at a high enough rate for that to have happened, and so the mass loss

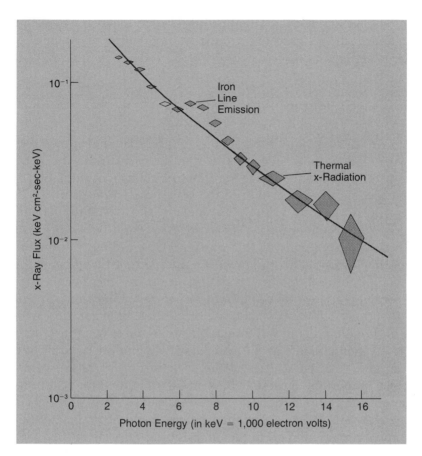

Figure 11.5 X–Ray Spectrum of a Galaxy Cluster
An x-ray spectrum of the Perseus cluster of galaxies was obtained from a satellite experiment in 1975. The flux of x-rays resembles that expected from a tenuous, hot gas at temperatures above 100 million degrees kelvin. This experiment was the first to detect iron line emission. The bump at 7000 eV is emission from almost completely stripped iron nuclei. Iron line emission is now known to be a common characteristic of x-ray emission from clusters.

must have occurred long ago. As galaxies age, they grow less luminous. In their early phases, many brighter and more-massive stars must have been present. These stars would have evolved rapidly. Shells of enriched matter would have been ejected as the lower-mass stars evolved into *planetary nebulae* and *novae* and the massive stars developed winds as supergiants and ultimately exploded as *supernovae* (Chapter 14). This enriched gas, especially that injected by supernovae, must be the source of much of the intergalactic gas.

GALAXY COLLISIONS

How was the gas removed from the galaxies? Here again we resort to speculation. One suggestion is that the pressure of the hot intracluster gas strips galaxies bare, both by virtue of the motion of the galaxies (when we refer to the pressure as *ram* pressure; see Chapter 14) and by virtue of squeezing and destabilizing interstellar clouds in the galaxies. This process seems to work efficiently in the central regions of rich clusters. One difficulty thought to limit this idea was that the process could occur only when clusters are newly formed; moreover, the resulting gas-free galaxies (type S0) were presumed to be similar in structure to the spirals. Specific observations dispute both of these properties: blue galaxies, which are undergoing recent star formation, are found to occur frequently in many rich clusters at the comparatively modest redshift of 0.5 or so, and detailed studies of the structure of S0 galaxies find a greater proportion of them to have more prominent nuclear bulges than do many spirals. Both the occurrence of so many blue galaxies in clusters and their accentuated bulges could be associated with the response of galaxies to recent stripping: the gas in the central bulge would undergo a final orgy of star formation. These distant blue galaxies are too remote for their morphology to be resolved by ground-based telescopes, but they are probably spiral galaxies that are not young but are still forming hot young stars, so that they appear blue. Since we see very few spiral galaxies in nearby rich clusters, some process must have removed their gas. Once the gas is removed, galaxies are likely to stay gas-poor. But how was the gas initially removed? Many gas-poor galaxies, including S0s, are found outside of rich clusters, and we must therefore seek a more universal mechanism that can account for their lack of gas and young stars.

One plausible possibility is that young galaxies lost much of their gas when they collided with other galaxies. The young galaxies were large systems that contained considerable amounts of gas. We observe extensive halos of stars around many galaxies; young galaxies, which were still contracting from their protogalactic clouds, must have been very extended indeed. Such large, diffuse systems would have a high probability of colliding with one another. Collisions between young galaxies would occur mostly in clusters, where the density of galaxies is highest, but would also occur even in groups, where a high frequency of smaller galaxies is found.

Collisions between young galaxies must have affected the gas and the stars within them. The diffuse interstellar gas in each galaxy suffered a tremendous shock as the galaxies collided with each other and the gas heated up. The energy imparted to the gas atoms provided so much heat that the gas blew away in a violent wind. Many of the stars in the halos of the colliding galaxies were then ejected, ripped out of the galaxies by the tidal forces that arose during the collision. The halo stars, bound to their parent galaxies by relatively weak gravitational forces, speeded up as a result of the tidal forces exerted during the encounter, and many would have escaped into the intergalactic medium. Finally, the denser clouds in either galaxy would be compressed by the passage of the shock wave through the diffuse interstellar gas, triggering the gravitational collapse of the clouds and the resulting fragmentation into smaller and smaller clouds. Eventually, star formation would occur in such fragments.

These dramatic events began when the clusters of galaxies were young, a few billion years after the big bang, or at a redshift of about 2 or 3. Collisions accompanied by gas ejection and star formation may have continued for several billion years, until even some of the outlying galaxies in rich clusters and in groups had participated in this activity. Once stripped of gas, many of the galaxies subsequently remained gas-free, to be identified as elliptical or S0 galaxies. In this manner, the bulk of the intercluster gas has accumulated by a redshift of about 0.5, corresponding to a distance of about 7 billion light-years, or a time about 7 billion years in the past. Only then would this hot gas efficiently strip the remaining spirals in the core regions of rich clusters of their gas. From this model, we might predict that intense x-ray emission produced by the newly shocked gas would be observable. Once the intergalactic gas was ejected from the galaxies, it expanded to fill the entire volume of the cluster. It remains hot. The gas is too diffuse to be able to cool effectively, and it continues to emit x-rays to the present. This x-ray emission may well be what is observed when x-ray astronomers study the great galaxy clusters. The numerous star-forming or recently star-forming galaxies found in rich clusters at a redshift of 0.5 (or larger) would be spirals and gas-rich ellipticals that have not yet been completely stripped of their gas; the stripping only occurs efficiently once a substantial amount of hot intracluster gas has accumulated.

DARK MATTER

The stars originally in the halos of galaxies in clusters must presently permeate intergalactic space. Tidal forces between colliding galaxies during the first few billion years of the cluster's existence stripped the outer halos of stars. Stripping was effective beyond a radius of about 100,000 light-years in a typical galaxy.

From observations of the Doppler shifts of their spectra, we infer that the cluster galaxies move at rather high random velocities. Because we can measure the dimensions of a cluster, we can compute how much mass must be present within the rapidly moving galaxies to contain the expansion. (If this mass were not present, the galaxies would simply fly apart—there would be no cluster.) The result is surprising: the required amount of mass per galaxy is several times larger than that inferred by other types of measurements, usually of nearby galaxies, whose dynamics we can study in sufficient detail to infer their masses. For example, by measuring the rate at which a nearby galaxy is rotating, we can infer its mass. We can also measure the velocities of nearby galaxies in a number of isolated close pairs to determine the average mass of the pair.

We can make these statements rather more precise by introducing the *mass-luminosity ratio*. We measure luminosity directly, and for every unit of luminosity (usually expressed in units of solar luminosity), we can assign a certain number of units of mass (expressed in solar masses). Thus, the sun has a mass-luminosity ratio of 1; the visible regions of the Milky Way galaxy, which consist for the most part of stars less massive and considerably less luminous than the sun, have a mass-luminosity ratio of 5. Rich clusters, however, appear to have a mass-luminosity ratio of between 200 and 400. Measurements of individual elliptical galaxies yield a mass-luminosity ratio of about 8, although this result is applicable only for the central, luminous regions.

By studying radio emission from neutral hydrogen, scientists have been able to measure the rate at which a spiral galaxy rotates. We can follow this rotation to the extreme outer parts of the disks of spiral galaxies, where few stars are visible. A surprising result has emerged: the rotation speed stays roughly constant. If most of the mass of the galaxy were in the inner regions, the outermost parts of the galaxy should be more weakly bound. They should therefore experience a weaker centrifugal force and be rotating less rapidly. But this is contrary to what is found. It appears from these measurements that spirals have larger mass-luminosity ratios than we would predict from studying their luminous inner regions (Figure 11.8). More mass must be present than we have previously realized. Their net mass-luminosity ratios must be about 30 or even larger. Precisely what form this nonluminous matter takes in the outer regions, or halos, is not known.

Rotation curves probe the outer regions of spiral galaxies, where there is little luminous matter. Two different techniques have been used to study ellipticals, which are gas-poor and therefore not amenable to rotation-curve studies at large galactocentric distances. X-ray emission has been discovered around ellipticals. The x-rays are produced by hot gas at about 10 million degrees kelvin, gravitationally confined in the halos of the elliptical galaxies. To confine the gas requires a considerable amount of mass: it is inferred that the ratio of total mass, including dark halo, to optical luminosity, which comes entirely from the inner regions, may be as large as 100.

Another discovery also indicates a considerable amount of dark matter in the halo of the elliptical. Elliptical galaxies reveal the presence of faint shells on deep photographic plates (Figure 11.6). These shells extend out to two or three times further than the bulk of the starlight. As many as 20 shells have been discovered around one bright galaxy. The shells appear to be fossil "splashes" remaining from a merger of a smaller satellite galaxy into the core of the elliptical. The spacings of the shells are a measure of the gravitational field, and computer simulations of the merger result

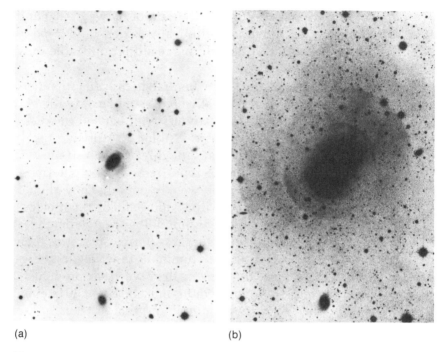

(a)   (b)

Figure 11.6   Shells Around an Elliptical Galaxy
A normal exposure (*a*) on this negative print of elliptical galaxy NGC 3923 revealed
some intriguing but barely perceptible fuzziness. A deep exposure (*b*) by David Malin
and David Carter on the Anglo-Australian telescope revealed the presence of an
extensive network of circular arcs extending over 10 arc-minutes (corresponding to a
linear distance of about 150,000 light-years) from the central image. Note that both
pictures are on the same scale, as may be seen by comparing the positions of stars.

in a similar array of concentric shells (Figure 11.7). Modelling of
the shells requires the presence of a massive dark halo.

Classical methods of mass determination, based on optical studies of the luminous inner regions (Figure 11.8), leave open the
possibility of galaxies having considerable amounts of mass in their
extended halos. Galaxies could be very extended indeed, conceivably filling most of space with exceedingly tenuous halos. In
clusters, the halos were stripped during collisions between the
galaxies. However, the excess mass should still be present in the
intergalactic medium. But the precise form of the dark mass poses
a great astrophysical puzzle. The mass cannot be very luminous,
or astronomers would be able to observe it directly. It cannot be

gaseous, for gas, whether hot or cold, ionized or neutral, is difficult to hide. Many searches have been performed for intergalactic gas. Some gas has been discovered in rich clusters, but not enough to account for the mass discrepancy.

Two hypotheses have emerged to account for the mass that is inferred to be present in clusters and in galactic halos. One hypothesis argues that the dark mass is baryonic. It might consist of stars of very low mass, which are so faint that they have escaped detection. Alternatively many collapsed remnants—perhaps white dwarfs or even black holes—of an early generation of massive stars constitute the hidden mass. A second hypothesis argues that the dark matter is nonbaryonic. It consists of one of the exotic-particle species that we earlier hypothesized could exist in sufficient quantity to yield a critical density for closure of the universe. The observed dark matter in halos and in clusters amounts to only about 10 percent of the critical density required to reverse the expansion of the universe, if we measure it in terms of the ratio of hypothesized mass to observed luminosity averaged over a suitably large region of space.

Black holes would have formed as a result of catastrophic stellar explosions, and the ensuing radiation should, in principle, be detectable. The current consensus is that if black holes account for the dark mass in clusters of galaxies, their formation must have occurred sufficiently early in the universe for the cosmological redshift to have hidden the associated optical emission from our observations. At a redshift of, say, 10, the protogalactic radiation produced when the massive stellar precursors of the black holes evolved and collapsed would now be visible only in the infrared region of the spectrum. In the infrared, observations are extremely difficult because of atmospheric emission (such as the terrestrial airglow) and attenuation resulting from absorption by ozone, water vapor, and other molecules.

White dwarfs or neutron stars are a more conservative choice than black holes for dark matter. They are the only dark-matter candidates that we can unambiguously state must exist, although whether enough actually exist is another matter. If they are to be numerous enough to account for dark matter, white dwarfs must have been produced by a large number of stars of moderate mass, formed early in the evolution of the galaxy. We cannot exclude

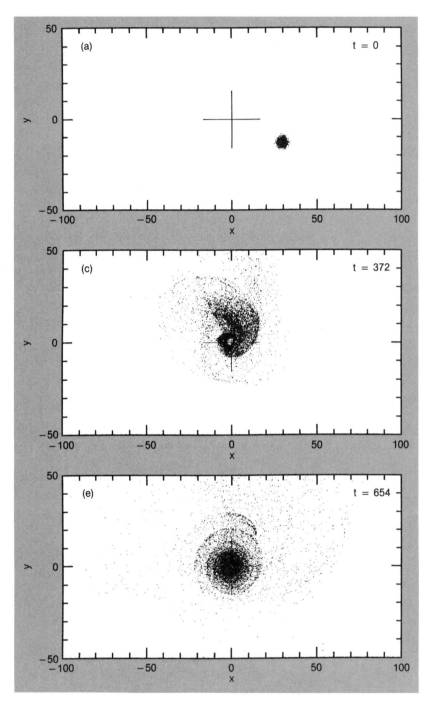

Figure 11.7 A Galaxy Merger

A high resolution computer simulation of a galaxy merger shows how the smaller galaxy orbits the massive companion before sinking into the middle, leaving behind a series of concentric shells of stars. (Courtesy of L. Hernquist and P. Quinn.)

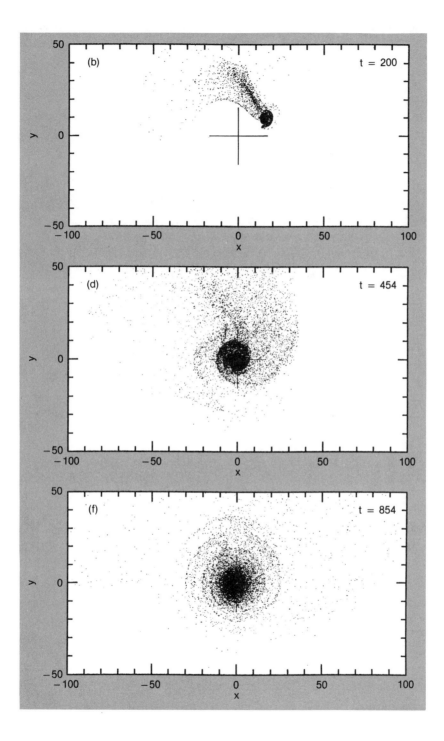

(a)

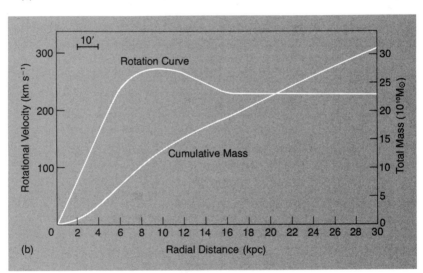

(b)

Figure 11.8 Rotation and Mass of The Andromeda Galaxy

Atomic hydrogen extends more than twice as far as the visible stellar component of the
Andromeda galaxy (a). The atomic gas (detected by line emission at a wavelength of
21 centimeters) provides a tracer of the mass distribution of the galaxy. From the
Doppler shift of the radio spectral line, a rotation curve (b) can be constructed, in which
rotational velocity is plotted (*left vertical axis*) against distance from the center along
the major axis of the galaxy (*horizontal axis*). At large distances, the rotational velocity
levels off at about 250 kilometers per second, which implies that the cumulative mass
distribution in the galaxy (*right vertical axis*) must continue to rise. In the outermost
regions of the galaxy, the mass-luminosity ratio is extremely high (in excess of 100 solar
masses per solar luminosity); in the inner regions, it is approximately 10. There is no
indication from the rotation curve that the mass distribution is falling off yet at 30
kiloparsecs; however, studies of binary galaxies suggest that massive dark halos are
confined to within about 50 kpc and typical masses do not exceed 10^{12} solar masses.

such a hypothesis, but we can seek ways to test it. For example, the white dwarfs would have cooled down, but they might still be dimly visible as reddish dwarfs. The ejecta produced when compact remnants, black holes, neutron stars, or white dwarfs formed would be chemically enriched and would show up in the composition of old stars that formed out of the recycled debris of this first generation of stars. Studies suggest that remnants of very massive stars, either black holes or neutron stars, are implausible candidates for the dark matter unless the black holes are much more massive than ordinary stars, but white dwarfs are a possibility.

Stars of low mass are also a possible source of the dark mass. Stars of very low mass populating the halo of our galaxy would occasionally pass close enough to the sun to be recognizable. They would appear as very faint nearby stars with appreciable apparent motions and the high velocities characteristic of their halo origin. Because few such objects are seen, the orbits of such stars must restrict them predominantly to the outer halo. Presumably, their orbits are mostly circular. The dynamical characteristics of these objects would make them distinct from the ordinary halo stars in our galaxy, which have appreciable velocities in the direction of the galactic center, move in highly elongated orbits, and occasionally penetrate the solar neighborhood. These ordinary halo stars are also believed to have formed in the early phases of galactic evolution, when the galaxy was collapsing, which accounts for their circular orbits. Alternatively, these outer halo stars could be "Jupiters"—essentially invisible giant planets that were not massive enough (less than 0.08 solar mass) to become normal stars.

If we possessed an adequate theory of star formation, we should be able to choose between these hypotheses of massive versus low-mass star formation. Even if low-mass stars predominate, there must also have been a considerable number of massive stars in the halo of a newly formed protogalaxy. The processed gases ejected during supernova explosions of the massive stars would account for the origin of the enriched intergalactic matter that is observed in rich clusters of galaxies. However, our knowledge of star formation is likely to remain so imprecise that direct observations will be required to determine the form of the dark mass if it is baryonic.

The nonbaryonic alternative for dark matter is in an equally unsatisfactory state. In Chapter 6, we described various terrestrial and astrophysical experiments designed to help verify the existence of nonbaryonic dark matter. None has been discovered, but particle physicists are sufficiently confident of the existence of such a particle to devote time and money to experimental searches. The reason, as we saw in Chapter 7, is that one of the great success stories of big bang cosmology is in accounting for the abundances of the light elements, one can account for the primordial deuterium abundance only if the baryonic density does not exceed about one-tenth of the critical closure value of the density that eventually causes the universe to collapse. Add inflation, which requires that the density of the universe equal the critical closure value, and we infer that a nonbaryonic dark-matter candidate is required. Perhaps such a candidate will come from the giant particle accelerators being built to probe the ultimate structure of matter. At present, there is no evidence from terrestrial experiments for such particles. However, cosmology tells us that if we adopt the simplest assumptions, they most likely do exist. If a stable, weakly interacting particle were discovered as a result of hurling protons against antiprotons or electrons against positrons at sufficiently high energy, then our hypothesis of an early hot and dense state in the big bang guarantees that some of these particles should still be around as viable dark-matter candidates.

ORIGIN OF THE GIANT CLUSTERS

The great clusters of galaxies developed inevitably after galaxies formed. The process of hierarchical clustering is well understood. Looking at the scale of masses throughout the early universe, the larger the scale, the smaller are the expected fluctuations, until, on the largest scale, there is complete homogeneity. From this assumption, we would predict that the smaller, less massive systems would develop first. At first, small galaxies formed in regions where the mean density of matter was slightly higher than average. This large-scale density enhancement exerted a slight gravitational attraction on surrounding matter, which was slowly pulled toward it. Smaller systems clustered and merged, and larger and larger systems formed. Ultimately, the great clusters of galaxies developed by this process of gravitational attraction (Figure 11.9).[16] According to this hierarchical-clustering theory, the peculiarities of

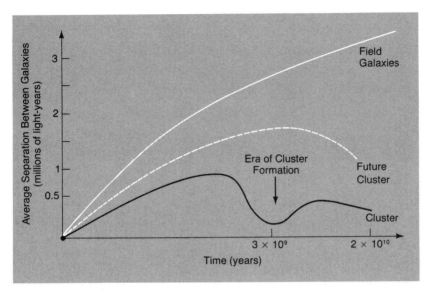

Figure 11.9 Formation of Clusters of Galaxies
The formation of an isolated cluster of galaxies can be visualized by looking at the typical separation between any pair of galaxies. For field galaxies, this separation continues to increase with time. For galaxies in a region where their mean density is higher than that of the field, the galaxies will lag behind the expansion of the universe and will eventually begin to fall together. The densest regions will experience a bounce; after the collapse, the galaxies will tend to miss one another and fly apart again. However, their motion will have become randomized during this process, and they will, on the average, end up only half as far away from the center of the cluster as they were initially. Other processes of relaxation that occur when galaxies pass close to other galaxies can further accentuate the development of a cluster of galaxies.

the initial distribution of primordial density fluctuations explain the distribution of galaxy masses.

A SYNTHESIS

Many of the preceding ideas can be combined into a unified cosmogonical theory of galaxy formation. A first step is to note that the light distribution in elliptical galaxies declines with distance from the center, as does the average matter density in galaxy clusters. If the ratio of mass to light is constant over the luminous regions of ellipticals, then the matter content of ellipticals also possesses a similar radial variation. Moreover, elliptical galaxies are smooth, round, and structureless. Such an amorphous distribution of matter that declines uniformly with no sharp boundaries results naturally from hierarchical clustering. One can understand this morphology if dynamical relaxation of stars, as opposed to gaseous dissipation, were the dominant process during formation. For example, one possibility is that perturbations of 100,000 solar masses and greater originally clustered and merged into larger condensations, forming stars throughout the duration of this process. These condensations merged into still larger condensations, until the elliptical galaxies formed. At any given stage, the structure would have formed from the merger of two or three substructures, in the manner depicted in Figure 10.11. If colliding substructures merge to form the next stage in the clustering hierarchy, considerable numbers of stars are ejected. This sequence of events leads to the formation of spheroidal galaxies surrounded by large diffuse halos. These elliptical galaxies would not be rotationally supported, and they consequently could not be very flattened; otherwise, they would be unstable and form central rotating stellar bars.

The first stars contained no heavy elements, and we must hypothesize that successive mergers incorporated successive generations of stars that formed from the enriched ejecta of previous generations of stars. As we have previously argued, these early stars are not presently detectable, but they may dominate the matter of the universe as dark stellar remnants to account for the extensive dark mass inferred from dynamical measurements. These stellar remnants were originally in the halos of the forming elliptical galaxies, but they were stripped and ejected as mergers occurred.

The flattened shape of spiral galaxies suggests that spirals formed from a single massive gas cloud that dissipated its energy, cooled, and contracted. Rotation supported the cloud in the equatorial plane, and the cloud collapsed into a thin disk. Tidal torques between neighboring protogalaxies are responsible for the acquisition of a small amount of angular momentum. The gas spins up as it contracts. This would not be sufficient to account for the observed rotation of spiral galaxies were it not for one additional ingredient. The luminous mass of a typical spiral galaxy is only 10^{11} solar masses. Studies of both rotation curves and isolated pairs of spiral galaxies orbiting about one another suggest the presence of extensive dark halos that contain about 10^{12} solar masses and exceed 100,000 light-years in radial extent (Figure 11.8).

The dark matter consists not of gas but of some relatively inert and weakly interacting form of matter that remains in a diffuse halo as the gas loses energy by radiation and contracts further. We have seen that what the dark matter consists of is uncertain: it is made up either of an exotic-particle species or of dark stellar remnants such as white dwarfs or black holes. However once formed, whatever its nature, the dark halo provides a dynamically inert backdrop for the unfolding drama of galaxy formation. While the gas contracts further, it continues to spin more and more rapidly, aided by the dark halo. The gas torques against the dark matter, just as a high diver somersaults off of a diving board. In the absence of any dark halo, the spin-up would have been insignificant. The gas contracts to form a thin, rapidly spinning disk that eventually fragments into stars. The disk is thin because centrifugal force operates in a plane perpendicular to the rotation axis: in a direction parallel to the axis of rotation, only thermal motions of the atoms provide a pressure that resists further collapse. A thin, self-gravitating disk is unstable. Gravitation causes it to fragment into small chunks, typically the mass of star clusters. These gaseous fragments in turn are gravitationally unstable and continue to collapse, finally fragmenting into stars. Remarkably, we can study the properties of this first generation of stars. Some low-mass members of the population have survived as extremely metal deficient stars and are still luminous today. Most have long since burned out, but during their lifetimes of nuclear burning, they produced heavy elements that were recycled into later generations of long-lived stars. Early star formation was responsible for the production of the heavy elements that are seen in the oldest disk stars.

Such stars have heavy-element abundances of less than one-tenth that of the sun. Here we can link theory with observation, for in rich clusters of galaxies, the enriched material would be swept out into the intergalactic medium as a consequence of galaxy collisions. Outside clusters, in low-density regions of the universe, the gas in a forming galaxy settles into a disk. It is a pleasing coincidence that the mass of (potential) disk material (about 10^{11} solar masses) available in forming each cluster galaxy can provide the right amount of enriched hot gas needed to account for the cluster x-ray observations.

As we have argued, we would not expect the intergalactic gas density to have built up significantly in galaxy clusters by this collisional stripping process until a redshift of about 0.5. Many galaxies could grow disks during this earlier phase. Once there is a significant gas density in the ambient medium, however, the pressure of the gas (because of both the motion of the galaxy and the high temperature of the intergalactic gas) removes the remaining gas rapidly (within 1 billion years). Disks stop growing, and gas-free disk galaxies with no continuing star formation are produced. These are S0 galaxies, and a natural implication of this scenario for their formation is that many S0 galaxies will have less prominent disks than do spiral galaxies. Observations of S0 galaxies indeed confirm this property.

What is the origin of the numerous dwarf galaxies, the globular star clusters, and the luminous halos of spiral galaxies? A natural implication of the preceding arguments is that these formed, like the elliptical galaxies they closely resemble, from the hierarchical clustering of density perturbations. Many globular star clusters and dwarf elliptical galaxies must be present in intergalactic space. These will be accreted by the great protogalactic gas clouds to form the luminous halos of spiral galaxies.

Dwarf irregular galaxies may have formed as gas-rich fragments of the large clouds from which the spiral galaxies condensed. Supporting evidence for this theory comes from the high abundance of dwarf galaxies (many are likely to be present in the Local Group and nearby groups, whereas spiral galaxies such as the Milky Way occur much less frequently); the fact that there are possibly some intergalactic globular star clusters 300,000 light-years or more from the Milky Way; and the occurrence of *hypergalaxies*, in which large spiral galaxies are accompanied by a host of small dwarf elliptical galaxies. These systems (the Milky Way and the Andromeda galaxies are examples) have been studied by the Estonian astronomer Jaan Einasto. The small companions seem to have clustered around the spirals. Globular star clusters could behave similarly. Because they are the least massive systems, they would be correspondingly more numerous, according to the general trend of the galaxy-luminosity function to increase toward lower luminosity and lower mass systems. Finally, the luminous spheroids of spiral galaxies could have formed by the mergers of many globular clusterlike systems early in the history of our galaxy. The observed globular clusters are likely to be the last survivors of a vast cloud of such systems that once surrounded our galaxy. Interestingly enough, giant elliptical galaxies such as Messier 87 possess an unusually large population of globular star clusters, perhaps two or three times more than expected, even though the number of globular clusters is ordinarily proportional to the luminosity of the parent galaxy. We may speculate that the mergers that long ago formed such giant ellipticals enhanced the rate of globular-cluster formation.

Primordial density fluctuations in the early universe are probably responsible for the globular star clusters, for dwarf galaxies, for the spheroidal-bulge components of galaxies, and generally for elliptical galaxies and the dark halos. The prominent disks of the great spiral galaxies are formed by secondary infall, a process suppressed in dense regions such as galaxy clusters. The infalling gas could also fragment into surrounding dwarf irregular galaxies, a process that could have occurred until relatively recently. In the higher-density regions, where the great clusters of galaxies develop, clustering and merging predominate, forming mostly elliptical and S0 galaxies. Spirals tend to form in more isolated, lower-density regions, where gas clouds can gradually dissipate and develop into a disk, unperturbed by interactions with neighboring systems of comparable mass.

According to hierarchical-clustering theory, young galaxies should be rare at very low redshifts or in nearby regions. At present, the evidence seems ambiguous. Some nearby galaxies appear to be young, but their appearance could be caused by a recent burst of star formation. The frenzied activity associated with active star formation, perhaps triggered by interaction with a neighboring galaxy, might obscure the presence of any older stars.

Confirmation of young galaxies in nearby regions of the universe would favor the *top-down*, or fragmentation, theory, thereby allowing the possibility of late formation of galaxies. Vast sheets of fragmenting gas could have formed from the anisotropic collapse and pancaking of large gas clouds or through the explosion-driven shells of compressed gas. Another possibility appeals to delayed infall of gas onto preexisting galaxies: the accretion provides the raw material for a new, vigorous phase of star-forming activity.

Detection of a number of gas-rich dwarf galaxies in the vicinity of our galaxy and in other nearby groups of galaxies indicates the presence of a few young galaxies. This could provide evidence for ongoing fragmentation or accretion of gas. The average density of matter in the universe appears to be sufficiently low at present that gravitational instability should occur only within regions of high density, such as galaxy clusters. There, cooling instabilities are likely to occur, and cooling gas may rain down on a few slowly moving galaxies; these are the likely candidates for nearby young galaxies. However, these objects should be rare; only in the gas-fragmentation theory would galaxy formation be a common process at present. This is certainly not the case, and as we have already seen, it is for this reason that the fragmentation theory, in its guise as hot dark matter, is in disfavor. According to the *bottom-up*, or hierarchical-clustering, theory, applicable to a cold-dark-matter-dominated universe, galaxies developed from small fluctuations in the early universe, and they must have become distinct entities relatively early in the expansion, perhaps at a redshift of 10 or more. Consequently, young galaxies should appear prolifically in searches for highly redshifted objects. By taking very long-exposure photographs of an apparently blank field with a large telescope, we might expect large numbers of faint, remote, young galaxies to appear. They would be identifiable by their colors, luminosities, and spectra as youthful and highly redshifted. This research program is now being vigorously pursued.

Another controversial issue is the emptiness of the great voids (Figure 11.3). Cold dark matter simulations do not evacuate these voids, but pancake-collapse models do leave behind vast empty regions. However, the voids could still be full of truly dark matter that has not formed any luminous galaxies. Perhaps only the highest-density peaks in the primordial-fluctuation spectrum actually formed galaxies. These were necessarily rare, like the highest mountain peaks on earth, and clustered, in locations like the Himalayas. We would be sampling only the tip of the iceberg when we study luminous galaxies if this were the situation.

Investigations of quasars and radio galaxies have also enabled us to accumulate evidence for considerable evolution in the remote past, when these objects were emitting the radiation we now observe. The available evidence suggests that there may have been a much greater abundance of these luminous objects in the past, at a redshift of 2 or even higher. If quasars, for example, could be identified as newly formed galaxies, we would indeed have found evidence for formation of galaxies predominantly at cosmological distances. Unfortunately, as we shall see in the next chapter, the interpretation of quasars poses as much of a puzzle as the existence of young galaxies.

· 12 ·

RADIO GALAXIES
AND QUASARS

Twinkle, twinkle, quasi-star
Biggest puzzle from afar
How unlike the other ones
Brighter than a trillion suns
Twinkle, twinkle, quasi-star
How I wonder what you are.
—*After George Gamow*

In our search for clues to evolutionary processes in the early universe, we next turn to an examination of radio galaxies and quasars. Radio galaxies and quasars appear to be centers of violently energetic emission, far surpassing other known energy sources in the universe. In this respect, these two phenomena may be compared to galactic nuclei, and it is indeed likely that quasars and radio galaxies represent active stages of galactic evolution. The greater abundance of quasars and radio galaxies at higher redshifts also indicates early formation and evolution over a cosmological time scale. However, as we shall see, we are far from a generally accepted theory of these objects, and their precise relation to ordinary galaxies has not been definitively established.

RADIO EMISSION AND RADIO GALAXIES

The vast majority of galaxies emit most of their radiation in the visible part of the spectrum, although emission can occur across the entire spectrum. The source of the light is, of course, the stars that make up the galaxy. Radio galaxies, however, emit a tremendous amount of energy in the form of radio waves. The output of a radio galaxy in radio waves is at least comparable to and may even considerably exceed the optical luminosity of its stars. By this standard, the Milky Way galaxy is an extremely feeble radio galaxy. Its emission of radio waves amounts to less than 0.1 percent of its optical emission.

The radio waves from radio galaxies are produced predominantly by cosmic-ray electrons traveling near the speed of light that spiral in tight loops around the interstellar magnetic field. The electrons radiate because they are accelerated by the magnetic force (Figure 12.1). The basic laws of electromagnetism require that accelerated charges must always lose energy by radiation. The characteristic wavelength of the *synchrotron radiation* by electrons moving at relativistic velocities is in the range of centimeters to meters, a band just below AM (around 300 meters wavelength) and FM (around 3 meters wavelength) radio transmission. Fortunately, terrestrial transmitters are restricted from broadcasting freely throughout this band, and we can listen to radio galaxies in certain narrow bands without distraction from our own electronic noise.

Synchrotron radiation is highly directional; the electrons radiate only in a narrow beam around their instantaneous direction of motion. Thus, if the magnetic field is reasonably uniform and aligned, synchrotron radiation will usually be polarized. This means that the electromagnetic oscillations, which constitute the radio (or any electromagnetic) wave, have a preferred direction. Polarization is a unique characteristic of synchrotron radiation that enables us to identify this process with the cosmic sources of radio waves in radio galaxies. Other forms of radio emission, such as emission from a hot gas, are not polarized.

Radio astronomers use gigantic radio telescopes, hundreds of meters or more in diameter, to focus as much of the weak signal as possible from a distant radio source onto the receiver. The receiver electronically amplifies the signal and analyzes it over a range of different wavelengths. The signal from a radio galaxy, if

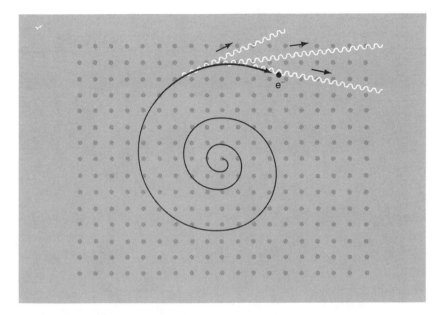

Figure 12.1 Synchrotron Radiation
When a relativistic electron spirals around magnetic lines of force, it emits intense radio signals. The lines of magnetic force (*dots*) are to be visualized as perpendicular to the page, and the electron emits a beam of radio waves in a direction that is tangent to its spiral orbit around the magnetic field lines.

played over a loudspeaker, would sound like a continuous hiss. It does not have any preferred wavelength but, like noise or radio static, contains a superposition of many different wavelengths broadcast simultaneously.

Because our galaxy, like most galaxies, is transparent at radio wavelengths, we can probe the very center of the galaxy by studying the longer-wavelength synchrotron radiation. At shorter radio wavelengths, atomic hydrogen and various molecules possess characteristic wavelengths at which radio emission can occur. These wavelengths are called spectral lines, just as in the optical part of the spectrum, which we described in Chapter 3. They are the radio analogues of the optical emission lines in the spectrum of a hot gas. A spectral line is emitted by atomic hydrogen at a radio wavelength of 21 centimeters.

Spectral lines are also produced by many different molecules, such as water vapor, carbon monoxide, and formaldehyde, at somewhat shorter wavelengths. These molecules, and many others, have

(a)

(b)

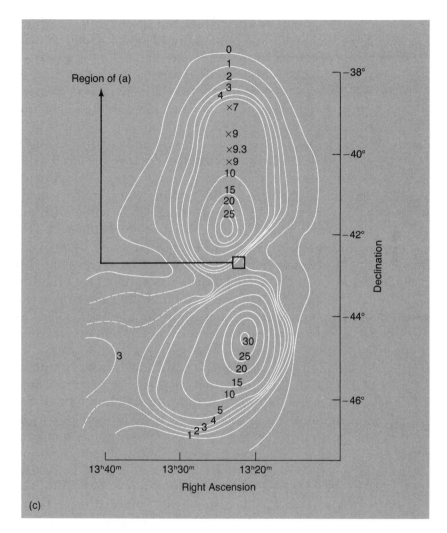

Figure 12.2 A Radio Galaxy, Centaurus A

A very strong radio source, Centaurus A may show evidence for a collision or a merger between an elliptical and a spiral galaxy (*a*). Part of the radio emission is produced in the nucleus of the galaxy, in a region that is less than 1 light-year across. The emission is highly variable with time. The bulk of the radio emission comes from two huge regions that extend for millions of light-years on either side of the galaxy. X-ray and infrared emission have also been detected from the nucleus of this galaxy, and its radiation spans almost the entire electromagnetic spectrum (*b*). Luminosity is equal to the absolute or intrinsic brightness per unit frequency interval multiplied by the mean frequency; it is plotted versus frequency in this figure. In the radio map (*c*), the contours (similar to altitude contours on a topographic map) mark different levels of intensity of the radio signals.

been discovered in dense interstellar clouds throughout the Milky Way. Thus, the molecular line observations have provided us with a unique probe of the structure of remote regions of the galaxy.

However, radio spectral-line emission is so weak that outside the Milky Way, we can detect it only in relatively nearby galaxies at low redshift. A radio galaxy emits much more intense synchrotron radiation which we can detect at even greater distances than we can detect visible light and can consequently utilize to study very remote galaxies. Radio surveys of the sky reveal a large number of radio sources that appear to have no optical counterpart. Closer examination of such specific regions where radio signals originate can be achieved by taking a deep CCD frame with the largest available modern telescopes. This procedure sometimes reveals an exceedingly faint image of a remote galaxy. In some cases, the galaxy is too faint to be detectable optically. Occasionally, if the light from one of these faint galaxies is sufficiently bright to pass through a spectroscope, the spectrum of the galaxy reveals a considerable Doppler shift, perhaps amounting to a redshift of as much as 1.0, or even in rare cases as much as 2. Thus, an emission line in the spectrum of light from the galaxy would be shifted to twice or three times the wavelength at which the line was emitted.

Often, we find that radio emission from such a distant source does not emanate directly from the optically visible region of the galaxy. Very frequently, a double source is found, which is centered on the optical object (Figure 12.2). It appears as though a gigantic explosion has liberated two vast clouds of radio-emitting plasma and ejected them in opposite directions. Occasionally, there is more than one pair of these *radio lobes*, indicating the possibility of a series of explosions in the central galaxy. These radio-emitting regions extend so far that the optically visible galaxy seems small by comparison.

A graphic representation of the presence of intergalactic gas has come from radio maps of cluster galaxies with extensive radio tails (Figure 12.3 and Color Plate 2). We have seen that radio galaxies are often surrounded by a pair of extensive lobes of radio-emitting plasma. Within the cluster, the rapid motion of the parent galaxy causes the lobes to be grossly distorted by the resulting pressure of the intergalactic gas and to trail behind the galaxy. The situation is reminiscent of atmospheric vapor trails left by jet airplanes. Be-

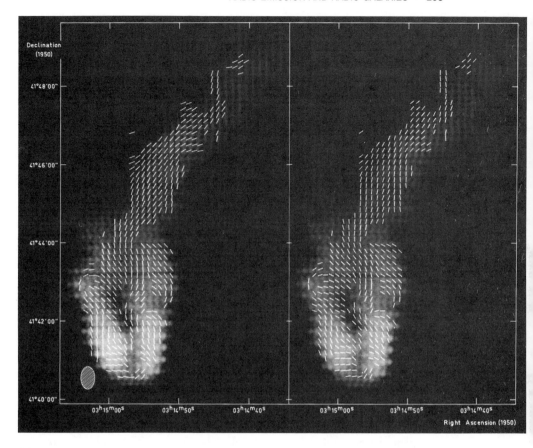

(a)

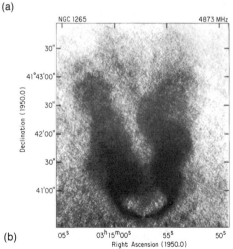

(b)

Figure 12.3 A Radio–Tailed Galaxy

The galaxy NGC 1265 in the Perseus cluster of galaxies reveals a prominent radio head and tail. (a) Superimposed on the radio photograph of the radio intensity, taken with the Westerbork Synthesis Radio Telescope in Holland, are bars that show the direction of the magnetic field. The radio emission is highly polarized. The magnetic field and radio emission appear to emanate from the head region. (b) A radio photograph at higher resolution taken with the Very Large Array, Socorro, N. Mex.

cause the radio-emitting plasma is confined to a narrow tail, we can directly infer the density of the ambient intergalactic gas. Not surprisingly, this density is similar to the value inferred from studies of the amount of gas required to produce the observed x-ray emission.

One of the dramatic discoveries of the new technique of radio interferometry has been to record the birth of a radio nucleus within an optical galaxy. A *radio interferometer* consists of two or more radio telescopes located far apart, perhaps in different continents, that observe the same source simultaneously. Unlike visible light, radio waves are not affected by atmospheric "seeing," and retain their coherence. This stability means that, by simultaneously measuring the same wavefront, the two telescopes are able to operate effectively as a single enormous unit (Figure 12.4). This technique greatly enhances our ability to resolve the structure of the remote radio galaxy. The resolving power of a telescope is directly proportional to its linear size, which in this case is the distance separating the radio telescopes. The technique of *very long baseline interferometry* has meant that with radio interferometers, we are able to study the nuclei of distant radio galaxies. Structures can be detected on a scale more than 1000 times finer than that detectable with the largest optical telescopes. In some cases, we have directly observed the radio-emitting regions to expand in the nuclei of radio galaxies which are undergoing tremendously powerful outbursts of radio waves. The galaxies are so remote that the expansion velocity must be very great for the changes even to be detectable. In fact, such explosions are so phenomenal that the apparent velocity in some cases actually exceeds the speed of light.

However, no actual physical velocities are known to exceed that of light. One explanation of this phenomenon appeals to what amounts to an optical illusion. A similar geometrical phenomenon occurs when a searchlight sweeps the sky. To us, the beam appears to have a very high apparent velocity relative to the stars. The stars are so distant that, no matter how great their velocities, we would barely notice any motion. Yet the searchlight beam traverses the sky in a matter of seconds. The apparent velocity of the beam, if it were sufficiently powerful to reach the nearest stars, would be much greater than the speed of light. Thus, apparent velocities can greatly exceed that of light, although the velocity of any material

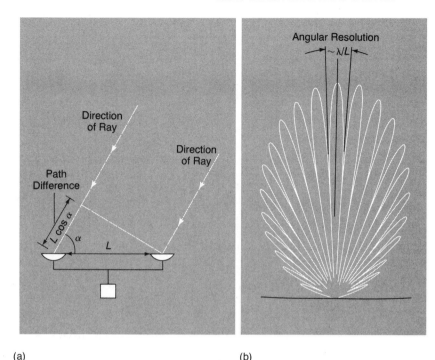

(a)   (b)

Figure 12.4   Very Long Baseline Interferometry
Radio telescopes in two different continents can simultaneously observe a distant radio
source (a). By keeping track of the phase of the radio waves and the exact times of
observation, the recorded signals can later be synthesized to simulate the effect of a
gigantic radio telescope, whose resolution is equivalent to that of a single telescope
spanning the two continents. An interferometer utilizes the principle that the radio waves
received by the two antennae have a slight difference in path length; when added
together, the two signals reinforce each other if the path difference is an even number
of wavelengths and cancel out if the path difference amounts to an odd number of
wavelengths. As the earth rotates, the path difference varies (because it depends on the
source elevation α). The interferometer has a multilobed directional pattern (b); the
angular resolution is the width of a single lobe or λ/L, where λ is the wavelength and L
is the separation of the two antennae. Because of the long intercontinental baselines
that can be used, the resulting resolution is more than 1000 times better than anything
attainable with an optical telescope.

object is always less than that of light. Another, more prevalent,
class of explanations appeals to relativistic jets emanating from galactic nuclei. Such jets of high-energy particles flowing at a speed
very close to the speed of light are postulated as a source of energy
in radio galaxies. If the beams occur very close to the direction in
which the observer views the sources, then an apparent superlu-

minal motion can occur in the transverse direction; it is, of course, only an illusion.

Exactly what initiates such explosions in radio galaxies remains a source of speculation. The size of the nuclear region where the explosion occurs is very small, perhaps only a few light-years across. It is located within the *galactic nucleus,* the innermost region where the light is centrally concentrated. This size rather strongly limits the possible energy sources. According to one theory, massive compact objects, weighing 1 million solar masses or more, are ejected at high velocity from the galactic nucleus. These objects serve as a source of high-energy electrons in the regions of radio emission. An alternative viewpoint argues that the explosion can be better understood as a sudden ejection from the galactic nucleus of many energetic particles. The particles stream through the gaseous medium in the central region of the galaxy (Figure 12.5). If the gas has a flattened distribution (because of its rotation), the energetic particles would tend to move along the path of least resistance, the minor axis of the gas distribution. Continuous streams of energetic particles could punch through the surrounding gas and penetrate any intergalactic medium present outside the galaxy. Eventually, these particle jet streams would terminate and accumulate in radio lobes.

This *twin-exhaust model* seems to provide a reasonable explanation of the morphology of observed radio galaxies, but the force that produces the twin beams of energetic particles remains very much a mystery. This is not because we have failed to construct a suitable theory. If anything, the opposite is true—we have a surfeit of theories to explain the nuclei of radio galaxies; we simply lack ways to test them. It seems likely that the mechanism for the basic energy input into the nuclei of radio galaxies is closely related to the quasar phenomenon, which, as we shall see, also occurs in a highly compact region.

QUASI-STELLAR RADIO SOURCES

Many faint radio sources have been discovered to be associated with optical objects whose images were indistinguishable from those of ordinary stars. Because the photographic image of a galaxy

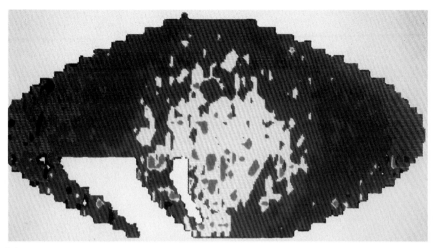

(a)

(b)

Color Plate 1 The cosmic microwave background.
An all-sky map in celestial coordinates (North celestial pole at top). The sky coverage was 85 percent at a wavelength of 3 mm. (a) All the data; the dipole (or 180°) variation is due to the Doppler shift induced by the motion of the earth relative to the cosmic background radiation. The effect amounts to a 0.1 percent modulation and indicates a velocity of 400 kilometers per second, which measures the vector sum of the motion of the earth around the sun, the sun around our galaxy, and our galaxy in the Local Group and in the Virgo Supercluster. (b) The residual fluctuations when the dipole signal is subtracted off. These are at a level of 0.01 percent and are due primarily to receiver noise, but impose an upper limit on any residual cosmic anisotropy that is expected to be present.

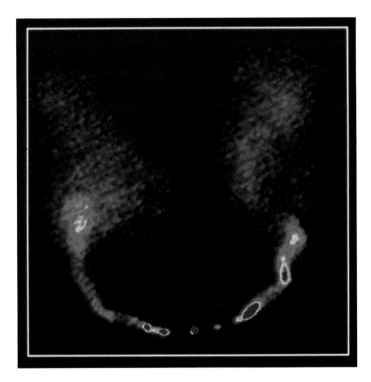

Color Plate 2   A false color radio image of the galaxy NGC 1265.

Distinct knots of radio emission can be seen in the twin jets that emanate from the central source. Because of the high velocity of the galaxy, the radio lobes trail behind and can be traced for hundreds of thousands of light-years.

Color Plate 3   A view of the Large Magellanic Cloud prior to Supernova 1987A.

Color Plate 4   The Large Magellanic Cloud shortly after the appearance of Supernova 1987A.

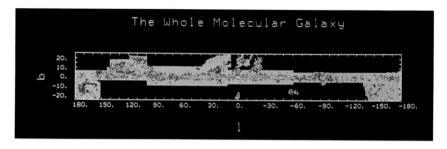

Color Plate 5   The Milky Way in molecular clouds.
A map in the microwave emission line of the carbon monoxide molecule.

Color Plate 6   The star-forming region NGC 6193.

(a)

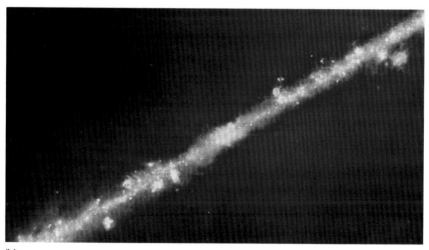

(b)

Color Plate 7   An infrared view of the sky from the IRAS experiment.

The IRAS (Infrared Astronomy Satellite) telescope was used to take images of nearly the entire sky, over six months of observing. Point sources are displayed in (a), and the diffuse infrared emission from the central part of our galaxy is shown in (b). The bright horizontal band is the Milky Way, with the galactic center located at the center of the image. Many hundreds of thousands of extragalactic point sources are visible. The colors represent infrared emission detected in three of the telescopic wavelength bands, with red being the cooler material and blue or white the hotter matter. Many of the galactic point sources are young stars, and the diffuse emission is radiated by dust that is being heated in regions of active star formation. Black stripes are regions of the sky that were not scanned by the telescope.

Color Plate 8   The spiral galaxy Messier 51.

This is also known as the Whirlpool galaxy because of its distinctive appearance. The companion galaxy may be responsible for driving the "grand design" spiral pattern by means of a density wave. The blue color of the spiral arms is produced by young, hot stars.

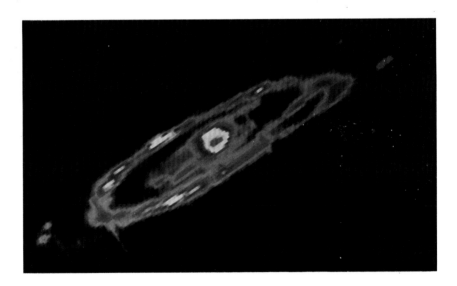

Color Plate 9   The Andromeda galaxy: an infrared view.

This image was obtained by the IRAS telescope, and emphasizes the warm, star-forming regions of the Andromeda galaxy.

Color Plate 10   The Eta Carina Nebula.

In the core of the giant HII region is the star Eta Carinae. It is a highly variable massive star; in 1843, it was surpassed only by Sirius in brightness. The nebulosity is produced by extensive mass loss from this supergiant star, which many astronomers consider a likely candidate for the next supernova in our galaxy.

Color Plate 11   The Sombrero galaxy.

The large central bulge of this galaxy contains old stars; the thin disk of this spiral galaxy, with many young stars, is seen almost precisely edge on, and so is largely obscured by dust. The outer parts of the bulge are progressively bluer than the inner regions; this is because the outermost stars are mostly metal-poor.

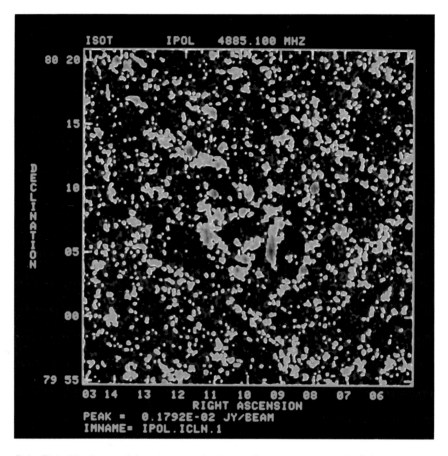

Color Plate 12 A map of the microwave background over a square patch of sky one-sixth of a degree across.

The fluctuations are at a level of 0.01 percent and are probably due to instrumental noise, although some very remote radio sources may be present. Any intrinsic fluctuations from which galaxies formed are hidden in the noise.

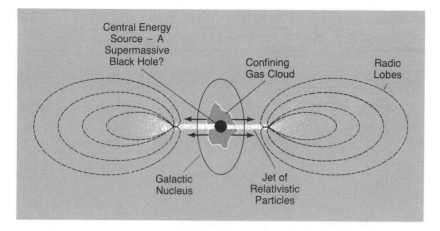

Figure 12.5 Twin—Exhaust Model for a Radio Galaxy
A powerful energy source at the nucleus of a galaxy emits a continuous stream of high-energy particles. These push their way along the lesser axis of the confining gas cloud, and two jets pierce through the surrounding intergalactic medium. The jets of particles eventually splatter to a halt, creating huge lobes of plasma that emit radio waves. The radio emission is due to synchrotron radiation by relativistic electrons that are being injected into the magnetized lobes from the nucleus.

is extended and not pointlike, it is easily distinguishable from the image of a star. Of course, stellar images do have a finite size, which is determined by the amount of scattering and refraction of the starlight in the earth's atmosphere. Atmospheric turbulence also causes the image to wander slightly, or scintillate. The turbulence is produced by small temperature irregularities in the atmosphere, on scales of a few centimeters or more. Scintillation accounts for the twinkling of stars. Planets tend not to twinkle, as their larger images make this effect much less noticeable.

It was soon realized that the starlike counterparts of radio sources were not ordinary stars (Figure 12.6). The faint optical sources were christened *quasars,* an abbreviation for *quasi-stellar radio source,* when spectra of their optical counterparts revealed a major difference from spectra of ordinary stars: the light of quasars is considerably redshifted. Recent surveys have found that the light from the most distant quasars can be shifted to the red in wavelength by, in some instances, as much as 5 times the wavelength at which the light was emitted. Consequently, the quasar has a redshift (rela-

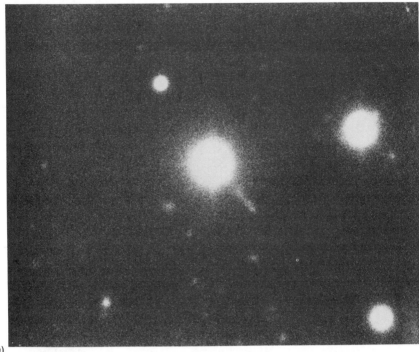

(a)

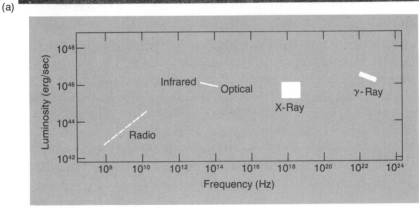

(b)

Figure 12.6 The Nearest Quasar

This photograph of quasar 3C 273, a fourteenth-magnitude starlike object, reveals a small wisp, believed to have been explosively ejected by the quasar some millions of years ago (*a*). Its electromagnetic spectrum (*b*) spans a wide range and is similar in many ways to that of the nucleus of Centaurus A (Figure 12.2). The absolute or intrinsic brightness per unit frequency multiplied by the frequency is a measure of luminosity and is plotted against frequency. One important difference is that the quasar is intrinsically 1000 times more luminous—it is one of the most luminous objects in the universe.

tive increment in wavelength) equal to 4. For such a star, the Lyman alpha line of hydrogen, normally an ultraviolet transition at a wavelength of 1216 Ångstroms, would be observed at 6080 Ångstroms in the red region of the spectrum.

The spectral lines identified from quasar spectra are quite unlike those of any star (Figure 12.7). They resemble more the spectra of light from the nuclei of certain (relatively unusual) types of galaxies known as *Seyfert galaxies*. Seyfert nuclei contain clouds of hot gas moving at velocities of thousands or tens of thousands of kilometers per second. Emission by the hot gas overwhelms emission from any stars that are present in the nucleus. Similarly, quasar spectra are characterized by emission from hot gas. Dark absorption lines are also seen. Absorption lines usually appear when one looks through cooler absorbing gas in front of a source of emission (such as much hotter gas.) In quasars, the absorption line spectrum has a different (usually lower) redshift than the emitting gas, which indicates that the absorption arises in a region that is very distinct physically from the emission line region. Often several different absorption line redshifts are measured in the same quasar.

Most astronomers agree that the origin of the quasar redshift is similar to the origin of the redshift of distant galaxies. The redshift in their spectra is essentially the Doppler shift, which results from their recession at high velocities over great distances. Controversy has arisen at times, and noncosmological interpretations of quasar redshifts have been proposed. The most plausible noncosmological interpretation of quasar redshifts appeals to the *gravitational red-shift*—light will be greatly redshifted, for example, if emitted sufficiently close to a black hole. To understand this phenomenon, consider the following analogy. Imagine a lift taking skiers up a very high mountain. If 1000 skiers per hour pass by a lift at the base, only 999 skiers per hour will climb off at the peak. What is wrong? Time is passing more slowly in the deeper gravitational potential at the mountain base, and clocks tick more slowly. So, each 1000 skiers take slightly longer to pass by the base than the peak.

Of course, this example grossly exaggerates the effect, but the frequency of light is similarly decreased when emitted in a gravitational field. This effect has been directly measured on the earth, where the redshift amounts to only 1 part in 10 billion. Near a black

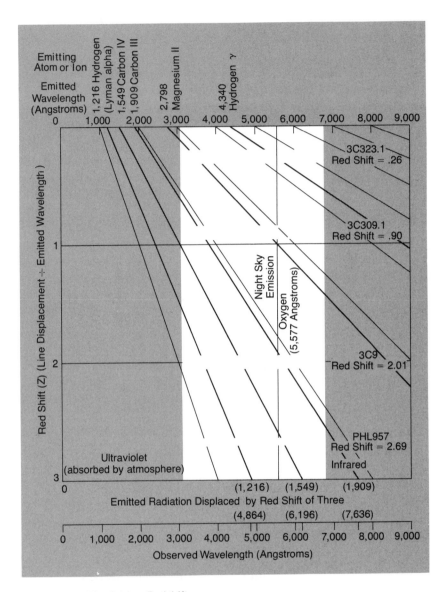

Figure 12.7 The Quasar Redshift

Four quasar spectra at different redshifts illustrate the range in wavelength that must be covered in order to identify quasar redshifts. Several emission lines are needed to establish any particular redshift, because one line could be redshifted by an arbitrary amount. The slanting lines correspond to radiation emitted by hydrogen, carbon, and magnesium. At a redshift (z) of 1, the strongest hydrogen line (Lyman alpha) is observed at 2432 Ångstroms, stretched in wavelength by a factor of 2 from its emission wavelength; at a redshift of 2, it is observed at 3648 Ångstroms. The photons that produced the spectrum of the quasar PHL 957 were emitted when the universe was only about 20 percent of its present age.

hole, a much larger redshift is attainable. For a quasar, the black hole would have to be very massive, weighing 10^8 solar masses or more in order to liberate enough energy from infalling gas, but redshifts of 3 or 4 (as are found for the most highly redshifted quasars) could be achieved in principle. However, one would expect enormous gravitational fields near a black hole to distort the emission spectrum produced there. Lines would be highly broadened by the motion of gas atoms at near-relativistic velocities encountered close to a black hole. This is not seen: the evidence of relatively narrow spectral lines in quasars has invalidated the gravitational redshift interpretation.

No compelling evidence has ever been presented to cast doubt on the cosmological nature of quasar redshifts. Since the cosmological redshift provides the simplest explanation of the quasars in terms of known physical laws, we shall generally assume that quasars are at distances inferred from their redshifts. Several types of observations have provided evidence that quasars are at cosmological distances. Quasars occasionally occur in clusters or groups of galaxies of similar redshifts. Also, in several quasars, absorption lines are found that coincide in redshift with other galaxies along the same line of sight. Finally, the host galaxies around a number of the nearer quasars have been identified. This evidence suggests that, if we accept the standard interpretation of galactic redshifts, most quasars must be extremely remote and, moreover, must be galaxies with exceptionally bright nuclei.

Quasar absorption lines that are well separated in redshift from the rest of the spectrum are probably caused by absorption in gas in the outermost parts of intervening galaxies. Intergalactic gas clouds may also cause absorption between earth and the quasar. The quasar is usually so bright and so remote that we cannot see images of other intervening galaxies along the line of sight to the quasar. However, if the intervening galaxies contain interstellar gas, their presence could be inferred by absorption of the light from a quasar. For a galaxy to absorb light from a remote quasar, one must assume that gas is present in the outermost parts of the galactic halo, a region that may extend considerably beyond the visible boundary of the galaxy. Several absorbing regions that are widely separated in distance could occur along the line of sight to a distant quasar. Thus, we would expect to find a number of different absorption redshifts in a quasar at any given (large) emission redshift. This is in fact more or less what is observed. There are several well-established cases of an intervening galaxy that produces absorption at the redshift of the galaxy in the spectrum of a more highly redshifted quasar. In addition to the strong absorption line systems that correspond to intervening galaxies, there are numerous hydrogen clouds that produce weak redshifted Lyman alpha line absorption in the quasar spectrum (Figure 12.8). These Lyman alpha clouds are metal-poor and are very common at a redshift of 2 or higher. They are presumed to be the counterparts of dwarf galaxies that have not yet collapsed to form stars. Apart from these discrete gas clouds, there is no distributed component of atomic hydrogen in the intergalactic medium; such diffuse gas would absorb a broad trough out of the general quasar spectrum, corresponding to continuously redshifted Lyman alpha lines, and this effect is not observed.

The energy output of a quasar exceeds that of the brightest galaxies. The average quasar is brighter than 300 billion suns. A few quasars are exceptionally luminous and exceed this value 100 times. Quasars are also intense sources of x-radiation, and as much energy is radiated in x-rays as in the entire optical region of the spectrum. Yet all this energy must be released within a volume with a diameter smaller than the distance separating the sun from its nearest stellar neighbor. We can be confident of this, because the light output varies over a time scale of years or even days in many quasars. Some quasars periodically erupt, brightening by several magnitudes over periods of a few months or less. Even if the source were exploding at the speed of light, it could still be no more than a light-day in diameter for its luminosity to be able to change sufficiently rapidly. Light emitted instantaneously by the quasar is received by us over a time interval that is determined by the time lag for light to reach us from the far side relative to the near side of the quasar (Figure 12.9). Consequently, the quasar lightburst is spread out over a distance that corresponds to the time it takes for light to travel across the diameter of the quasar.

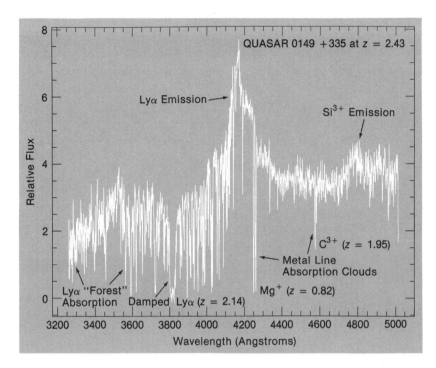

Figure 12.8   Spectrum of a Quasar

This plot of light intensity versus wavelength displays the spectrum of a quasar at redshift $z = 3.299$. The emission lines (peaks) are systematically shifted in wavelength toward the red by a factor $1 + z = 4.299$: the first two fundamental lines of atomic hydrogen Lyman alpha and Lyman beta are seen at 5240 and 4440 Ångstroms, respectively. Also visible is the "forest" of Lyman alpha absorption lines (dips: the many small dips are noise, but the deeper features are real) produced in numerous gas clouds along the line of sight to the distant quasar. The absorption lines are at lower redshift and seen at shorter wavelengths, the rest of unredshifted wavelength of Lyman alpha being 1216 Ångstroms. On the extreme left of the spectrum, a dense cloud at a redshift of 3.22 produces a trough of continuum absorption; the incident quasar photons are sufficiently energetic to ionize the hydrogen and be absorbed in doing so at all wavelengths below the Lyman limit value of 912 Ångstroms (corresponding to the energy required to ionize the hydrogen atom), which is redshifted to where the trough begins at 3850 Ångstroms.

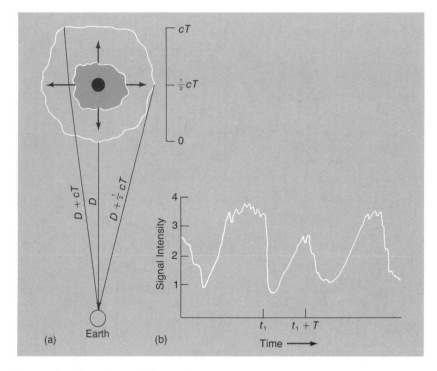

Figure 12.9 Time Scale of Quasar Light Variability

The minimum span of time during which we can expect to observe any light variations from a quasar is determined by the size of the emitting region (*a*). The apparent magnitude of the quasar 3C279 (at a redshift of 0.5) varies by almost 7 magnitudes over 2 or 3 years (*b*). No matter how rapidly the source might actually be fluctuating, the light we receive will be spread over this time span, light taking slightly longer to reach us from its back than from its front. By measuring the variability time scale (*T*), we infer the maximum size for the quasar light-emitting region to be the corresponding light travel distance *cT*.

GRAVITATIONAL LENSES

The universe is full of bizarre events, but one that ranks high on any list is surely the *gravitational lens*. Astronomers discovered this effect when they found a pair of quasars separated by a few seconds of arc. The pair had identical redshifts and similar emission spectra. The odds against such a situation occurring by chance are huge: the typical separation between quasars (with arbitrary redshifts) is several thousands of arc seconds. One clue that might explain the double image is the set of absorption lines common to the quasar pair. These lines were produced in an absorbing cloud whose redshift is about half that of the quasars, which means that about halfway along the immense distance to the quasars, some tens of billions of light-years away, there is an intervening cloud. When we detect such intervening clouds in other quasars, we interpret them as clouds in the halos of ordinary disk galaxies. Such galaxies occur frequently enough and are large enough to account for the occasional coincidences in direction with background quasars.

If a sufficiently massive disk galaxy intervenes, it can behave like a gigantic lens, bending the quasar's light in its gravitational field. Just like an ordinary glass lens, it produces another image. This double image of a single quasar is responsible for the observed quasar pair. If this explanation is correct, we ought to be able to detect the luminous lensing galaxy. Some seven examples of lensed quasars have been found, and in several cases, astronomers think they have identified the lens. In other cases, with no luminous candidate galaxy for the lens, we have to speculate about other intervening objects that may be responsible, perhaps a burnt-out galaxy or something more exotic, such as a supermassive black hole or more mundanely, a foreground galaxy cluster. In fact, examples are known of an entire cluster of galaxies which conspires to form a collective gravitational lens. If a background galaxy is suitably oriented with respect to the intervening cluster, the lensed image is distorted into an arc, or even ideally a ring; at least three such luminous arcs have been discovered.

A further complication of a gravitational lens is that, unlike a glass lens, it should produce an odd number of images. In at least one case, a triple quasar image has been found, confirming that a gravitational lens indeed is responsible. In other cases, we expect only a double image because the third image will generally be too faint to see. Yet a further complication arises because the light paths of the two images are slightly different. Light may take several months or even years longer on one of the light paths than on the other to traverse the ten or more billion light years to the quasar. This means that if the quasar is variable in its light output over the course of a few months, as many quasars are, one would expect a time delay between the image pair. Confirmation of this time delay should eventually provide further evidence for the gravitational-lens hypothesis.

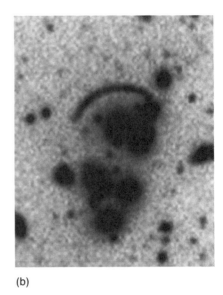

(a)   (b)

Figure 12.10 Gravitational Lenses
The luminous arcs imaged in these negative CCD images are in the central regions of galaxy clusters. The fuzzy objects are all cluster galaxies, and the arcs are caused by the clusters' lensing remote background galaxies. In (a), the cluster is Abell 370 at a redshift of 0.17 and the redshift of the arc has been measured to be 0.725; in (b), the cluster is CL 2244-02 at a redshift of 0.328.

THEORIES OF QUASARS AND RADIO GALAXIES

Several competing theories have been advanced to account for quasars and the related phenomena that occur in active nuclei of radio galaxies. According to one viewpoint, quasars are a phenomenon associated with galaxy formation. The matter shed by early generations of massive stars during the collapse of the protogalaxy collects in its nucleus, where, in a brief but intense burst of star formation, it attains considerable luminosity.

According to an alternative viewpoint, a quasar represents the final stages of the evolution of the nucleus of a galaxy. The density of stars in the central regions of a galaxy is very high; stars are separated by only one-hundredth of a light-year (or 1000 astronomical units). The stellar distribution tends to segregate into massive stars and binary stars, which fall toward the center, and less massive stars, which populate the outer regions of the nucleus. The central stars begin to collide with one another. Collisions between stars are catastrophic events that may lead to supernova explosions. A high rate of supernovae occurring in a compact, dense system of stars could provide the energy source for quasars.

These alternative viewpoints in fact may be identical. According to bottom-up theories of galaxy formation, the central regions of a galaxy develop first, and the outer regions fall in subsequently. The nucleus therefore has time to evolve rapidly while surrounded by the protogalaxy. What does it evolve into, what provides the quasar powerhouse? A popular viewpoint argues that a quasar is a supermassive object, perhaps a giant black hole of 100 million solar masses or more, at the dense nucleus of a galaxy (Figure 12.11). (Some theorists have favored a model consisting of a massive but compact cloud, supported by some combination of magnetic field, rotation, and turbulence. However, it seems difficult for such a system to remain stable.) The massive black hole disrupts and ultimately swallows the stars whose orbits pass too close to the hole. As matter accretes onto the black hole, it becomes very hot. The ensuing x-radiation would provide the energy supply for the quasar. One of the strongest arguments for such a compact source comes from the variability of quasars, which has been reported to be as short as one day. As we saw previously, the light emitted by a quasar must be produced within a region traversed by light during the period of observed variability. One light-day equals 300 astronomical units, and we infer that in a quasar, the light equivalent of 100 Milky Way galaxies is generated within a region less that the size of the solar system. The *Schwarzchild radius*, which delineates the imaginary surface around a black hole within which light is trapped, is equal to about 10 astronomical units or 1 light-hour for a black hole of 1 billion solar masses. It is this scale which crudely represents the size of the black hole. Any particle venturing within this sphere of influence cannot escape, even if it travels at light speed. On the way in, however, such particles collide with ambient gas particles and emit substantial amounts of radiation. The intense x-ray emission from quasars can best be explained if a massive black hole is the underlying energy source, because gas that falls into such a massive compact object heats up sufficiently to radiate x-rays. The gaseous debris infalling onto the central black hole spirals in and forms an accretion disk.

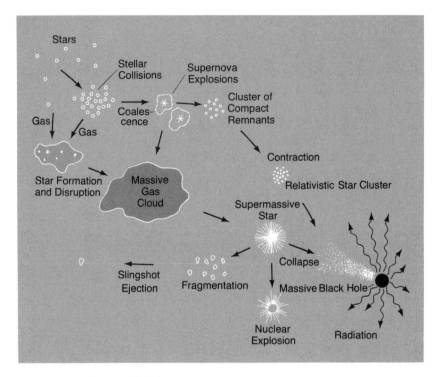

Figure 12.11 A Quasar Model

The power supply in quasars and active galactic nuclei is likely to derive from an extremely dense stellar system that is undergoing catastrophic and violent evolution. We know that the star density in the nuclei of many galaxies exceeds 1 million solar masses pc^{-3}; theory suggests that it will continue to increase as stars evolve and shed matter, and new stars will form. Stellar collisions will likely occur, leading to coalescence and massive star formation. The massive stars would explode as supernovae, other stars would collide disruptively, and the resulting pool of gaseous debris would fall farther in toward the galactic nucleus, thereby increasing the density. New stars would form, and the cycle would repeat itself. The supernovae would leave compact black hole or neutron star remnants behind that would evolve into a denser and denser cluster. So much energy is produced that star formation is eventually inhibited, and an amorphous supermassive cloud results. Such a cloud would be unstable, collapsing first into a supermassive star and then into a massive black hole of perhaps 100 million solar masses. The stellar remnants would be successively swallowed by this hole, releasing vast amounts of energy in the process. Alternatively, the supermassive star may undergo a violent nuclear explosion as it fragments into a few massive clouds, which are occasionally ejected like a slingshot from the nucleus of the galaxy. Although we cannot yet strongly favor any one of these possible scenarios, it seems clear that the fate of a dense stellar system in the nucleus of a galaxy is to yield prolific amounts of energy.

Some recent evidence favors the massive black hole theory of active galactic nuclei and quasars. Such a hole may be present in the nucleus of the giant elliptical galaxy

The radiation emitted by this disk of hot gas—held out from the black halo by its angular momentum—is largely responsible for the quasar luminosity. Gas in the disk gradually loses angular momentum as gravity drags it into the black hole, where it finally disappears into invisibility. For any specified black hole mass, there is a maximum luminosity that can be generated by infalling gas being swallowed by the hole. Exceed this luminosity, and the radiation pressure halts and even reverses the inflow. In this way, we estimate the minimum hole masses required to account for the observed quasar luminosities. These hole masses turn out to be between 0.1 and 1 billion solar masses. These massive holes may also be present in galaxies that manifest a less vigorous state of nuclear activity, such as Seyfert galaxies; however, in this case, the rate of fueling by gas or star infall into the central engine is greatly reduced, and the luminosity is relatively low (typically about 10 billion suns) compared with that of a quasar.

Of the various theories of quasars, the *massive black hole* model appears to be the most plausible. If we adopt this model, we can conceive of black holes lurking in every large galaxy. Only intermittently would such massive black holes become activated by accreting gas or stars. In particular, Seyfert galaxies exhibit phenomena that require a powerful central energy source, one that is also capable of producing intense x-radiation. Many optical characteristics of Seyfert galaxies, such as the emission lines from the nuclear regions, are similar to quasar properties, although the total luminosity is considerably lower.

It is likely that the Seyfert phenomenon is a more quiescent version of quasar activity. When the fueling rate onto the central black hole is large, we get a quasar. Eventually, the fuel supply becomes depleted, and a Seyfert galaxy is the result. Finally, in the centers of ordinary galaxies, we may have a passive black hole,

---

M87. There is an unusually high concentration of stars at the very center of the galaxy, and the stars are moving more rapidly, as if their random motions are responding to the intense gravitational field of a massive central black hole. For quasars, the evidence is much more tenuous, but observations of quasar emission can be modeled by a thick disk of accreted hot gas orbiting a massive black hole.

the only remnant of what might have been a much more violent and luminous past phase in galactic activity.

The statistics of the distribution of quasars with redshift suggests that quasars become increasingly frequent at earlier eras of the universe. They seem to occur about as commonly as we might expect large galaxies at a redshift of 3 to occur. Evidently, quasars have burned out by the present era. However, some trace of a quasar phase may remain embedded deep within the nuclei of nearby galaxies and possibly even in the center of our own galaxy. We actually know little about the nuclei of galaxies.

The evidence for a central black hole in our own Milky Way galaxy is exceedingly controversial. Some observations suggest the presence of a massive object of about 1 million solar masses; others favor a dense star cluster at the center of our galaxy. One intriguing property is worthy of note. Both the Andromeda galaxy and its dwarf elliptical companion have a central light distribution that peaks at the very center of each galaxy. Spectroscopic measurements of the nucleus of Andromeda suggest that about 10 million solar masses are lurking in the central parsec, and this cannot be in the form of ordinary stars that would otherwise be visible. Perhaps this is a black hole. A more extreme case may be the giant elliptical galaxy M87 in the Virgo cluster, which also has a central cusp. In this case, spectroscopic observations have recently shown that the stellar distribution has a higher velocity dispersion toward the cusp. That is, stars are moving more rapidly, as might be expected if an extreme concentration of mass is at the center of the galaxy. We could conceivably be looking at a distribution of stars that has accumulated around a very massive and compact central object, such as a giant black hole. The inferred mass at the center of M87 exceeds 1 billion solar masses. A less radical alternative is that the remnants of massive stars, perhaps neutron stars or black holes, would have become more centrally concentrated than less massive stars as a consequence of dynamical relaxation in the galactic nucleus. The resulting mass concentration could attract sufficient ordinary stars to account for the central light cusp.

A similar phenomenon may have occurred in several globular star clusters, in which the light distribution also possesses a central cusp. One suspects that massive stellar remnants have concentrated in the central cores. Here, the phenomenon has been made more intriguing by the discovery of associated x-ray emission. Periodic bursts of x-ray emission occur, and there is an underlying component of steady x-ray emission as well. The locations of the x-ray sources in globular clusters indicate that, on the average, the x-ray source is not at the very center, as we would expect for a massive black hole that weighed more than 100 solar masses; instead it is displaced. From this finding, we can infer a typical mass, several solar masses for the counterpart of the x-ray source. It has been suggested that we are witnessing the infall of debris into a black hole of modest mass, weighing in this instance perhaps 3 solar masses. The debris would be generated by a wind from a close companion star that is in its red-giant phase of evolution. Other models are possible, of course, and the interpretation of the globular cluster x-ray sources is currently a source of controversy. The possibility remains, however, that we are viewing accretion of gas onto a black hole.

How do these massive black holes form? Black holes of a few solar masses, as in the case of a globular cluster, would have formed at the endpoint of normal stellar evolution. As we shall see in the next chapter, the early stages of galactic evolution are marked by the rapid evolution of an early generation of many massive stars of up to 100 solar masses, which produced the bulk of the heavy elements now found within such stars as our sun. These massive stars leave black hole remnants, some of which would have lower mass stellar companions that eventually evolved into red giants and fed their associated black hole. We may speculate that, in the dense nucleus of a galaxy, these black holes eventually coalesced into a much larger black hole. As a result of continuing growth by swallowing nearby stars and gas, the central black hole ultimately achieves its present dominance in quasars and, perhaps to a lesser extent, in the less luminous active galactic nuclei.

· 13 ·

FORMATION OF THE STARS

I don't pretend to understand the Universe—
it's a great deal bigger than I am.
—*Thomas Carlyle*

An observer present 10 billion years ago, prior to the birth of the sun, would have witnessed a spectacular visual display. Many stars would have been forming, and the Milky Way galaxy would have been perhaps 100 times more luminous than it is at present. Other stars would have been erupting in supernova explosions, which marked their violent death. The births and deaths of stars are commonly observed events in the universe today, and astronomers have a good understanding of processes of stellar evolution. In this chapter, we shall examine these processes to complete our emerging picture of the evolution of the early universe.

BIRTH AND DEATH OF STARS

When a star first condenses, it is very distended. It begins to shine as a very red, intensely luminous object. Gradually, the star contracts, becomes hot, and emits radiation at shorter wavelengths and at a less prolific rate. If the star is fairly massive, perhaps 20 times the mass of the sun, it will rapidly exhaust its supply of nuclear fuel. For a star, life is a precarious balance between the inward pull of gravity and the outward pressure exerted by the hot gas in the interior as it tends to expand and cool. Once the nuclear fuel has been depleted, the center of the star can no longer maintain the pressure difference to support the star against the gravitational force. A catastrophic collapse occurs in the core of the star, and the result is a supernova explosion—the outer layers of the star are blown off in a brilliant flash of light.

Many stars died a violent death by supernova explosions during the early years of our galaxy, but a supernova is now a rare event in the Milky Way. It has been said that the next bright supernova will herald the birth or death of a famous astronomer. The last observed supernovae in our Milky Way galaxy were named for the astronomers who first recorded them, Tycho Brahe and Johannes Kepler.

Much of our galaxy is obscured by interstellar dust, and it is entirely possible that supernovae have exploded more recently. We can obtain a better idea of the rate of supernova explosions by studying other galaxies (Figure 13.1). A star is observed to brighten suddenly and within a matter of weeks to become brighter than a billion suns. The supernova makes a significant contribution to the light from the entire galaxy. Within a year, the supernova rapidly dims and soon becomes indistinguishable from other stars in the galaxy.

On February 23, 1987, a great commotion arose among astronomers when the brightest supernova in 300 years was discovered. It was in the Large Magellanic Cloud, our nearest galactic neighbor, first recorded by deep underground neutrino detectors in Japan and in Cleveland. Shortly after the neutrino flash, astronomers in Chile and in Australia saw a brilliant star, of about magnitude 3 (Color Plates 3 and 4). It had increased a millionfold in brightness. It slowly faded over the next 12 months but would be studied for many years to come with large telescopes. SN1987A, as this supernova was christened, was destined to go down in astronomical annals as the first supernova for which a neutrino burst was measured, confirming the theory that the inner core of the star had collapsed and released its final store of energy. Supernovae are markedly more luminous and violent than novae, which occur far more frequently. A star often undergoes repeated nova outbursts, but a supernova explosion is truly catastrophic—the star

(a)   (b)

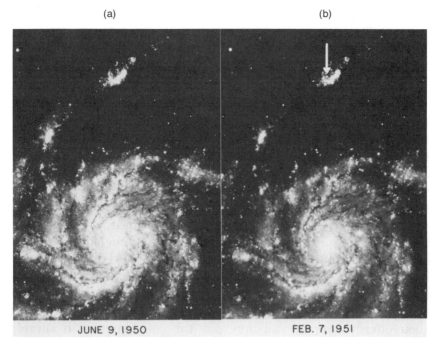

JUNE 9, 1950   FEB. 7, 1951

Figure 13.1   A Supernova in Another Galaxy
Photographs of the spiral galaxy M101 without (a) and with (b) a supernova.

has died. Such supernova events are found to occur in other spiral galaxies similar to our Milky Way galaxy approximately once every thirty to fifty years. During the early years of our galaxy, we believe that supernovae occurred far more frequently, perhaps as often as once a year. Our galaxy must have been immensely more active when it was young.

It takes a certain amount of detective work to be able to make this rather remarkable inference about the early evolution of our galaxy. First, we assume that the heavy elements that are found in even the oldest stars must have been synthesized in supernovae. Elements as heavy as iron could have been produced only under the extreme densities and pressures of matter attained during the final phases of stellar evolution. These heavy elements are blown out by the supernova explosion into the interstellar medium. The enriched gas mixes with older gas and eventually is recycled by a later generation of star formation. A large number of supernovae must have exploded to account for the abundances of heavy elements that we now observe. Most of these supernova explosions must have occurred during the very early evolution of our galaxy, when the Milky Way was relatively large and distended.

The oldest stars that astronomers have studied are found in the galactic halo. They evidently formed during the initial collapse of the galaxy. Determining the age of a star is a complex procedure. It is necessary to compute the rate at which the star is burning its supply of nuclear fuel. The less massive a star is, the cooler its central temperature will be, because less pressure is needed to support the star against gravitational collapse. As a result, less massive stars are cooler and radiate much less effectively. The luminosity of a star is found to depend only on its mass, at least during the hydrogen-burning phase. The lower rate of nuclear energy release required by the lower luminosity leads to longer lifetimes for such low-mass stars. Hydrogen-burning stars of 1 solar mass live about 10 billion years, but stars of 20 solar masses may live only a few million years. Massive stars rapidly consume their nuclear energy supply, and they pay the consequence in shorter lifetimes.

Stars that have exhausted their supply of hydrogen as a nuclear fuel will next begin to burn helium into carbon. After the helium is exhausted in the stellar core, heavier atoms burn, until the star finally collapses. Low-mass helium-burning stars are found in globular clusters, which consist of the oldest stars in the galaxy. Their ages are believed to be about 15 billion years. Compared with the sun, these old stars are often deficient in heavy elements. As we shall see in Chapter 15, the origin of the heavy elements provides an intimate link with the earliest development of our galaxy.

We are fairly confident that massive, short-lived stars will become supernovae. By contrast, stars like the sun have very long lives. Perhaps the early galaxy contained predominantly massive stars; otherwise, we would observe many stars of about 1 solar mass with few or no heavy elements. We find instead a surprising paucity of stars that are totally deficient in heavy elements. Even the oldest stars seem to contain about 1 percent of the abundance of heavy elements found in the sun. Short-lived, massive stars must have supplied the heavy elements for the gas from which later generations of stars condensed.

A second type of observation that allows us to draw conclusions about the early phases of galaxies is examination of the spectra of distant galaxies. Galaxy spectra often vary in character between the inner and outer regions. Many billions of stars are contributing to the light from these galaxies, and the stellar population appears systematically to have a greater abundance of heavy elements toward the central region of a galaxy. This result suggests that as stars evolved and shed enriched matter, gas rich in heavy elements fell toward the inner regions of a galaxy, where it mixed with unprocessed material and condensed into a successive generation of stars. This process may have been repeated a number of times (Figure 13.2). The end result is that the innermost stars are progressively more enriched in heavy elements than stars farther away from the center. Evidently infall of gas and continued enrichment through massive-star formation was an important process in the early evolution of the galaxies.

THE FIRST STARS

We can now return to our attempt to reconstruct the evolution of the galaxy by considering how the first stars may have formed.[17] An early collapsing protogalactic gas cloud would have consisted predominantly of hydrogen, with about 10 percent of its atoms in the form of helium. Essentially, no heavier elements were present at first. The gas, which may initially have been rather cold, heated as the cloud collapsed. During the collapse, turbulent eddies formed, ran into one another, and dissipated their energy into heat. Thus, the gross motions of the collapse, which ordinarily would tend to grow as the collapse proceeded, would gradually be dissipated by turbulence and formation of shock waves.

However, the collapsing gas soon became sufficiently dense to begin to radiate energy from atomic collisions. The gas heated to about 10,000 degrees kelvin, where hydrogen is partially ionized. Because of a fundamental property of the hydrogen atom, the gas remained at this temperature while continuing to collapse. As the temperature rose, the atoms and free electrons began to move more rapidly, thereby ionizing more of the hydrogen atoms and freeing more electrons. These newly liberated electrons collided with

Figure 13.2 Enrichment of the Galaxy

Supernovae and evolving stars produce enriched matter in the outermost regions of the galaxy, where the earliest stars must have formed. This ejected matter falls inward, eventually accumulating into clouds that form stars. These stars in turn produce more highly enriched material. In this way, a gradient of enrichment builds up, so that the stars that are richest in heavy elements are found closest to the galactic nucleus.

other hydrogen atoms, to which they transferred energy. In excited atoms that have just suffered such a collision, the bound electrons are in a higher energy level than they were prior to the collision; they cannot retain their excess energy, however, and almost immediately radiate it away as a discrete quantum of radiation (Figure 13.3). Hence, when its ionization level increases, the gas is more readily able to radiate away some of its internal energy. In other words, its rate of cooling increases.

But the process is self-limiting. If the gas cools very much, all the free electrons will disappear. Cooling means that the free electrons lose energy. As they slow down, they can more easily be captured by protons, and they form atoms. If excessive recombination into atoms were to occur, cooling would no longer be possible. The continuing release of gravitational energy as the cloud continues to collapse would now heat the gas and reionize it. The gas evidently maintains a delicate balance between ionization and cooling. The temperature stays at about 10,000 degrees, because hydrogen begins to be partially ionized from collisions between atoms at this temperature.

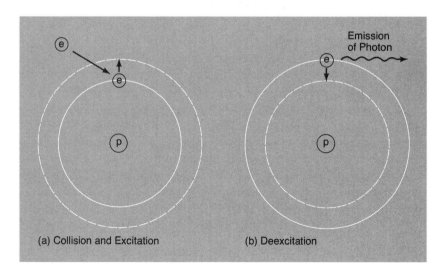

Figure 13.3 Cooling by Radiation
The energy of rapidly moving free electrons is converted into radiation when electrons collide with hydrogen atoms and excite the bound atomic electrons (a). The result is a net cooling (loss of energy) of the gas by radiation (b).

We seem to have discovered a natural thermostat. The temperature is prevented from dropping or increasing appreciably as the collapse proceeds. Because the temperature does not rise, the pressure forces, which tend to inhibit breakup of the cloud, fail to increase as fast as the gravitational forces as the cloud compresses. As a result, the cloud fragments into smaller and denser pieces. In each of these fragments, gravity and pressure forces are initially in balance. The fragments themselves collapse, however, as the cloud collapses. The fragments will continue to subfragment as long as the gas can lose energy freely by radiation. As long as radiation can be freely emitted, pressure forces can never recover sufficiently to stabilize or counteract the collapse.

Eventually, when the density has risen sufficiently, even the weakly ionized hydrogen gas becomes opaque. Once this occurs, radiation can no longer leave freely. The gas temperature begins to rise, thereby raising the pressure. Under sufficient compression, electrons attach themselves to hydrogen atoms to form negative hydrogen ions (H^-). This ion, unlike the hydrogen atom, absorbs light strongly. Consequently, when the abundance of H^- ions has risen to a certain level, the radiation emitted by the hydrogen atoms becomes trapped, and the gas can no longer cool effectively. The cloud continues to collapse, however, because gravity still exceeds any opposing pressure force, and the gas once again heats up. Further fragmentation eventually ceases, when pressure forces support the fragments against their own self-gravity. Once they become supported by pressure, the protostellar fragments enter a stage of extremely slow contraction. They radiate at a greatly reduced rate because of their newly acquired opacity, which helps to trap the radiation. No star can be perfectly opaque. The center must always be hotter than the exterior to supply the pressure difference that supports the star against its own self-gravity. Radiation from the center must eventually diffuse out, and it will be highly degraded in energy if it has to undergo much absorption and reemission.

During the early stages of star formation, a compact, opaque *protostellar core* forms, which grows in mass as matter is accreted from the more diffuse and transparent surroundings. The cloud was initially rotating, and the accreting matter forms a flattened, spinning, protostellar disk of dense gas. The infalling matter spirals in and releases energy in the form of heat as it crashes onto the core. The hot gas radiates its energy away, as gravitational contraction of the core continues. The temperature of the core steadily rises as it contracts. The protostellar contraction phase terminates only when the central region becomes so dense and hot that nuclear reactions occur.

Several reactions occur prolifically. At sufficiently high energy, a pair of protons are thrown together with sufficient violence to merge, by ejecting a neutrino and a positive charge in the form of a positron (positively charged electron). The resulting product of the fusion is a nucleus of deuterium (heavy hydrogen), which consists of a proton bound to a neutron. When the temperature has risen to some millions of degrees, the protons are moving sufficiently fast to merge together, overcoming the repulsive force that ordinarily exists between equally charged protons. Deuterium synthesis provides the first injection of nuclear energy: a free proton and a neutron weigh slightly more than a deuterium nucleus, and this mass difference is released as nuclear energy. In return for expending this energy, the deuterium nucleus has a binding energy, or energy required to break it up into a free neutron and proton, of 2.25 million electron volts, about 0.1 percent of its rest mass. This is not very much by nuclear standards, and in fact deuterium is very easily disintegrated by the hot radiation field at the center of the protostar. Once destroyed, fragile deuterium nuclei do not reform, and helium nuclei, each consisting of two protons and two neutrons, are eventually synthesized out of hydrogen nuclei. Unlike deuterium, helium is a relatively stable nucleus, and survives quite happily inside the stellar core at a temperature of about 10 million degrees.

This process of fusing hydrogen into helium releases considerable energy. A small fraction (0.007) of the initial rest mass of the hydrogen is converted into pure radiation in the form of highly energetic photons and neutrinos. These nuclear reactions provide a steady, continuing source of heat and pressure in the core of the star. Gravitational contraction ceases, and the luminosity of the star becomes constant: the star has now evolved into the hydrogen-burning *main sequence* (Figure 4.3). The sun is a very ordinary, hydrogen-burning, main-sequence star. When stars exhaust their supply of hydrogen in the core, they are said to leave the main sequence. Subsequently, such stars evolve more rapidly, collapse further, heat up, and burn heavier elements, until their nuclear fuel supply is ultimately exhausted.

The distinguishing characteristic of the first stars that formed was their initial lack of any elements heavier than helium. The primary result of this absence of heavy elements was to affect the ability of primordial matter to cool and eventually to acquire opacity. The presence of heavy elements is crucial to the efficiency of both of these processes. The electrons of elements such as carbon have low energy levels that can be excited in atomic collisions in a fairly cold gas, at about 100 degrees kelvin. Molecules such as carbon monoxide have still lower energy levels (which are associated with the rotation of the molecule). These energy states occur in quantized amounts, just as is the case for the energy levels of bound electrons in atoms. Excitation of the rotational energy levels can enable a molecular gas cloud to cool to temperatures of only a few degrees kelvin.

Hydrogen protostars formed at a relatively high temperature (about 1000° K) because no heavy elements were present to cool the gas.

Even a temperature of 1000 degrees kelvin would be surprisingly low for a metal-poor gas, were it not for the fact that some hydrogen molecules are present. Hydrogen atoms cool by excitation of electronic energy levels. The excitation is by collisions, and kinetic energy is radiated away, resulting in a cooling of the gas. The lowest level in atomic hydrogen is excited at a temperature of about 10,000 K and a metal-poor atomic gas could not cool below this temperature. However, hydrogen molecules have lower-lying energy levels, which are associated with the rotation of the molecule. These molecules are normally formed by a catalytic reaction that uses the surface of a dust grain to bring two hydrogen atoms together. Although a primordial gas without heavy elements has no dust grains, nature fortunately provides an alternative channel for forming hydrogen molecules: the primordial hydrogen is predominantly atomic but contains trace amounts of electrons and protons. This ionization is a relic of its past history; once a fully ionized plasma, the hydrogen recombined during the decoupling epoch, but because of the cosmic expansion, the recombination was not quite complete. Perhaps one electron in a thousand avoided finding a partner proton. Much later, when the radiation temperature had decayed, with expansion, down to a few hundred degrees, the residual electrons could combine with hydrogen atoms to form negative hydrogen ions. Collisions between hydrogen gas clouds generate shock waves that also result in some hydrogen ionization. Once the negative hydrogen ions are formed, they rapidly react with atomic hydrogen to form hydrogen molecules. The net effect is that as much as one part in a thousand of the hydrogen becomes molecular. This is enough to cause strong cooling as gas clouds contract and consequently fragment.

The molecular hydrogen cooling becomes even more dominant once high densities are reached because three hydrogen atoms can react together to form a hydrogen molecule (and an atom). The three-body reaction converts essentially all the atomic hydrogen into molecular form. Eventually, the density becomes so high that the molecules are dissociated by collisions. By then, however, extensive fragmentation has occurred. The cooling of molecular hydrogen ensures that fragmentation occurs on scales as small as one-tenth or even one-hundredth of a solar mass (Figure 13.4). Massive stars also must form, but we cannot deduce the actual distribution of stellar masses with any confidence. One clue is that molecular hydrogen can be destroyed, as might happen if a cloud undergoes extensive stirring that results in shocks. If the molecules fail to reform in the shocked gas, cooling can occur only down to about 10,000 degrees kelvin, which means that greater mass would be required to contain the high pressure than would be the case for colder protostars containing hydrogen molecules. Thus, the resulting fragments would have been correspondingly massive, and we conclude that in this case the first stars generally must have been more massive than the sun. A typical first-generation star might have been 10 times more massive than the sun. It would have been extremely bright, perhaps 10,000 times brighter than the sun, and relatively short lived, lasting perhaps 10 million years.

Theory is not sufficiently precise to distinguish the primordial formation of low-mass stars from that of high-mass stars. The possibility of molecular-hydrogen cooling suggests that primordial stars spanned precisely the same mass range (0.1 to 100 solar masses) as stars that form in the interstellar medium today. The second possibility, of cooling by atomic hydrogen in the absence

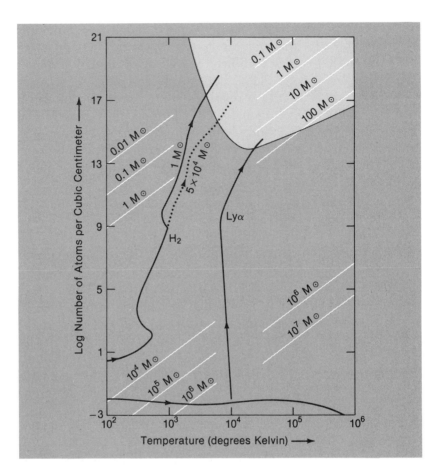

Figure 13.4 Formation of the First Stars

Star formation proceeds in the direction of the arrows. Because the first stars contained no heavy elements, the gas cloud could not cool if it consisted only of hydrogen and helium atoms. If the atoms were partially ionized, the free electrons could collide with atoms, and radiation would result. In the figure, this radiation is labeled Lyα, for the Lyman alpha line radiation from the ground state of the hydrogen atom. The cloud would remain partly ionized at about 10,000 degrees kelvin as it collapsed and fragmented. This process halted when the fragments became opaque due to absorption by molecules that formed at high densities when electrons could attach to hydrogen atoms. Because of the relatively high temperature and pressure, the initial fragments were much more massive, perhaps 10 solar masses or larger, than they would be with dust grains present to provide strong cooling, as is the case for later generations of stars. An alternative possibility is that some hydrogen molecules (H_2) form. These act as a thermostat that keeps the cloud at about 1000 degrees kelvin until a very high density is attained. In this case, much lower stellar masses can form, just as they can when heavy elements are present. Whether or not H_2 cooling is significant in primordial clouds cannot be decided with any certainty.

of any molecular gas, results in relatively massive stars. One piece of empirical evidence supports the second hypothesis and is a resolution of the "G-dwarf problem." The sun is a G dwarf star, with a hydrogen-burning lifetime of about 10 billion years. Had the first generation of stars included a significant number of stars of solar mass, we would observe a large number of surviving metal-poor G-dwarfs today. However, studies of the metallicities of stars in the solar neighborhood show that there are proportionately fewer and fewer stars of lower and lower metallicity. This fact is most simply understood if the vast majority of primordial stars were much more massive than the sun and therefore have long since been extinct. While this is not the only solution to the G-dwarf problem (for example, the early disk of the galaxy may have been much less massive than the present disk and therefore have formed few metal-poor stars), we can be confident that whatever the initial distribution of stellar masses, the very fact that a protogalaxy was predominantly gaseous means that the star-formation rate would have been enhanced several fold relative to a present-day spiral galaxy. The early evolution of a galaxy must have been a very spectacular display indeed.

These first-generation stars provided the source of the first heavy elements. The hydrogen in the stellar cores burned to helium, and the helium burned to carbon. In the innermost core, the carbon eventually ignited to form oxygen and silicon. The final product, iron, is the most stable element, as we shall see in Chapter 15. Eventually the stellar core, exhausted of nuclear energy, collapsed in a violent supernova explosion. The debris, consisting of enriched material that had undergone nuclear processing, was ejected into the interstellar medium and recycled into gas clouds, which collapsed to form the next generation of stars. This cycle of star birth and death may have been repeated many times in the young galaxy.

CURRENT STAR FORMATION

The gradual addition of heavy elements to the interstellar gas from the deaths of the earliest massive stars at first made no discernible difference to the star formation process, because the enriched material was immediately diluted by primordial gas. The lifetimes of massive stars that eject large amounts of matter are short, perhaps only a few million years—only a small fraction of the dynamic time scale necessary for any gross characteristics of a galaxy to change. Eventually, after several generations of massive hydrogen stars had evolved and died, the average level of heavy-element enrichment would have risen considerably. The point was reached where heavy elements could play a greater role than hydrogen in radiating energy from fragmenting gas clouds. Heavy elements affect protostar formation primarily through the presence of tiny interstellar dust grains. These solid particles are like sand grains, although they are much smaller. (Typical grains are only about 0.0001 millimeter in radius.) At first, the dust grains condense from the hot gases that flow from evolving stars. Most of the heavy elements in a typical interstellar cloud appear to be in the form of dust grains.

Dust grains are believed to be made of a combination of refractory, rocklike materials, including silicate minerals such as quartz. The hard cores are surrounded by a volatile layer of ice, such as water ice or possibly ammonia or methane ice. When dust grains are present, as they are in ordinary interstellar clouds, the degree of transparency of the cloud is determined by the amount of solid material in grains along the line of sight. When the cloud collapses, it becomes highly opaque. A ray of light is almost entirely absorbed or scattered many times by the dust grains in the cloud.

As a cloud collapses, the presence of the grains enables the gas to cool very effectively. The grains stay cold and emit far-infrared radiation, at a wavelength much longer than that of the radiation they absorb. Recall that the small particles can interact effectively only with radiation whose wavelength is not too much longer than they are. Far-infrared photons have wavelengths between 0.1 and 0.01 millimeter, greatly exceeding the sizes of interstellar grains. Thus, the far-infrared radiation escapes freely and provides an efficient mechanism for cooling a dense cloud. Any remaining heavy elements in the gas will probably freeze out to form icy mantles around the cold grains. It should be apparent that the number and size of the grains will determine both the gas temperature and the opacity of the cloud.

One consequence of the presence of dust grains is clear: if more than a certain critical fraction of heavy elements relative to hydrogen is present, the collapsing cloud will begin to be cooled by the grains rather than by the hydrogen, and the temperature will drop dramatically, to perhaps only 10 degrees kelvin. The cloud will continue to collapse and cool, and it will remain cold as long as the grains can radiate freely. Smaller and smaller fragments form, until the smallest fragments become dense and opaque enough for their temperature to rise and for protostellar cores to form. These cores have a relatively low characteristic mass, just as we saw for the primordial protostars, where molecular-hydrogen cooling played a critical role, despite the fact that dust grains and heavy elements were lacking. This low mass is a direct consequence of the low temperatures (and correspondingly low pressures) at which cores can form when molecules and dust grains are present in a gas cloud. Typical masses of the opaque protostellar cores are found to be about 0.1 solar mass. The protostellar cores can grow further by accretion, and the process of star formation proceeds much as described for first-generation stars (Figure 13.5).

MOLECULAR CLOUDS

The raw material for star formation is interstellar gas. The Milky Way galaxy contains about 2 billion solar masses of diffuse gas, some 5 percent of its stellar mass. The density of a typical interstellar gas cloud would be similar to that of a perfect vacuum on the earth, perhaps with a hundred atoms in every cubic centimeter. Individual clouds extend for many light-years, however; cumulatively, they give a detectable signal in different wavelength bands. Interstellar gas was first discovered in the optical part of the spectrum, when binary stars were found to have absorption lines that resulted from intervening sodium and calcium atoms and whose strength did not vary with the binary period. Evidently the absorption was not circumstellar but interstellar, or produced in intervening gas clouds. Hydrogen was discovered to pervade interstellar space in great quantity as a consequence of its 21-centimeter radiation emitted because of the spin-flip transition; the electron and the proton can have their spin axes either parallel or antiparallel in the hydrogen atom, and there is a tiny energy difference corresponding to a photon of wavelength 21 centimeters between

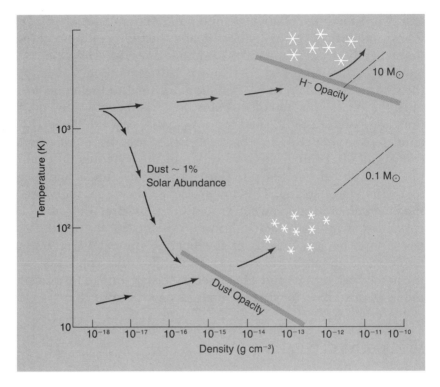

Figure 13.5 Current Star Formation

Once the heavy-element abundance builds up sufficiently to permit strong cooling to occur (primarily by dust grains), stars of lower mass can form. The heavy-element abundance must rise to about 1 percent of the solar value before primordial star formation (primarily of massive stars) is inhibited. The greater the abundance of heavy elements, the sooner can the onset of strong cooling occur and lessen the likelihood of massive star formation. Arrows show the path of first-star formation and current star formation.

these two states. Hydrogen atoms form in both states, but prefer to end up in the state of lowest energy, the transition to which requires emission of a photon of 21-centimeter wavelength. This transition occurs so infrequently that it was measured for the first time in interstellar space, where in a cloud containing some thousands of solar masses of hydrogen, enough hydrogen atoms radiated at 21 centimeters to give a strong signal. The main constituent of these gas clouds turned out to be hydrogen, which was first detected in the 1950s in interstellar clouds at 21 centimeters. The

1970s heralded a new discovery: about half of the interstellar gas was found to be in the form of molecular hydrogen (Color Plate 5). High-resolution mapping of interstellar clouds at microwave frequencies during the subsequent decade led to the discovery of many complex molecular species that coexisted with the molecular hydrogen (Figure 13.6). Such molecules as carbon monoxide, formaldehyde, water, and ammonia are abundant; indeed, there is enough ethyl alcohol in a single molecular cloud to serve a party of enough human beings to populate the entire galaxy. We know that the interstellar medium is a complex mix of gas in several phases: ionized, atomic, and molecular. There are giant molecular cloud complexes and small globules, tenuous very hot gas and wisps of cool atomic gas, and shell-like remnants from ancient explosions of supernovae or winds from massive stars (Color Plate 6).

Molecular clouds are of especial interest because they are the sites of ongoing star formation. Molecular clouds are denser than clouds of atomic hydrogen; the molecules form as a consequence of this high density. About 1 percent of the mass of an interstellar cloud is in the form of small, solid grains of dust. The concentration of these grains in a cloud is dense enough to block out light from background stars and be visible as a black cloud silhouetted against the Milky Way. Many dark clouds, often globular shaped and containing some ten or hundred solar masses of gas have been mapped. These regions are especially rich in molecular species: only when ultraviolet radiation is blocked can molecules form in great abundance. The molecules permit the gas to cool effectively because energy is radiated in microwave photons by excitation of molecular states as well as by the dust grains. Loss of thermal energy means that pressure forces cannot play much of a role in cloud support, and the molecular clouds are destined to collapse and form stars.

STELLAR MASS DISTRIBUTION

One of the crucial clues to the nature of star formation is the mass distribution of newly formed stars. Stars forming today range from 0.1 solar mass to about 100 solar masses. Relatively few massive stars are being born, however, and the vast majority of stars being formed in the Milky Way today have relatively modest mass and

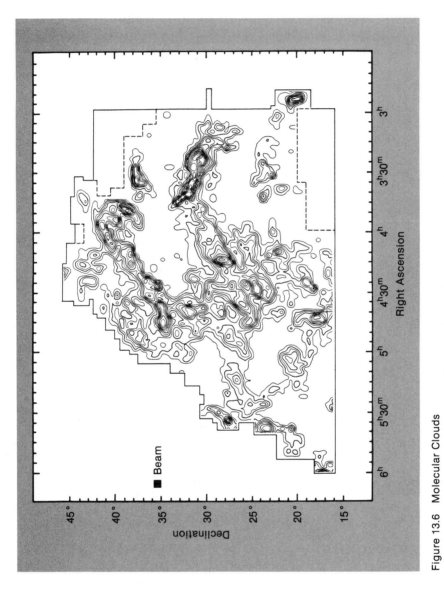

Figure 13.6   Molecular Clouds

A radio map of molecular clouds in the Taurus region made in radiation emitted by carbon monoxide. These clouds are at a distance of about 500 light-years; the scale of the map is about 200 light years.

luminosity. These stars are long lived; stars of 1 solar mass or less have ages comparable to that of the galaxy. The first stars appear to have had a different mass distribution, for we observe few stars with very low abundances of heavy elements. Moreover, the oldest stars that we see, which populate the halo, have heavy-element abundances that average about 1 percent of that of the sun (Figure 13.7). Thus, the virtual absence of extremely metal-poor stars is a clue to the nature of the first generation of stars.

As we have already seen, the first stars were predominantly massive, luminous, and short lived. The debris from supernova explosions that marked the first stars' deaths caused the average heavy-element abundance to build up to about 1 percent of the solar abundance. Once this buildup occurred, a new mode of star formation, dominated by grain cooling, took over. This resulted in a dramatic change in the mass distribution of newly formed stars: less massive, longer-lived stars could form. These stars survived to form the stellar population that constitutes the halo of our galaxy today. Matter shed during the course of evolution of these stars accumulated in the disk of the galaxy, where it was recycled into successive generations of stars that were progressively more abundant in heavy elements. This gradual enrichment was important during the first billion years or so of the galaxy, when much gas was present and the star-formation rate was correspondingly high.

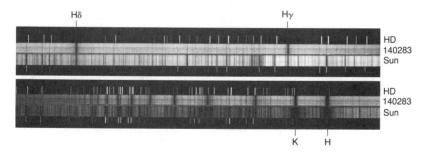

Figure 13.7 Normal and Metal-Poor Stars
Spectrograms are shown of two stars in our galaxy, a high velocity halo star (HD140283) and a normal disk star of similar luminosity and temperature. Spectra at top and bottom provide reference lines for calibration purposes. The high velocity star shows weak absorption lines for all elements except hydrogen. The disk star, our own sun, shows many strong absorption lines characteristic of such heavy elements as carbon and iron.

Star formation has continued since then, but the rate of star formation has been relatively slow. In elliptical galaxies, almost all the interstellar gas has been lost, and star formation has practically ceased. In highly flattened galaxies, such as spirals, enough gas remains that stars continue to form.

Imagine a collapsing gas cloud that breaks up into a large number of protostellar fragments. These fragments will have dense, opaque cores surrounded by a larger, more diffuse region of infalling gas. At first the protostellar fragments will participate in the overall collapse of the cloud. Very soon, however, fragments will start to collide. Can we imagine what may happen next?

It seems clear that just as the motions of planets are perturbed by the presence of their neighbors (precisely this effect led to the discovery of Neptune), nonradial motions will be acquired by fragments as they suffer near collisions with nearby fragments. At first, the fragments will be very extended, which implies that many direct collisions between fragments will occur. The relative velocities at which encounters take place are quite low, compared with the velocities required to disrupt a fragment. We can imagine that the ensuing collisions will be very sticky. For an analogy, imagine dropping a marble onto a thick rug rather than a hard floor—the marble will lose much of its energy in the first bounce. If much energy is dissipated in a collision, the colliding fragments will become gravitationally bound to one another, and mergers will eventually occur. Many of the smaller fragments will be swallowed by the larger ones. Eventually, collisions will become less frequent, because fewer fragments remain and the remaining individual fragments have contracted and become more condensed. In this way, it seems plausible that a certain mass distribution of fragments will arise that may not depend directly on the masses of initial fragments.

We can try to test such a theory by examining the observed distribution of stars by mass (Figure 13.8). There are many small stars and few large stars. Of course, massive stars have shorter lifetimes, but we can allow for this effect because we know how long stars of a given mass have lived. Knowing the lifetimes of stars of different masses enables astronomers to compute the mass distribution with which the stars were born. Very roughly, it is found that the number of stars of any given mass born in a particular region

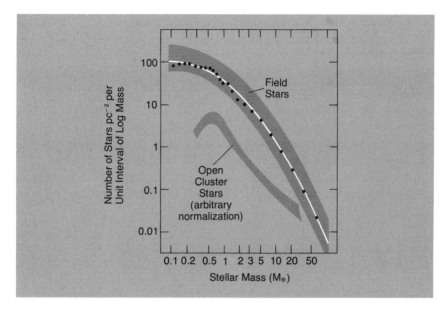

Figure 13.8 The Initial Mass Function of Stars

Progressively fewer massive than low-mass stars are observed. The number of stars that are born in a given region varies approximately as the inverse of the square of the stellar mass. At low masses, this relation is uncertain, since we cannot readily detect these stars. In some dark clouds, massive stars appear to be deficient, and some young stellar associations appear to be lacking in low-mass stars.

is proportional to the inverse of the square of that mass. This relation is known as the *Salpeter function,* after the American astrophysicist Edwin Salpeter. We find relative to the frequency of occurrence of stars of 1 solar mass, only one-quarter as many stars of 2 solar masses and only one-sixteenth as many stars of 4 solar masses. We find a similar mass function in a variety of different star clusters. These systems are usually studied because the stars within any cluster often appear to be nearly coeval. However, there are exceptions to the general rule; some regions, such as dark clouds, appear to be deficient in massive stars; other regions, notably those where many young stars are present, are deficient in low-mass stars.

By developing this theory, we can hope to understand variations in the mass function with which stars are born. It seems likely that fragment collisions may cause a similar mass distribution eventually to develop in any dense interstellar gas cloud. However, where this distribution begins and ends on the scale of mass must be largely determined by local conditions. For example, in a quiescent, cold cloud, the initial fragments might have very low mass and might form protostars of only moderate mass. However, if the cloud were initially warm or turbulent, perhaps because of the influence of nearby stars, the initial fragments would have a greater pressure and could be relatively large. In this case, massive protostars could form, and the cloud might lack low-mass stars.

The distribution of stellar masses provides a vital clue to the stellar birth process. We observe variations in our galaxy, both in regions of active star formation and in regions such as the globular star clusters, where star formation has not occurred for tens of billions of years. Although our understanding of the theory of star formation is limited to the crudest outline, we can relate observed star formation to the prolific star formation that occurred when the galaxy was young.

Let us now try to incorporate our knowledge of star formation into the context of a forming galaxy. It seems that galaxies began with a tremendous burst of activity. Most stars were massive and short lived, exploded as supernovae, and supplied the residual amounts of heavy elements that we find even in the oldest visible stars in our galaxy. Throughout this phase, the galaxy may still have been in a state of dynamic contraction from the gas cloud out of which it condensed. The free-fall time of such a cloud to condense into an object the size of the galaxy is about 100 million years, ample time for many cycles of massive star formation. A star of 30 solar masses exhausts its nuclear fuel in only a few million years. Once the heavy elements had accumulated to a critical level of about 1 percent of the heavy-element abundance in the sun, the first generation of star formation was essentially complete. At this stage, dust grains played a crucial role in enabling gas clouds to cool and fragmentation to proceed, down to scales of mass well below 1 solar mass. Long-lived stars like the sun formed, and the galaxy began to acquire its present stellar distribution. The theory of star formation by fragmentation enables us to understand the range of stellar masses that we observe today.

Star formation is an ongoing process. There are regions in our Milky Way galaxy where we can see many young stars, known as

T Tauri stars, after the prototype in the Taurus region. These are low-mass stars which are cool and relatively distended compared to hydrogen-burning stars, and are believed to have recently condensed (some within the past 10^5 years) from the interstellar medium and not yet have settled onto the main sequence of stellar evolution (Figure 13.9). Many of these stars are still shrouded by dense dust and molecular gas, which is also an intense emitter of spectral lines. A still earlier stage in the star formation process may be represented by Bok globules, after Dutch-American astronomer Bart Bok (Figure 13.10). These compact dark clouds are so dense that they completely obscure the light from background stars. Bok globules are believed to be on the verge of collapse into protostars, and they may be at the final stage of fragmentation. Molecular line observations enable radio astronomers to peer into even denser gas clouds; here also they find observational evidence that stars are forming via fragmentation into protostellar-sized clumps of gas. Mapping molecular clouds with far-infrared telescopes, both on the ground and airborne or in space (in balloons, in a converted airplane observatory, and in a recent satellite experiment), has enabled us to probe deep into the densest cores of molecular clouds and view the earliest stages of star formation (Color Plate 7). Within these cores, protostellar outflows have been discovered, a phenomenon associated with formation especially of massive stars when a vigorous stellar wind is generated. This wind clears away much of the accreting gas still in a disk around the protostar, and marks the termination of the protostar phase. We conclude that star formation is a vigorous and continuing process throughout our Milky Way galaxy; low-mass stars are being born in the coldest and most quiescent molecular-cloud cores, such as Bok globules, whereas massive stars are being born in relatively warm, turbulent regions in the giant molecular-cloud complexes.

Theory and observation evidently merge to support the idea that stars form by continuous fragmentation of interstellar clouds. The clouds themselves are parts of massive cloud complexes extending for hundreds of light-years; each contain some 10^5 solar masses of gas. In our Milky Way galaxy, gas shed by evolving stars collects into these molecular cloud complexes, so that there is an approximate equilibrium between the supply of gas and its consumption through star formation.

Figure 13.9 A Region of Active Star Formation

Within the Cone nebula are a number of young stars, known as *T-Tauri stars*, that are still embedded in the dense cloud of interstellar gas from which they formed. The brightest diffuse patch is a gaseous nebula (or HII region) being ionized by very massive young stars. Some nebulosity is seen in reflection against the light from these stars. Cool dusty matter is also visible in silhouette against the background star field and is an intense source of molecular spectral line emission.

Figure 13.10 Dark Globules
Dark *Bok globules* are so dense that they completely obscure the light from background stars. The mass of these two visible globules is estimated to be about 20 solar masses. They may be cold clouds on the verge of collapse into protostars.

But this was not always so. At early epochs, star formation was much more vigorous. Some spiral galaxies are even now far more active than our own galaxy. Yet in elliptical galaxies there appear to be no young stars. The history of star formation must surely have undergone significant variations to produce systems as diverse as spiral and elliptical galaxies. We must now place star formation in the broader context of galactic evolution to ascertain the origin of the morphology of galaxies.

· 14 ·

THE MORPHOLOGY
OF GALAXIES

When star formation is going on in an area,
it spreads in some way like a disease.
—*Walter Baade*

Our theories of collapse of protogalactic clouds, fragmentation, clustering, and star formation have brought us past the era of formation of galaxies and into the era of galactic evolution. In Chapter 13, we examined the fundamental properties of stars for clues to our theory of star formation. In this chapter, we examine the properties of galaxies for clues to the dynamic processes that occurred during their early evolution. We will formulate hypotheses that explain stellar evolution in spiral galaxies, the unique characteristics of elliptical galaxies and globular star clusters, and the observed gas content in the galaxies. By examining the dynamical interactions that must have existed in this era, we can hope to get a clearer picture of early galactic evolution.

ROTATION AND DENSITY WAVES

Spiral galaxies are spinning. The sun will complete an orbit around the center of the Milky Way galaxy every 250 million years. Rotation accounts for the highly flattened shapes of spiral galaxies.

Elliptical galaxies are not spinning as rapidly, and they are much rounder. The origin of the galactic rotation must be intimately related to the process of formation of the galaxies. According to a popular viewpoint (Chapter 11), galactic rotation originated in the mutual gravitational interactions between neighboring protogalaxies. Because of the recession of galaxies from one another, protogalaxies were once much closer to one another than they are now. They were also considerably larger, and at their era of formation, they must have been practically touching one another. Just as the gravitational pull exerted by the sun and moon is responsible for tides on the earth, the proximity of a protogalaxy would cause immense gravitational tides in its neighbor. These tides would cause the protogalaxies to acquire a certain amount of spin (Figure 14.1). The galaxies would be set spinning in opposite directions, so that the net amount of rotation does not change as a result of their interaction. The quantity that is conserved is called *angular momentum,* and it is a measure of the product of the mass of a body, its linear size, and its rate of rotation. The principle of *conservation of angular momentum* explains why a high diver will curl up tightly to maximize the number of somersaults he or she can make

Figure 14.1 Tidal Interactions and the Origin of Galactic Rotation

The tidal interactions between any pair of neighboring protogalaxies causes an immense torque to be exerted on each system, and each acquires a small amount of spin. The rotations of the two galaxies are equal and opposite, since angular momentum must be conserved. As the protogalaxies collapse into galaxies, they rotate more rapidly.

(Figure 14.2). Similarly, infall of the gaseous disk material toward the center of a galaxy enables the disk, by conserving angular momentum, to spin faster and faster, until it acquires the rotation speed that we now observe. The infall is halted by the onset of star formation. This generally accepted hypothesis seems to account for the rotation of the spiral galaxies, provided the gas has collapsed from a large enough radius. Collapse from a radial distance of about 300,000 light-years seems required if dark halos are present.

Spiral galaxies are conspicuous for their spiral arms, which carry newly formed stars along with much of the gas and dust from which they condensed. The frequency of young stars is much higher in spiral arms than in other regions of galaxies. Many interstellar clouds in these regions have recently contracted to form stars. If a cloud is compressed suddenly, it will begin to collapse and will eventually fragment. The sudden compression that appears to have occurred in spiral galaxies manifests itself as a *spiral density wave* (Figure 14.3). Density waves may be triggered by two possible phenomena. Passage of a nearby companion galaxy causes tidal forces that can drive a density wave. If the central bulge of the galaxy is not spherical but bar shaped, the asymmetry leads to outward propagation of density waves as the bar rotates. Both phenomena—spirals with close companions and galaxies with central bars—are common.

Just as a sound wave propagates, alternately compressing and expanding the air at a given point, a density wave can travel through stars and gas. Because of differential galactic rotation, the wave propagates in a spiral. The spiral density wave is actually carried by the gravitational forces exerted by the stars in the disk, because they contain most of the mass of the galaxy. The gas present in clouds and as interstellar gas throughout the disk experiences the extra gravitational force in the peaks of the wave. As a result, the clouds aggregate together and coalesce, and are also compressed because of the increase of interstellar gas pressure. Unlike stars, whose motions are weakly perturbed, gas clouds respond dramatically to a density wave: they collide, lose kinetic energy of motion, and pile up in a cosmic traffic jam. Consequently, the clouds become gravitationally unstable and are induced to collapse gravitationally. Once the gas is over the threshold of collapse, fragmen-

(a)

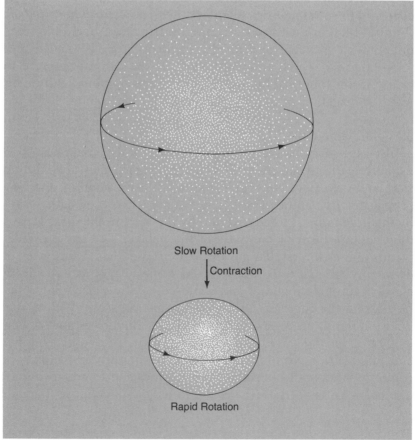

Slow Rotation

Contraction

Rapid Rotation

(b)

Figure 14.2 Conservation of Angular Momentum

A high diver (*a*) will curl up tighter in order to somersault during a dive. Conservation of angular momentum allows the diver to increase the spin rate and the possible number of somersaults. In much the same way, a contracting gas cloud spins faster as it collapses (*b*).

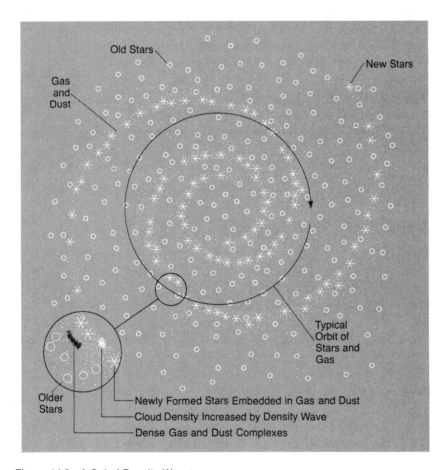

Figure 14.3 A Spiral Density Wave

Clouds moving in circular orbits enter the spiral density wave peak and are compressed by the locally increased gravitational field, which triggers the process of star formation. The newly formed stars and nearby regions of gas and dust provide a tracer of the spiral density wave pattern. Beyond the region of greatest compression, more time has elapsed since star formation was initiated, and older stars are found there.

tation continues as the cloud radiates away the energy that the gas acquired during the compression. Collapse and fragmentation do not cease until the fragments become opaque enough to slow their radiation losses. Protostars form at this stage.

The compression induced by the spiral density wave triggers star formation along the spiral pattern delineated by the wave peak. The stars and gas pass through the wave peak in a few million years. No further star formation subsequently occurs. The wave peak is therefore associated with bright young stars. It stands out dramatically from the rest of the disk in photographs of spiral galaxies (Color Plate 8). The young stars and associated gas are much more luminous than the older disk stellar population, and so they delineate the structure of the arms.

Not all spirals have "grand design" patterns. Floccular spirals possess bits and pieces of spiral waves as though the star-formation process were much more chaotic. Elliptical and SO galaxies seem to have lost most of their gas outside their nuclei. Consequently star formation does not presently occur in these systems.

THE ROLE OF MAGNETIC FIELDS

Galactic rotation has another interesting consequence. We have seen how the gas cloud from which the galaxy formed was also rotating. This gas initially may have been very weakly magnetized. Where this weak primordial magnetic field originates is uncertain. The field may be spontaneously generated by a turbulent dynamo. Tiny magnetic fields are generated naturally in an ionized, turbulent plasma, and if there also is a certain degree of ordered motion, the fields grow by dynamo action. Rotation of the gas would have the effect of greatly amplifying any seed magnetic fields (Figure 14.4). The magnetic field of the earth may originate in a similar way, as the earth's molten metallic core rotates. We can think of the magnetic lines of force as behaving like elastic strings; the tension along these strings builds up because of the motions of the clouds at the ends of the strings, until the tension is comparable to the dynamical forces of the rotating gas cloud.

Interstellar clouds are permeated by magnetic fields. The magnetic field is strong enough to align interstellar grains, which consequently polarize starlight. It also provides support in dense

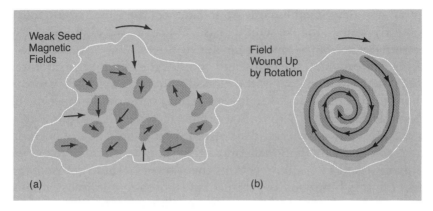

Figure 14.4 Origin of the Galactic Magnetic Field
Rotation of a collapsing gas cloud amplifies any primordial seed magnetic field (*a*). Turbulence may also play a role in coupling the energy of the rotation into magnetic energy as the galaxy forms (*b*).

clouds against gravitational collapse. Clouds below a critical mass of about 10,000 solar masses are magnetically supported. The magnetic field slowly leaks out, because only the charged particles (ions and grains), which are rare, are coupled to the field, and the neutral particles eventually become more and more concentrated. This process takes time, perhaps tens of millions of years, and it helps to explain why the interstellar gas is not aglow everywhere with forming stars. Massive molecular clouds collapse despite the magnetic field; one current view is that these clouds grow as they sweep up gas and accrete mass while orbiting around the galaxy, eventually loosing their gravitational stability.

Magnetic forces may play a crucial role in star formation and in the formation of planetary systems, because of the magnetic pressure that tends to inhibit small clouds from collapsing. Since only large clouds of 10,000 solar masses or more can collapse and overcome the average interstellar magnetic pressure, we infer that stars preferentially form in stellar associations and groups.

Once the clouds collapse, the magnetic field continues to play a crucial role. Like the entire galaxy, interstellar clouds are in a state of differential rotation. If a cloud were to conserve its angular momentum as it collapsed, it could not contract very far before the centrifugal forces became important. Once the centrifugal force and the gravitational force were in balance, the cloud would attain a steady state: collapse would cease. What saves a cloud from this terminal state far below stellar density is that the magnetic field threads throughout the cloud and couples the cloud to the ambient gas. The field winds up as the cloud rotates and transmits a torque that despins the cloud. Finally, the cloud can now contract to sufficiently high density to form a protostar, which is surrounded by a centrifugally supported disk. A protostar is an object whose energy source is gravitational contraction, not the thermonuclear energy released by fusion reactions that powers a star. Protostars are exceedingly luminous and diffuse; most of their energy is radiated in the infrared region of the spectrum. The terrestrial atmosphere is fairly opaque in the infrared; it required the launch of the first infrared astronomical satellite (IRAS) in 1984 to probe deep into nearby molecular clouds and reveal hordes of newly forming embedded stars, still immersed in their womblike shrouds of protostellar nebulosity (Color Plate 9).

Studying infrared radiation is not the only way we learn about protostars. One surprising finding, to which we briefly alluded in the previous chapter, emerged from microwave studies of molecular emission: that the formation of a protostar often triggers energetic outflows of molecular gas, apparently beamed along the path of least resistance: the rotation axis of the protostellar disk. Astronomers speculate that these outflows are triggered by release of substantial amounts of magnetic energy. The magnetic field is compressed and sheared in the contracting protostellar disk and eventually annihilates, energizing the molecular outflows.

Spiral galaxies contain a considerable amount of diffuse gas that has not condensed into stars. This gas is not left over from the formation of the galaxy but is continually produced as stars evolve, shed mass, and die. Some stars expel shells of gas, after they complete their hydrogen-burning phase, and undergo a core-contraction phase; such stars are known as planetary nebulae (Figure 14.5). During the late stages of nuclear burning, massive stars burn shells of helium and hydrogen simultaneously, becoming supergiants which undergo prolific mass loss in the form of a stellar wind (Color Plate 10). In the final stages of evolution, the inevitable supernova explosion must also lead to ejection of considerable amounts of material. Very luminous younger stars seem to lose mass prolifi-

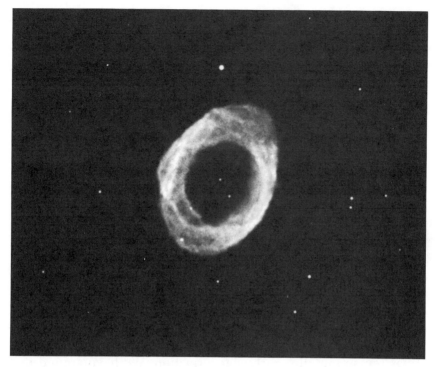

Figure 14.5 A Planetary Nebula
The Ring nebula in the constellation Lyra is one of the best-known planetary nebulae. It consists of a small, hot, evolved star that is ionizing a surrounding shell of gas that was previously blown off by the star.

cally by means of an energetic wind during their protostellar phase. The net effect is that the mass shed by stars accumulates in star-forming clouds as well in the more diffuse interstellar medium. As we have seen, the passage of a spiral density wave can trigger gravitational collapse in the clouds. The explosion of a nearby supernova or the mass outflow and ionizing radiation from a massive, hot young star can similarly trigger star formation. (Supernovae or ionizing radiation are likely to be found in spiral arms, because one of their common precursors, a massive, short-lived star, is found only there.) Nearby clouds collapse and produce massive stars, which initiate further collapse. In this way, star formation, like an influenza epidemic, can be contagious and can almost spontaneously (with respect to the much longer dynamic time scale of the galaxy) break out over large regions. The flocculent spirals, in which the star-forming regions are relatively disorganized, are believed to be dominated by processes of this sort, with star formation being initiated at more or less random locations, but then self-propagating over scales of hundreds of light-years.

ELLIPTICAL GALAXIES AND GLOBULAR CLUSTERS

Elliptical galaxies generally do not appear to contain young stars. We usually find gas in these systems only in the innermost nucleus, where it is ionized and diffuse. Gas appears to be distributed throughout the disk in spiral galaxies, where many young stars are being formed. We may infer that the early gaseous dissipation of a collapsing protogalactic gas cloud caused the flattened shape of the galaxy. Although the first generation of star formation occurred when the galaxy was still round, *most* of the matter remained gaseous and continued to collapse toward the disk, being held outward radially by rotation. The disk stars formed *after* most of the gas had collapsed. However, elliptical galaxies cannot have had a great deal of gas remaining after early star formation. We have argued that they formed by mergers of smaller, predominantly stellar subsystems. Efficient early star formation consumed most of the gas and left the stars in a relatively round configuration; however, *some* gas should also be left. There should be an even greater amount of gas shed by evolving stars over the tens of billions of years of stellar evolution. But elliptical galaxies are relatively gas-free, and they certainly do not contain newly formed stars. What has been the fate of all this gas?

This gas must have somehow been removed. To understand how this may have happened, let us first consider the simpler system of a globular star cluster. These clusters are nearly spherical systems containing a million or so old stars and no detectable gas. Globular clusters are miniature versions of elliptical galaxies, and they present a similar puzzle. Their stars are continually evolving and shedding mass. If this gas accumulated over billions of years, it should be easily observable. We are again faced with the question of what has happened to the gas.

The oldest stars in our galaxy are found in globular star clusters, and the first outburst of star formation must have occurred either in the clusters or before the clusters formed. Our galaxy probably has close to 200 globular clusters, some of which are in highly eccentric orbits that take them far out of the plane of the galaxy (Figure 14.6). Globular clusters have been observed even at over 10 times the distance of the sun from the galactic center. Perhaps these clusters are more properly considered to be intergalactic globular clusters.

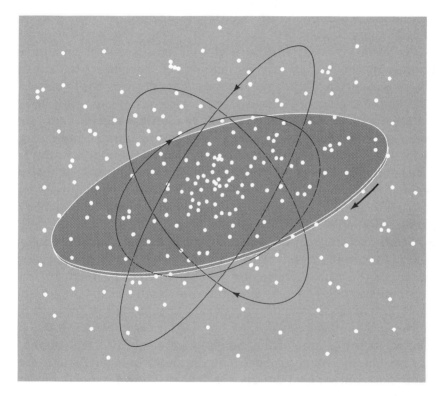

Figure 14.6   Orbits of the Globular Clusters
A schematic view of our galaxy illustrates how the globular clusters (*dots*) form a halo around the flat disk of the galaxy. This halo has a diameter of about 100,000 light-years. The clusters have eccentric, elongated orbits. The halo actually consists of tens of billions of stars of low luminosity in addition to the globular clusters. Because the halo stars are deficient in heavy elements, compared with the disk stars, they are older than most disk stars.

Globular clusters have another remarkable property—their distribution according to their heavy-element content. The more remote their average position from the galactic center, the lower, on the average, is the abundance of heavy elements found in the globular cluster stars by direct spectroscopic observations. This relation holds true for clusters which are only a factor 3 less abundant than the sun in heavy elements as well as for clusters with as little as 0.1 percent of the solar abundance. This strongly suggests that the most distant globular clusters must have been the first to form. From this hypothesis, we can explain why the abundance of heavy elements in globular clusters increases toward the central regions of the galaxy.

As the most massive stars in globular clusters evolved and became supernovae, the gaseous stellar ejecta became systematically more enriched with heavy elements. This gas was not retained in the globular clusters, because these systems have too little gravitational attraction by themselves to contain gas ejected by the violence of a supernova explosion. Instead, the enriched gas mixed with other interstellar material and gradually spiraled in toward the central regions of the galaxy. Massive clouds of gas that became enriched by the addition of some of this matter eventually collapsed to form new globular clusters with a higher abundance of heavy elements.

We find a similar phenomenon in elliptical galaxies: the heavy elements are progressively more abundant toward the central regions of the galaxy. Even the youngest stars observed in these systems are believed to be older than the sun. Thus, the progressive enrichment must have occurred very early in the evolution of the ellipticals. Any gas remaining from star formation would have collected in the center of the galaxy, where it may have formed a dense, massive nucleus of gas. The gas then would have condensed into stars, and some of the gas was perhaps blown out of the galaxy by supernova explosions from the deaths of more recently formed massive stars.

We are compelled to conclude that gas is continuously being swept out from both elliptical galaxies and globular clusters. The problem is least severe for globular clusters, which are rather loosely bound systems, compared with our galaxy. Because they are loosely bound, we can readily conceive of processes that will remove any gas within them. Globular clusters occur mostly in the galactic halo, but they have eccentric orbits that take them through the galactic disk every 100 million years or so. For at least two reasons, their passage can be violent: the cluster may intercept a cloud of interstellar material, and the cluster will certainly interact with diffuse interstellar matter. The resulting ram pressure will tend to sweep gas out of the globular cluster. The crossing of the galactic disk also exerts a strong tide that induces a severe reaction in any gas in the globular cluster. Any gas present will be strongly compressed and heated and consequently will evaporate. In this way, clusters can be periodically cleansed of any gas.

If the processes of sweeping and evaporation do not entirely account for the extremely small amounts of gas that we observe in clusters, a third process should suffice. Globular clusters contain many luminous blue stars. These hot stars generally have exhausted the hydrogen in their cores and have evolved to an advanced stage, at which the principal nuclear fuel is helium. The blue stars ionize and heat any gas present. Once a gas atom is ionized, it will be moving so rapidly that the weak self-gravitational attraction of the cluster cannot retain it, and the gas is driven away in a continuous wind. A similar process occurs in the outer layers of the sun, which continually drive off the corona in the form of the *solar wind*.

For elliptical galaxies, our theories are less clear or certain. A wind could be present, but the great mass of such galaxies requires an extremely strong wind. Gas that is merely ionized by stellar radiation would not be hot enough to escape from an elliptical galaxy. We have already hypothesized that supernova explosions could provide a sufficiently powerful energy source to drive a galactic wind. Another possibility is the effect of intergalactic gas. Elliptical galaxies are mostly found in clusters of galaxies, where intergalactic gas is also present. Ellipticals are moving at high velocity through the intergalactic gas in their orbit around the galaxy cluster. One consequence of this rapid motion (typically of the order of 1000 kilometers per second) is that the intergalactic gas exerts an enormous ram pressure on any gas in the galaxy. The galactic gas is continually bombarded by rapidly moving intergalactic gas ions, which maintain a steady pressure. This ram pressure does not affect the stars, but it can drive the gas from the galaxy (Figure 14.7).

One test of this theory is to examine the spatial distribution of galaxies of different types within rich clusters. As we would expect, spiral galaxies appear to be absent from the central regions of the great clusters, where the intergalactic gas density, as detected by its x-ray emission (Chapter 11), is greatest. At the same time, we would not expect to find any elliptical galaxies entirely free of gas in isolated regions, away from clusters or any other source of intergalactic gas. A few elliptical galaxies are observed to be relatively isolated in space. However, we cannot yet tell whether these systems contain significantly more gas than comparable galaxies within rich clusters.

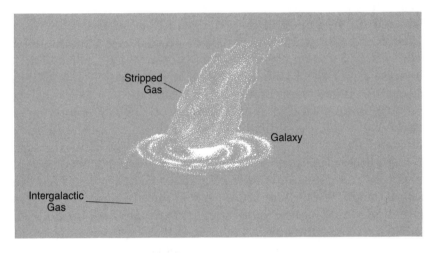

Figure 14.7   Ram-Pressure Stripping

The high velocity of a galaxy in a rich cluster (thousands of kilometers per second) results in an enormous ram pressure exerted by the intergalactic gas on any gas in the galaxy. The ram pressure is sufficient to eject the gas from the galaxy because it greatly exceeds the interstellar gas pressure. This process may explain why elliptical galaxies are gas-free and spiral galaxies are not observed in the central regions of rich clusters: the interstellar gas, which produces the young stars that largely account for the spiral arms, has been swept out. A spiral galaxy with all its gas and young stars removed resembles an S0 galaxy, and many of these are indeed observed in clusters. According to an alternative viewpoint, S0 galaxies acquired their characteristic properties when they formed, and ram pressure only helps to keep them gas-free at later times.

COLORS OF GALAXIES

Galaxies exhibit a wide range of colors and luminosities. Elliptical galaxies are red, and spiral galaxies are blue. Color is a property we can understand in terms of the different populations of stars within these galaxies. Spiral galaxies are undergoing active star formation in the spiral arms, and their blue color comes from hot, relatively massive and luminous stars. Massive stars are blue because they develop a greater central pressure to withstand gravity than do less massive stars. Thus, they are very luminous and hot. Elliptical galaxies generally reveal no traces of current star formation and contain predominantly old, cool, red stars.

We can understand this difference in terms of our previous discussion. Because elliptical galaxies contain little gas, they have no recent opportunity to form stars. There are, however, intriguing variations in color among elliptical galaxies. The more luminous the elliptical galaxy, the redder is its general appearance. Moreover, the central regions of elliptical and spiral galaxies alike are redder than the outermost regions (Color Plate 11). Color differences of this sort can be interpreted in terms of variations in the abundance of metallic elements relative to hydrogen.

To understand how this effect may occur, let us consider a star of 1 solar mass that contains only one-tenth of the solar abundance of heavy elements. This metal-poor star would have a more transparent atmosphere than the sun, because the electrons from the ionized metallic atoms contribute significantly to the atmospheric opacity. Radiation can therefore be more freely emitted from the metal-poor star, and the star would be more luminous than the sun. Yet it would have the same mass and radius as the sun, and consequently it would be hotter and have a slightly bluer color; the temperature of its outermost layers would be higher than that of the sun. The sun has an effective temperature of 5770 degrees kelvin (this is the temperature that a perfect blackbody radiator would have of the same radius and luminosity as the sun). The metal-poor star would have an effective temperature several hundred degrees hotter.

To explain the variations in color among elliptical galaxies requires variations of less than a factor of 10 in heavy-element abundance. The origin of this variation has not received a completely satisfactory explanation. Since this effect is observed mainly in clusters of galaxies where ellipticals are most frequently found, one possibility might be that galaxy collisions triggered star formation early in the evolution of the cluster, when all the galaxies were still largely gaseous. Because the most massive galaxies would have undergone the greatest number of collisions, they probably also would have undergone the greatest amount of star formation and consequently experienced the greatest degree of enrichment. During an encounter between two galaxies, any diffuse interstellar gas would have been heated and ejected into the intergalactic medium. Dense interstellar clouds would only have received a small net acceleration; however, they would have been compressed as a result of the shock wave, which would then have triggered their collapse. They would have fragmented and formed stars in each of the colliding galaxies. The more massive newly formed stars would then rapidly evolve and eject enriched material in the form of stellar winds, planetary nebula shells, and supernova remnants. The less massive galaxies would be progressively less able to retain the enriched matter produced by evolving stars and supernova explosions because of their relatively weak gravitational fields. Consequently, the less massive (ultimately less luminous) galaxies would undergo less enrichment than massive (luminous) galaxies. This sequence of events is highly speculative, but some combination of these effects seems needed in order to account for the apparent correlation between luminosity and metal abundance we infer from observations of elliptical galaxies.

MORPHOLOGY AND GALAXY FORMATION

The preceding arguments enable us to understand how the morphological distinctions between spiral and elliptical galaxies may have arisen. A galaxy forms when a collection of gas clouds condenses out of the expanding universe and merges together (Figure 14.8). The relevant parameter is the ratio of the time scale for star formation to the time scale for the protogalaxy to collapse. If star formation occurs more slowly than overall collapse, the galaxy remains gaseous. The gas radiates, loses energy, and is able to flatten; rotation supports it in the radial direction. Eventually, star for-

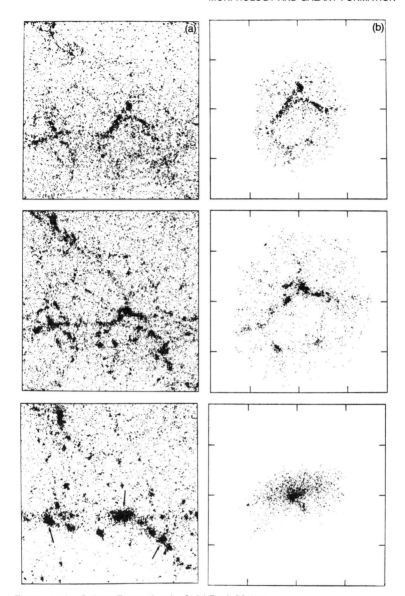

Figure 14.8 Galaxy Formation in Cold Dark Matter
Galaxy formation according to a numerical simulation of cold dark matter collapse. The time sequences shown is from early in the universe (*top*) to the present (*bottom*): (*a*) a 3-million-light-year cube of the expanding universe contains three developing galaxies: (*b*) a close-up (in nonexpanding coordinates) of one such protogalaxy, showing the many blobs that merge together to form a galaxy. (Courtesy of M. Davis, G. Efstathiou, C. Frenk, and S. White)

mation occurs in a highly flattened, disklike system, and we have what is potentially a spiral galaxy. However, if most star formation occurs before the collapse into the final configuration has proceeded very far, little flattening occurs, because a distribution of stars cannot easily lose energy. In this case, the galaxy is likely to remain spheroidal if it was spheroidal initially. Large spheroidal systems very likely result as the consequence of mergers of smaller stellar subsystems.

Spiral galaxies are spinning relatively rapidly, whereas elliptical galaxies are spinning slowly. We have seen that gaseous infall from a greater radial distance accounts for the spin-up. In general, the rotation has a tidal origin, typically induced by interaction with the nearest neighboring galaxy of comparable size. The spiral pattern delineates a region of recent star formation, triggered by the passage of a density wave through the gas-rich disk. Elliptical galaxies are almost devoid of gas and show no spiral pattern; it is likely that they are continuously being swept clean of any gas that accumulates from the debris of ongoing stellar evolution. The sweeping mechanism is likely to be either a wind (generated by supernovae in the central bulges of elliptical galaxies) or ram pressure exerted by the ambient gaseous medium through which the system is moving if it is in a rich cluster of galaxies. Ellipticals are slowly rotating because they probably formed as a consequence of mergers of many smaller clouds and protogalaxies. Spirals form less violently, from the collapse of a large, isolated cloud that spins up as it shrinks.

Ellipticals are generally found in rich clusters; spirals, in the outer regions of galaxy clusters and in the field, where the density is low (Chapter 11). Astronomers suspect that galaxy collisions and tidal interactions played an important role in stimulating star formation, especially for the formation of the ellipticals (Figure 14.9). Evidence that galaxy interactions stimulate star formation has come to us with the discovery that rare galaxies seem to be unusually strong sources of far infrared radiation. The IRAS satellite discovered that these galaxies emit up to 100 times more strongly than galaxies such as the Milky Way at infrared wavelengths. The infrared radiation seems in many cases to be produced by a burst of star formation, and the mechanism that concentrates and destabilizes the interstellar gas appears to be tidal interaction with a nearby

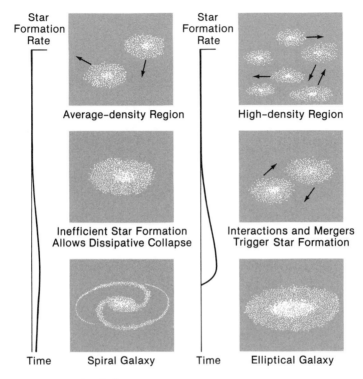

Figure 14.9 Galaxy Morphology
Elliptical galaxies (*right*) most likely form as a consequence of mergers of smaller subsystems, they are therefore likely to form in dense environments where star formation is efficient. The violent dynamical interactions of the forming stellar systems result in smooth, gas-poor elliptical galaxies. Disk galaxies (*left*) form in low-density, relatively quiescent, gas-rich regions, where star formation remains relatively slow and inefficient; the continuing infall of gas and gaseous dissipation result in formation of a galactic disk. The graphs show the rates at which stars form in each case.

companion galaxy. Collisions, particularly with one of the numerous dwarf galaxies, may have instigated the sweeping process by initially helping to disrupt and remove the interstellar medium. The gas shed in early star formation finds a natural resting place in the vast intergalactic spaces of the clusters.

One unifying factor underlies our account of how spiral and elliptical galaxies have evolved: the origin of the heavy elements in the debris from early generations of stars. Heavy elements are crucial to the formation of all the presently visible stars in all galaxies, and we now look more closely at their origin.

· 15 ·

ORIGIN OF
THE HEAVY ELEMENTS

We have found it possible to explain, in a general way,
the abundances of practically all the isotopes of the
elements from hydrogen through uranium by synthesis
in stars and supernovae.
—*E. M. Burbidge, G. R. Burbidge,*
W. A. Fowler, and F. Hoyle

We are children of the stars. Ten billion years ago, practically every atom now in our bodies was once near the center of
a star. The first stars that illuminated the collapsing galaxies consisted mostly of hydrogen, and 10 percent of their atoms were the
helium that was created in the first moments of the big bang. But
there was no carbon, no oxygen, no iron; none of the elements
necessary to form the earth and sustain life were present. A nuclear
stumbling block—the inability of a helium nucleus to combine
with a proton or with another helium nucleus to form a heavier
stable nucleus—prevented the synthesis of heavier elements in
the big bang.

A considerable fraction of the heavy elements was produced by
the first stars. Dramatic confirmation of this has come from the
discovery that the intergalactic medium in many of the great clusters of galaxies is practically as rich in iron as the sun. The intracluster gas must have been enriched by stellar ejecta from the clus-

ter galaxies, and this gas could have been produced only when the galaxies were very young; there is far too much iron to be explained by current rates of stellar evolution and mass loss (page 226).

A closer look at individual galaxies (including our own) reveals a systematic trend: heavy-element abundance declines away from the central regions of the galaxy. This distribution tells us that enrichment must have been a gradual phenomenon. Heavy elements were first produced in the outer regions of the galaxy, and the enriched gas fell inward and was consumed by formation of new stars. Successive cycles of star formation led to greater degrees of enrichment, and the greatest enrichment occurred where most of the gas accumulated—in the central regions of elliptical galaxies and in the disks of highly flattened spiral galaxies (Chapter 14).

The production of heavy elements is therefore a key link in the chain of evolutionary events that led to the formation of the sun. We must examine this link more closely to understand how crucial it is: the absence of heavy elements would mean that life could not exist. As we shall see in this chapter, the process was not inevitable; the production of heavy elements depended on the timely formation and evolution of massive stars and on their self-destruction in violent supernova explosions.

NUCLEAR EVOLUTION OF STARS

Inside stars, hydrogen nuclei fuse to form helium. This nuclear reaction, or sequence of reactions, releases energy, because the helium nucleus weighs slightly less (0.7 percent less) than the four protons that fuse to produce it. The mass deficit is released mostly in the form of intense gamma radiation and neutrinos. The radiation is trapped in the center of the star, which remains hot, creating an enormous pressure difference, or *pressure gradient,* between the hot interior and the cool outer layers of the star. The pressure exerts an outward force that withstands the gravity of the star, as long as there is sufficient hydrogen present in the stellar core to produce helium.

Once the hydrogen supply begins to be exhausted and the core becomes mostly helium, the nuclear reactions that fuse hydrogen into helium become far less frequent. Less energy and heat is produced, the central pressure drops, and the precarious equilibrium between pressure and gravitational forces finally breaks down. Gravity becomes dominant, and the star begins to collapse. The collapse results in heating of the central core, just as compression always results in heating of a gas.

Eventually, the core of the star becomes hot enough for helium to burn. Helium requires a higher temperature than hydrogen before it can undergo nucleosynthesis into heavier elements, because it has a greater nuclear charge. Two helium nuclei repel one another more strongly than two protons. Penetration of this repulsive electrostatic force field, the *Coulomb barrier*, requires higher velocities. The helium nuclei speed up and acquire greater energy when they are heated to a higher temperature. Moreover, two helium nuclei fuse to form an unstable isotope of beryllium: it requires mediation of a third helium nucleus in order to result in a nucleus of carbon. Only because of a resonance in the nuclear structure of carbon, which allowed the carbon to form before the beryllium decayed, did this work: remarkably, the British astrophysicist Fred Hoyle predicted in 1954 that the resonance had to exist; otherwise carbon would not have been synthesized. Shortly afterwards, this prediction was verified in the laboratory. Helium burning again releases a certain amount of energy, because three helium nuclei weigh slightly more than the carbon nucleus that they synthesize. Somewhat less energy is produced by helium burning than is produced by hydrogen burning, but it suffices to sustain the star in gravitational equilibrium during its helium-burning phase.

A rather curious phenomenon occurs as the core of the star heats up: the rate at which the star is radiating rises drastically, and the outer envelope expands by 100 times or more. This puffing up of the stellar atmosphere results in cooling of the star's outer layers. The radiation being emitted becomes extremely red. The star has now evolved into a luminous *red giant*.

In sufficiently massive stars, helium burning is followed by carbon burning, which produces oxygen and other elements. When the carbon core is exhausted, burning of heavier elements occurs, until the stellar core consists largely of iron. Iron is the most tightly bound of all atomic nuclei, which means that iron nuclei weigh *less* than their constituent parts. Energy is released when iron is produced. However, this result no longer tends to be true for nuclei heavier than iron, and energy has to be *supplied* to synthesize elements heavier than iron. No further energy can be extracted by burning iron. Fusion yields energy for elements lighter than iron, but only nuclear fission of elements heavier than iron can yield energy. (Fission is the basic principle at work in the atomic—as opposed to the hydrogen or fusion—bomb.) For most elements heavier than iron to form generally requires a much more violent energy source than is available in the gradually evolving stellar core. Once the star develops an iron core, its fate becomes inevitable. After the supply of nuclear energy runs out and fails to provide adequate heat and pressure, gravitational collapse must ensue.

WHITE DWARFS

The star continues to collapse until the matter becomes so tightly packed that it becomes *degenerate matter.* A fundamental principle of quantum theory, the uncertainty principle, implies that individual particles have uncertain locations. Only the region of space they occupy can ever be ascertained. In a very crude analogy, we can conceive of an underlying motion, and therefore a pressure, possessed by elementary particles. A more precise definition requires us to consider the exclusion principle of quantum mechanics, which limits the number of electrons in a particular atomic level. At sufficiently high densities, this limit is approached, which results in a pressure that is purely a quantum mechanical concept. Unlike ordinary gas pressure, it depends not on temperature but on atomic random motions. Under normal circumstances, this *degeneracy pressure* is negligible, compared with ordinary pressure. When the matter is compressed to enormous densities, however, degeneracy pressure becomes increasingly important. The first particles to be affected are the electrons, which provide additional outward pressure when they become degenerate; they therefore resist being squeezed any further. When the stellar core reaches a state in which it is supported by degenerate electron pressure, the star has become a *white dwarf.* A star of 1 solar mass shrinks 100 times in size to become a white dwarf (Figure 15.1). Such stars have cores of carbon and oxygen or even of helium if less than

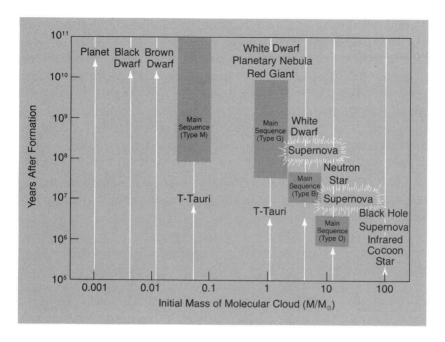

Figure 15.1   The Fates of Stars of Differing Initial Mass

White dwarfs contain less than about 1.4 solar masses. More-massive stars may become white dwarfs, because a considerable amount of mass can be lost during their evolution. It is currently believed that stars below about 6 solar masses become white dwarfs, and more massive stars explode as supernovae and leave neutron star cores behind. Very massive stars (above 30 to 50 solar masses) may form black holes. Protostars of less than 0.08 solar mass never develop enough heat in their cores to initiate nuclear reactions; these objects are called brown dwarfs (and eventually black dwarfs) because they emit feebly in the infrared. The planet Jupiter may fall into this category; it radiates 70 percent more energy than it receives from the sun.

about 0.5 solar mass. The dwarf star is white, because it is initially very hot as a result of the enormous compression. It gradually cools and disappears from sight over some billions of years. The maximum mass of a white dwarf is about 1.4 solar masses, although rapid rotation can somewhat increase this mass limit.[19] Stars shed a considerable amount of mass during the course of their evolution, and it is believed that stars whose initial masses were up to about 6 solar masses will evolve into white dwarfs.

Occasionally, a white dwarf in a binary system can accrete mass from a companion star that has evolved into a giant star. The addition of hydrogen-rich fuel results in an explosive mixture, which astronomers observe as a nova. After the explosion, the process can repeat; novae are recurrent objects that increase in luminosity by a factor of 100 or more over the course of a month or so, and may repeat every 10 or 20 years. Much more rarely, a binary pair of white dwarfs will result that are so close that they eventually lose orbital energy via frictional drag in a common envelope, and spiral together. In this case the mass limit of 1.4 solar masses is exceeded, and the consequence is dramatic. Collapse occurs, followed by a violent explosion; the result is a supernova. A supernova initiated by a pair of white dwarfs is said to be of Type I: it is characterized by a hydrogen-poor spectrum, and is associated with the older stellar population of galaxies.

NEUTRON STARS

More-massive stars face a far more catastrophic fate. The initial mass of such a star was sufficiently high to ensure the strong gravitational forces and high central temperatures that result in formation of an iron core of mass slightly greater than 1.4 solar masses. Once it exhausts the nuclear fuel supply, the gravitational force is too strong to be supported by degenerate electron pressure. The star collapses violently until, at far higher densities, the electrons and protons combine to form neutrons. These neutrons in turn eventually become degenerate. If degenerate neutron pressure can provide sufficient support to halt the collapse, a *neutron star* is formed. A neutron star is typically only a few miles across, yet such a star may contain more mass than the entire sun.

The neutron star forms from the core of the collapsing star, which must have been quite massive. As the star collapses, an enormous amount of energy in the form of x-rays, gamma rays, and neutrinos is released suddenly; this energy helps to blow off the outer envelope of the star, already greatly enriched by nucleosynthesis, in a violent supernova explosion. A supernova resulting from a massive star death is said to be of Type II: its spectrum is characteristically hydrogen-rich and it is associated with regions of star formation. In this way, elements heavier than helium are returned to the interstellar gas. SN 1987a is the best studied example of a Type II supernova (page 274).

A neutron star is essentially a giant atomic nucleus. The protons and electrons are squeezed together so tightly that they combine to form neutrons that are practically in contact with one another. The density within a neutron star is almost equivalent to the density of an atomic nucleus.

Neutron stars have been discovered by radio observations, x-ray observations, and even gamma-ray observations. Radio astronomers observe *pulsars*, which are characterized by the remarkably regular radio signals they emit. A period of 1 second or so is often maintained to better than one part in 10^{12}. The pulsar in the Crab nebula has a period of one-thirtieth of a second (Figure 15.2). It is slowing down at a rate that suggests an origin about 900 years ago. The Crab pulsar has been associated with the supernova of A.D. 1054 recorded by Chinese astronomers, which left behind the Crab nebula. Astronomers have observed the Crab pulsar to maintain the same period (or radiate in phase) at radio, optical, x-ray, and gamma-ray wavelengths. The fastest known pulsar (p. 128) has a period of only 1 millisecond.

Pulsars are believed to be rapidly spinning neutron stars formed during a supernova explosion. Only stars as compact as neutron stars could spin so rapidly. The regularity of the pulsar emission can be explained by a beam of radiation from the neutron star that sweeps around like a gigantic searchlight during each rotation period. The beaming effect is likely to be due to the funneling of radiating relativistic particles into the intense magnetic field of the neutron star. Because of this directionality (we have to be in the path of the beam to receive it), pulsars are not always detected wherever a supernova is known to have occurred.

Neutron stars have also been observed under more exotic circumstances. For example, many stars have close companions, and when one member of such a pair of stars becomes a supernova and leaves a neutron star behind, the evolution of the companion star will often be greatly accelerated. The companion may shed a considerable amount of mass from its atmosphere, enveloping the neutron star in a dense cloud of gas. This envelope has the effect of quenching the radio emission from the pulsar but produces intense x-ray emission as the gas is accreted and heated in the strong gravitational field around the neutron star. The best-known case of a binary x-ray source is Hercules X-1 in the constellation Hercules

(a)

Figure 15.2 The Pulsar in the Crab Nebula
Within the gaseous remnant (*a*) of a supernova observed by Chinese astronomers in
A.D. 1054 lies the Crab pulsar, a rapidly rotating neutron star. Radiation is emitted from
the pulsar (*arrow*) 30 times each second, as is depicted in the stroboscopic television
pictures taken at maximum and minimum light (*b*). Filaments of the nebula appear
white in this photograph; the amorphous gray areas are produced by synchrotron
radiation emitted by relativistic electrons moving through the magnetic field of the
nebula.

(Figure 15.3). In this case, the x-ray emitting star is not observed
optically. The companion of the x-ray source is a star called HZ
Herculis, which had long been known as a variable star. Hercules
X-1 emits x-ray pulses every 1.24 seconds. The precision of the x-
ray pulsar is such that, as the x-ray source moves in orbit every 1.7
days around HZ Herculis, the pulses are observed to speed up and
slow down slightly because of the Doppler shift associated with
the periodic orbital motion of the x-ray source. Because we know

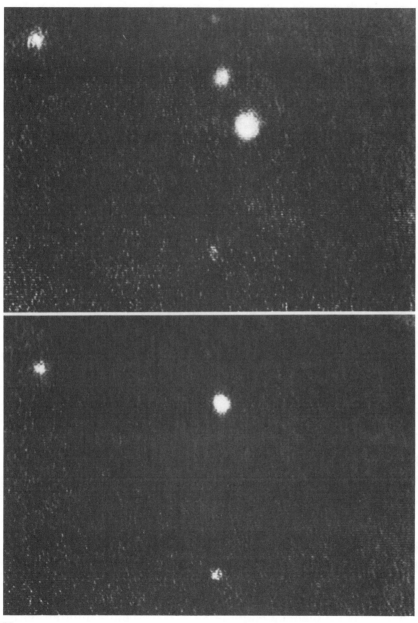

(b)

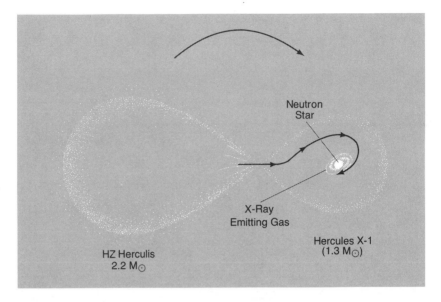

Figure 15.3 Hercules X–1

The x-ray source Hercules X-1 is a neutron star of about 1.3 solar masses orbiting a more massive companion star, HZ Herculis, which is still very luminous. HZ Herculis may once have been slightly smaller, in which case it would not have suffered any severe disruptive force. As it evolved, however, the core heated and the outer envelope swelled. At some point, the proximity of the neutron star caused the outer part of HZ Herculis to become distended, and matter was gradually dragged into the neutron star by tidal gravitational forces. The material lost by HZ Herculis spiraled into Hercules X-1, forming a hot disk that prolifically emits x-rays.

the mass of the optical companion from studies of its spectrum, we can infer the mass of the x-ray emitting star. Its mass and size suggest that Hercules X-1 is a neutron star.

BLACK HOLES

Many stars undergo considerable mass loss during their final stages of nuclear evolution. One manifestation of this is the planetary-nebula phase. A planetary nebula consists of a central helium-burning star surrounded by a glowing shell of recently ejected gas. If their masses are initially less than about 6 solar masses, such stars eventually become white dwarfs; they consist mostly of carbon because all the hydrogen-rich material has been ejected. Stars between 6 and 8 solar masses may undergo a final explosive collapse and leave no core behind. Because the pressure of degenerate matter does not rise as it is heated, more and more energy can be injected until a core temperature high enough to achieve catastrophic detonation of the carbon core is attained. In the mass range between 8 and 50 solar masses, the gravitational forces are such that the nuclear evolution of the core proceeds until an iron core of about 1.4 solar masses is formed. Exhaustion of the nuclear energy supply results in collapse of the core to form a neutron star, with the outer envelope being ejected in the ensuing supernova explosion. The reaction $p + e \rightarrow n + \nu$, where ν is a neutrino, signals the onset of neutron-star formation, and the energy of compression to nuclear density is carried away by the weakly interacting neutrinos. This theory was confirmed with the detection of the neutrino blast from SN1987a.

Stars more massive than about 50 solar masses are believed to collapse into black holes. We do not know the precise mass limit above which black-hole formation is inevitable. However, it is clear that for stars of sufficiently large mass, even the ultimate degeneracy pressure of the neutrons cannot restrain the collapse. The gravitational field becomes so intense that no light can escape. The situation is reminiscent of the grin of the Cheshire cat—when a star collapses to form a black hole, only its gravitational force is left behind. As with formation of a neutron star, the sudden release of nuclear and gravitational energy associated with the final collapse may result in the violent ejection of the outer envelope of the star.

One of the most exciting searches of modern astronomy has been the attempt to discover a black hole. Although no light can escape from a black hole, it is possible that a black hole could be one of a close binary pair that accretes matter from its companion star. Accreted matter would be heated as it spiraled into the black hole, forming a disk of material that emitted intense amounts of x-radiation before finally disappearing into the hole. X-ray astronomers have generally agreed that a few particular sources, such as Cygnus X-1, are likely to be black holes (Figure 15.4). The principal ar-

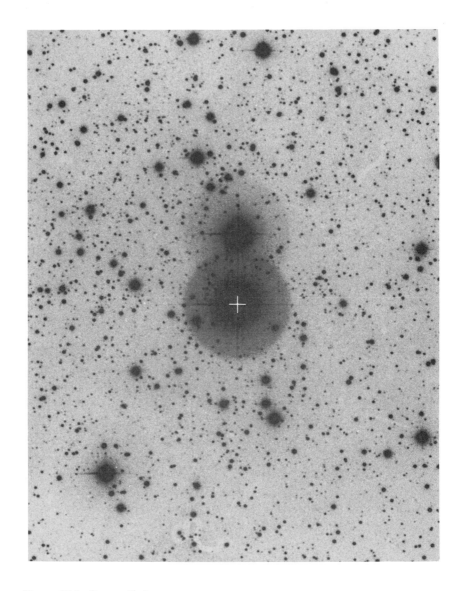

Figure 15.4   Cygnus X—1

The largest dark object at the center of this negative photograph is the star HDE 226868, associated with the x-ray source Cygnus X-1. Superimposed on the photograph is the location of a source of radio emission (*cross*), which coincides with the x-ray source location. The radio source has been located accurately, and variations in the radio intensity, coincide with changes in the x-ray intensity, thereby enabling astronomers to identify HDE 226868 with Cygnus X-1. The x-ray source itself is a compact object, probably a black hole, which orbits this star and accretes matter from its outer layers.

gument favoring Cygnus X-1 is that this x-ray source probably contains more than 8 solar masses. This result comes from analyzing the period and light variations of the optical companion star, which enables us to compute the orbit and the mass of the black hole. The x-ray source must itself be a compact object, either a neutron star or a black hole, according to any plausible model explaining the x-ray emission. However, theories of gravitational collapse and stellar stability indicate that no neutron star can contain more than 4 solar masses. A less definitive but more realistic upper mass limit for a neutron star is only 2 solar masses. Thus, the identification of Cygnus X-1 and of even stronger candidates, such as LMC X-1, as black holes seems reasonably convincing.

EXPLOSIVE NUCLEOSYNTHESIS

Supernova explosions are triggered by the energy released during the implosion of the central iron core to form a neutron star or, in the case of a very massive star, by the implosion of the entire central region to form a black hole. The outer layers of the star are ejected into the surrounding interstellar medium, which redistributes the material among a new generation of stars.

A supernova explosion is a very violent and energetic event. The sudden release of energy, as the central iron core implodes to form a neutron star, sends a blast wave through the surrounding layers of silicon, oxygen, carbon, helium, and, at the outer edge, hydrogen. The material briefly heats up to a temperature that far exceeds anything previously attainable during the history of the star. At the same time, prolific numbers of neutrons are produced in the collapsing core and in the surrounding hot, shocked layers. These help to *irradiate* the material of the stellar envelope. Neutron bombardment and transient high temperatures provide the setting for *explosive nucleosynthesis* to occur. Elements heavier than iron can now be produced, as can many of the rarer isotopes that find no nook in the early stages of steady nuclear burning. Supernova explosions thus provide the unique environment where energy is available to synthesize the heavier elements.

ELEMENTAL ABUNDANCES

How confident can we be that heavy elements originated in the debris of ancient supernovae? One clue comes from the abundances of the elements (Figure 15.5). When considered by increasing atomic mass, the nuclei of *even* atomic weight are more abundant than those of *odd* atomic weight. This odd-even variation is a direct consequence of the nuclear properties of the elements. It can be easily understood, if the heavy elements have formed by means of nuclear reactions, in terms of the stability of atomic nuclei or the degree to which the nuclei are tightly bound together (Figure 15.6).

Elements of even atomic weight tend to be more stable than those of odd weight. A nucleus containing an odd number of constituent particles (protons or neutrons) can be visualized as having

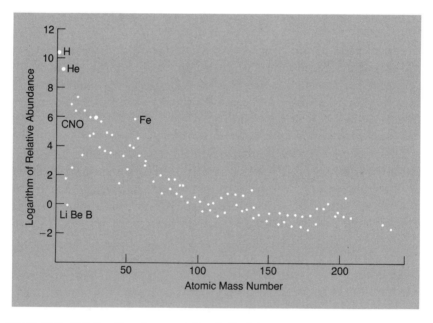

Figure 15.5 Relative Abundances of Atomic Nuclei

The abundances of naturally occurring atomic nuclei are highest for hydrogen and helium. Lithium, beryllium, and boron evidently form an anomalous grouping. Like deuterium, these three elements are very fragile, being destroyed rather than created in stars, and lithium may have been produced in the big bang. Note the relative prominence of iron among the heavier elements.

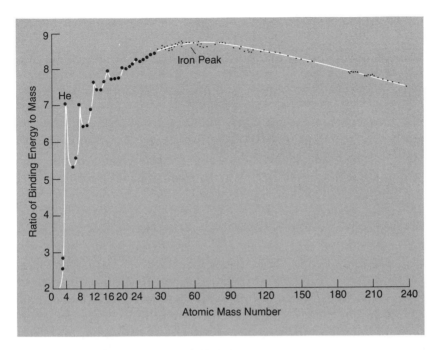

Figure 15.6 Binding Energies of the Nuclei

We can gain some insight into the relative abundances of the elements by examining the amount of energy that is stored in each nucleus. The nuclear forces bind the nuclei tightly together and release a certain amount of energy (the binding energy) when nuclei are formed. The binding energy per particle in the nucleus is plotted against the number of particles in the nucleus (mass number). The relative prominence of helium indicates the great amount of energy that is provided by its synthesis. After helium is made, extracting energy by synthesizing heavier nuclei becomes a downhill battle; the relative peaks become smaller and smaller. Energy is still released by fusion, but the available binding energy decreases as we move to heavier elements, becoming less and less until finally the end comes with nickel and iron. In synthesizing elements heavier than iron, we cannot extract energy; instead, energy must be supplied. Iron has the highest binding energy and is the most stable of all elements; it is consequently favored by the inevitable tendency to attain nuclear equilibrium. This is why the iron peak occurs in the graph of abundances (Figure 15.5).

one unpaired proton or neutron, whereas a nucleus of even atomic weight has no unpaired particles. It is necessary for the pairings to be of like particles (proton-proton or neutron-neutron). An unattached particle indicates lower stability, because spare bonds are available to attach to an additional particle. Nuclei with even numbers of neutrons and protons are accordingly more stable than nuclei of odd mass, and nucleosynthesis will therefore lead to a greater abundance of even nuclei. The odd-even alternation of the naturally occurring elements is a direct clue that the heavy elements must have been formed by nuclear fusion, wherein protons, neutrons, and helium nuclei combined with heavier nuclei. Stellar cores are the most likely environment for this process of nucleosynthesis. As we have seen, only catastrophic supernova explosions appear capable of disrupting these cores sufficiently to release the heavier elements locked within them. Thus, we conclude that previous generations of star birth and death must have occurred before our sun formed. The abundance of heavy elements observed in the sun and other stars provides strong evidence to support this theory.

The abundances of various common elements are remarkably suggestive of the composition of an evolved star. A star of 20 solar masses produces approximately the same ratios of carbon, nitrogen, oxygen, silicon, sulphur, and iron that we find on earth (Figure 15.7). A star of 20 solar masses would be more than 10,000 times as luminous as the sun. Such prolific expenditure of energy would mean that the star rapidly exhausts its supply of nuclear fuel. Consequently, massive stars are destined to evolve rapidly as they synthesize these common heavy elements.

We can now integrate our knowledge of the birth and death of stars into our theory of the evolution of the early universe. Within about 100 million years after the initial collapse of protogalactic gas clouds, opaque fragments began to form massive stars. These stars had lifetimes of several million years; they then collapsed and became supernovae. As the central cores of the stars imploded into compact neutron stars, the bulk of their mass was blown off in supernova explosions. The debris from each explosion formed a rapidly expanding gaseous nebula, or *supernova remnant*.

The Crab nebula is the best-known supernova remnant. Another, the Vela nebula, is visible from the southern hemisphere (Figure 15.8). Pulsars, identified as spinning neutron stars, have been discovered in each of these remnants. The Crab nebula is expanding at a velocity of about 1000 kilometers per second, and its radius is consistent with the time elapsed since its formation in A.D. 1054. The Vela nebula is larger than the Crab and is expanding at less than half of its rate. The Vela remnant is believed to be about 10,000 years old.

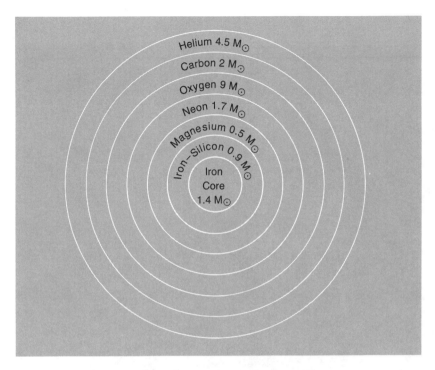

Figure 15.7 Heavy Elements Produced by a Massive Star
A star of 20 solar masses is depicted prior to its final stages of evolution. An iron core of 1.4 solar masses is surrounded by layers of processed material that have been produced in earlier phases of the star's life. At each stage of nuclear burning, a shell of lighter, unprocessed material is left behind, because the temperature was not high enough for it to burn.

The young supernova remnants decelerated as they swept up interstellar matter; eventually they became thoroughly mixed with interstellar gas. In this way, the heavy elements that formed in the first stars were distributed throughout the galaxy in interstellar clouds, which subsequently collapsed to form new stars.

In the early galaxy, supernovae occurred at a much more prolific rate than at present. There was perhaps one supernova per year for the first 100 million years of the galaxy's life. These supernovae were responsible for the heavy elements that permeate the disk population of our galaxy. The sun, which is near the periphery of our galaxy, formed some 5 billion years ago, when the galaxy was about 10 billion years old. At the time the sun formed, the super-

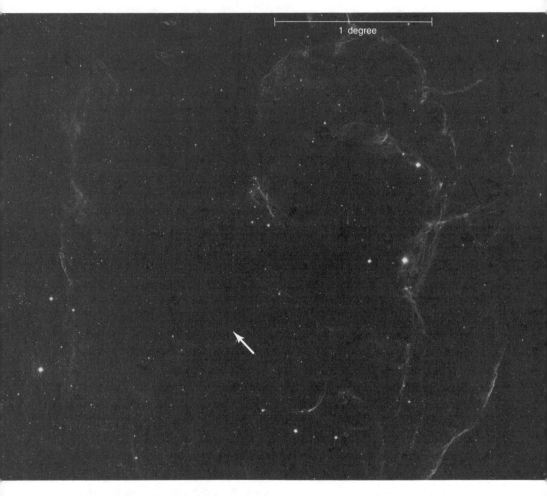

Figure 15.8 The Vela Supernova Remnant

This rapidly expanding supernova remnant is left over from an explosion that occurred about 10,000 years ago. The Vela remnant is also an intense source of low energy x-rays, and it contains a radio pulsar at the indicated position. The irregular shape of the nebulosity is produced by the interaction of the supernova blast wave with many interstellar clouds and filaments of gas. These glow after the explosion reaches them, and the visible complex of bright areas and filaments delineates the current extent of the remnant, which is still expanding at several hundred kilometers per second.

nova rate had declined to approximately its present rate of approximately one every 100 years, on the average, for each of the two types of supernova. Observations of the Crab and other supernova remnants demonstrate that the process of nucleosynthesis and enrichment of the interstellar gas is continuing to this day.

SUPERNOVAE AND THE SOLAR SYSTEM

It is possible that the collapse of interstellar clouds is triggered by the shock wave from adjacent supernova explosions. A fascinating clue points to the interaction of a supernova with the interstellar cloud from which our solar system formed. Studies of samples of a few large meteorites, which have yielded enough material for a thorough chemical analysis, have revealed the presence of anomalies in the abundances of certain isotopic species. The anomalies are found in small embedded regions, or inclusions, where the composition differs significantly from the surrounding meteoritic material (Figure 15.9). The ratios of certain isotopes in the inclusions differ markedly from terrestrial ratios. One of the key results has been the discovery of trace amounts of a rare isotope of magnesium (Mg^{26}) in aluminum-rich meteoritic inclusions. The magnesium isotope does not occur naturally on the earth, but in the inclusions, the greater the degree of aluminum enrichment, the more Mg^{26} is found. Radioactive decay converts an unstable isotope of aluminum (Al^{26}) into Mg^{26}; the conclusion seems inescapable that the Mg^{26} has been produced in this manner. What is intriguing is that the decay time (or half-life) of Al^{26} is only 1 million years. The unstable radioactive aluminum isotope has no long-lived progenitor; it is indisputably the parent species of the observed rare magnesium isotope.

The aluminum must have been produced in a supernova explosion, according to our present understanding of the origin of the elements. Radioactive aluminum-rich debris evidently was produced in the supernova explosion and solidified within less than 1 million years after the supernova event, before any appreciable dilution of the aluminum-rich material with interstellar matter could occur. The supernova would have immediately preceded the formation of the meteorite and thus also the formation of the solar

(a)

(b)

Figure 15.9 Chondrules

The chondrite type of stony meteorite contains embedded marblelike beads of material (chondrules) that appear to have undergone considerable melting when they formed at high temperatures. The chondrules are the oldest rocks known; they predate the formation of the sun. Isotopic abundance anomalies discovered in certain chondrules provide evidence for contamination, possibly by a near-by supernova, of the cloud from which the solar system condensed. The best-studied samples are from the Allende meteorite (a), a carbonaceous chondrite that fell in Mexico in 1969. A close-up view of a chondrule is also shown (b) from a chondrite that fell in Czechoslovakia.

system. One million years is a brief instant of time by cosmic standards. In fact, it is more or less equal to the time it would take a dense cloud, once its collapse was triggered, to fragment into stars.

Thus, we may reasonably conclude that some meteoritic debris found on earth has come undiluted from a nearby supernova explosion that immediately preceded the formation of the solar system. According to one hypothesis, the supernova debris condensed into tiny solid grains as it expanded and cooled. The grains acted rather like shrapnel, blasting their way into the collapsing interstellar cloud from which the solar system was destined to develop. The grains became incorporated into meteorites as the solar system formed.

The sun and the billions of other similar stars are the end products of this sequence of events. During the first 500 million years of the galaxy, massive stars evolved rapidly and created most of the heavy elements, which were distributed around the galaxy by supernova explosions. Billions of years later, the solar system formed from the ashes of the ancient stars that had been swept up into an interstellar cloud. About 5 billion years after the collapse of the interstellar cloud destined to form the solar system, one end product of this evolutionary scenario, the meteorite, fell upon another end product, the planet earth, and was examined by yet a third end product, intelligent life. In Chapter 16, we shall present our theory of the formation and evolution of the solar system and the unique objects that are found within it.

· 16 ·

FORMATION OF
THE SOLAR SYSTEM

This transition from the inanimate to the living was an almost incredible sequence of highly improbable events, but in a universe containing more than 10^{20} stars capable of building planetary systems and over a period of at least 10^{10} earth-years, almost anything can happen repeatedly. And obviously this unlikely sequence did happen at least once in this solar system, for here we are, the timid descendants of some rather nauseating gases and sundry flashes of lightning!
—*Harlow Shapley*

In our discussions of the big bang, the evolution of galaxies, and the evolution of stars, we have seen how astronomers have observed and classified systems deep in space and have constructed theories of origin and evolution for these systems. In this chapter, we shall survey our local neighborhood, the solar system, and present a theory of how the unique objects of the solar system formed and evolved. As we saw in the last chapter, it seems likely that the solar system formed from a cloud of dust and gas that was triggered to collapse by a supernova explosion. Although the cloud from which the solar system originated no longer exists, traces of evidence remain in the characteristics of the solar system objects that we observe today. We shall first survey the features that a theory of origin must explain. Among these unique features—besides the sun, planets, satellites, comets, asteroids, and meteorites—is life itself.

THE SOLAR SYSTEM

Perhaps the most striking feature of the solar system is that the orbits of all the planets are nearly circular and lie close to the equatorial plane of the sun (Figure 16.1). Not only do all the planets

Solar System Object	Relative Size (approximate)	Distance from Sun (millions of miles)	Diameter (thousands of miles)	Mass (earth masses)
Sun	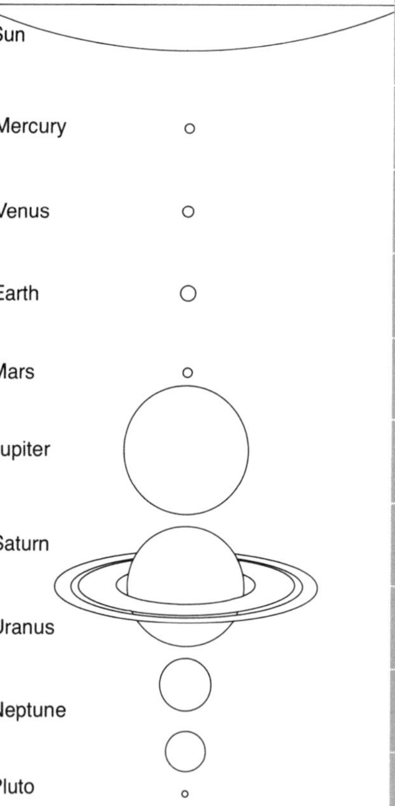	—	867	343,000
Mercury	o	36	3.0	0.1
Venus	o	67	7.6	0.8
Earth	O	93	7.9	1.0
Mars	o	142	4.2	0.1
Jupiter		486	89	317.8
Saturn		892	76	95.2
Uranus		1,790	32.5	14.5
Neptune		2,810	31.4	17.2
Pluto	o	3,780	1.3	0.002

Figure 16.1 The Planets

The planetary data are displayed beside drawings that show the relative sizes of the planets. The minus sign in front of the rotation periods of Venus and Uranus indicates that they rotate in a retrograde direction.

revolve around the sun in the same direction that the sun rotates, but practically all the planets are also spinning in the same direction. The rotation axes of the planets are nearly parallel to the axis of the sun, except for Venus, which is antiparallel, and Uranus, which is tipped about 90 degrees. The satellite systems, for the

Number of Moons	Density (g cm^{-3})	Rotation Period (days)	Revolution Around Sun (years)	Average Temperature (K)	Inclination of Equator to Orbit	Inclination of Orbit to Ecliptic
–	1.4	27	–	5,800	–	–
0	5.4	55	0.24	~600	<7°	7°
0	5.2	–244	0.62	750	177°	3.4°
1	5.5	1	1.00	180	23.5°	0°
2	3.9	1.03	1.88	140	24°	2°
20	1.3	0.41	11.86	128	3°	1.3°
20	0.7	0.43	29.48	105	27°	2.5°
15	1.2	–(0.6–1)?	84.01	70	98°	0.8°
2	1.5	0.66	164.79	55	29°	1.8°
1	0.7	6.4	248.4	53	50°	17.2°

most part, show a similar regularity. The inner planets (Mercury, Venus, Earth, Mars) are relatively small, metal-rich, dense systems that spin slowly, have rotation periods of one day or more, and have few satellites. The outer, giant planets are large, light, hydrogen-rich systems that often have many satellites.

We are all familiar with practical applications of Newton's law of motion: bodies in a state of uniform motion conserve their momentum. We can quantify the amount of rotation a body has by calculating its angular momentum. The sun rotates slowly and has rather less angular momentum than would be expected for an interstellar cloud of comparable mass that has collapsed into a star. Most of the angular momentum of the solar system turns out to be stored in the orbital motions of the planets—in particular, in the motion of the largest planet, Jupiter.

Besides the planets, the solar system also contains many smaller orbiting bodies, including asteroids, meteors, meteorites, comets, and the rings of Saturn, Uranus, and Jupiter. *Asteroids* range up to 1000 kilometers in size and lie primarily between Mars and Jupiter (Figure 16.2). There are only about 200 asteroids exceeding 100 kilometers in diameter, but there must be tens of thousands that are greater than 1 kilometer in diameter. A missing planet was once conjectured to exist in this *asteroid belt.* The basis for this prediction is *Bode's law,* which postulates an empirical relation between planet location and distance from the sun that indicates preferred locations for planet formation in the solar system. Bode's law can probably be understood on the basis of the gravitational pull of the sun and planets.[20]

The wide range of orbits of the asteroids appears to preclude the possibility of their being the disrupted remnant of a former planet. The chemical history of meteorites, presumed to be similar to that of asteroids, also differs from any terrestrial or lunar rocks and suggests an origin in several different parent bodies. Rather, asteroids and meteorites seem to be primitive debris that failed to condense into a planet. Periodic brightness variations suggest that asteroids are rapidly spinning and irregularly shaped, the projected area of the asteroid varying as it spins. Spectroscopic observations have shown that many asteroids closely resemble meteorites in composition. Many asteroids are black, with low reflectivity, and are similar to the stony meteorites in composition. Others seem to be rich in iron and nickel.

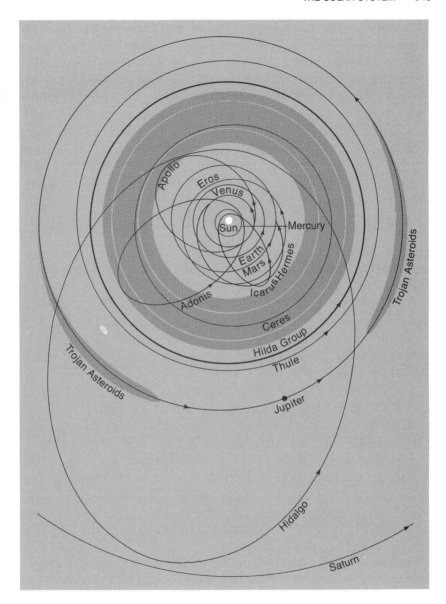

Figure 16.2 Asteroid Orbits

The asteroids lie predominantly between the orbits of Mars and Jupiter. The Trojan asteroids precede and follow Jupiter by about 60 degrees, occupying preferred locations where the gravitational pull is least. Hidalgo travels farther from the sun than any other known asteroid.

Meteorites can be broadly classified into two varieties, iron and stony. There is also a hybrid category of stony-iron meteorites. From radioactive dating, the stony meteorites (*carbonaceous chondrites*) are known to be as old as 4.6 billion years. Meteorites contain glassy inclusions, suggestive of a fiery past at temperatures high enough for melting to have occurred. Studies of meteorites have also found small traces of rare isotopes, which are produced by radioactive decay of unstable parent nuclei. We have previously inferred that some of these unstable nuclei were probably synthesized during a supernova explosion that immediately preceded the formation of the solar system.

Comets are massive chunks of frozen matter that orbit the sun in highly elliptical paths, which can range from beyond Pluto to within the orbit of Mercury (Figure 16.3). Hundreds of years may elapse before the comet leaves the cold, bleak outer regions of the solar system and approaches the sun. The comet is gradually heated, and it begins to evaporate as it reaches the inner solar system. A stream of hot gas that emanates from the sun, the solar wind, blows away the icy vapors, resulting in the formation of a cometary tail of hot glowing gas that always points away from the sun. Some comets (the short-period comets) appear rather more frequently; their orbits reach to the asteroid belt. These comets do not exhibit any tails, presumably because their supply of frozen matter is becoming exhausted.

Eventually, comets decay into a loose assemblage of tiny particles, visible as a *meteor shower* at certain times of the year when they pass near the earth (Figure 16.4). The prominent meteor showers have usually been associated with individual comets. Each year, when the earth crosses the comet's orbit, it intersects debris left behind by the comet. These fragments burn in the upper atmosphere and are visible as *meteors.*

Saturn's rings extend in diameter about 171,000 miles, but their thickness is only 2 miles. The ring system is so thin that observers lose sight of it when it appears edge-on for a few days every fifteen years. The rings are not solid but are inferred to consist of swarms of small icy particles roughly a few centimeters or less in size that move in circular orbits around the giant planet. Both Jupiter and Uranus have also been found to possess dark rings inside their satellite systems; in the latter case, these rings were discovered during the Voyager fly-bys, although ground-based observations originally detected the Uranus ring during the transit of Uranus in front of a bright star. Unlike the wide rings of Saturn, the rings of Uranus appear to be narrow, and they are separated by wide gaps. We do not yet know the composition of the faint rings of Uranus.

The aim of any satisfactory theory of the origin of the solar system must be to explain all these characteristics. From the giant planets to the asteroids and meteorites and the tiny specks of interplanetary dust, our theory must explain the evolution of the unique objects in the solar system that we observe today. Our current understanding of the origin of the solar system is a result of contributions by many different astronomers, and we can gain some insight by examining the historical evolution of theories about the solar system. We can distinguish three broad classes of theories, whose development can be traced over the past three centuries.

TURBULENCE THEORIES

Modern cosmogony began with the *vortex theory* of René Descartes (1644), who was the first writer to attempt to apply scientific reasoning to cosmology. He argued that motions had to be in closed orbits in a space filled with a mysterious "ether." This conception led him to postulate a system of vortices that extended over all scales. Erosion of the smallest particles led to infall and eventual accumulation into the sun, while larger pieces were captured and retained in the vortices. One of the early objections raised by other scientists to the vortex theory was that it could not account for the symmetry of the solar system, where all planetary motions are in the sun's ecliptic, or equatorial plane.

Modern versions of vortex cosmogony have refined the early ideas of Descartes to postulate the existence of a rotating and turbulent atmosphere of gas around the newly formed sun. Turbulence can be described in terms of a hierarchy of vortices or eddies that coexist over many different length scales. Dissipation of the smallest eddies would then result in the formation of denser regions, which act as condensation nuclei for planet formation. These planetary nuclei can grow in size by gravitational attraction of surrounding matter. One outstanding question is, what drives the turbulence? At present, this question has not been satisfactorily answered, and turbulence theories have consequently been rejected.

Figure 16.3 Comets
(a) The head of Brooks' comet, observed in October 1911; (b, page 348) Halley's comet, photographed on June 6 and 7, 1910, returned in 1986.

Figure 16.4   A Meteor Shower
Meteor showers are spectacular displays of shooting stars that are produced when a shower of meteorites burns up in the earth's atmosphere. Large clusters of these meteorites, believed to be the debris of short period comets, orbit the sun and periodically intersect the orbit of the earth. The engraving shows the Leonid shower of November 12, 1833, which reappears every 33 years.

TIDAL THEORIES

The first *tidal theory* was expounded by Georges Louis de Buffon (1785), who proposed the possible collision of a comet with the sun. At that time, it was thought that comets were as massive as the sun. We now know that comets have insufficient mass to be able to cause any appreciable tides on the sun. Modern versions of the tidal theory, developed in the twentieth century by James Jeans and Harold Jeffreys, invoked the close approach of another star to the sun, which would tear a large tidal protuberance out of the sun. The gaseous debris condenses into separate fragments that eventually cool and become the planets (Figure 16.5).

The principal theoretical difficulty with a tidal theory is that the close approach initially must produce very hot gas. The hot gas

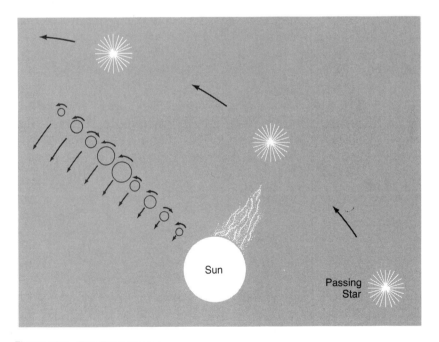

Figure 16.5 The Tidal Theory
According to the tidal theory, a passing star created enormous tides on the surface of the sun, gravitationally disrupting huge filaments of material, which condensed into the planets. This theory cannot be correct, because the gaseous debris would be far too hot to condense: it would not form planets but simply evaporate into interstellar space.

would expand and dissipate before condensation into a planet could occur. Consequently tidal theories have largely been abandoned. One implication of a tidal theory is that formation of the solar system requires the occurrence of a catastrophic event and is therefore likely to have been a unique event in the universe. This viewpoint has been superceded by the evolutionary theory, in which planetary formation is a natural and common phenomenon.

NEBULAR THEORIES

The third and ultimately most successful type of theory of the solar system began with the nebular hypothesis of Immanuel Kant (1755) and Pierre Simon de Laplace (1796). This hypothesis forms the basis of most modern *nebular theories* of origin of the solar system. According to this hypothesis, the *protosolar nebula* was initially a diffuse, slowly rotating gas cloud. As it gradually contracted by self-gravitation the centrifugal forces at the equator caused the ejection of rings of material. We now regard this description as highly oversimplified. For one thing, if centrifugal force was acting alone, the contracting gas cloud would simply become flatter and flatter rather than develop rings.

Current ideas have drastically modified the nebular hypothesis. We calculate that the observed angular momentum of the solar system is probably insufficient to have been responsible for the ejection of enough material in the form of a continuous disk to form planets. Nevertheless, it is attractive to try to account for planet and satellite formation from rings of orbiting particles.

THE ACCUMULATION THEORY

The modern view of the formation of the solar system postulates a slowly rotating gas cloud that maintained a pressure balance with the surrounding medium. The nebula existed as an ordinary interstellar cloud for tens of millions of years. Perhaps a massive star was born nearby as a consequence of compression produced by passage of a spiral density wave, and perhaps death of the star resulted in a supernova. The impact of the shock wave from the supernova explosion may have caused the cloud to collapse.

The presence of a magnetic field in the rotating, contracting gas cloud played a crucial role in the cloud's collapse. As the cloud spun faster and faster, it wound up the magnetic lines of force, which tended to respond like elastic strings (Figure 16.6). The magnetic stresses resulted in formation of a slowly rotating central core, with rapidly spinning matter remaining in a ring on the periphery. We appealed to the identical process in our discussion of star formation (chapter 14) to explain how a star condenses out of a rotating, magnetized interstellar gas cloud. The rapidly spinning

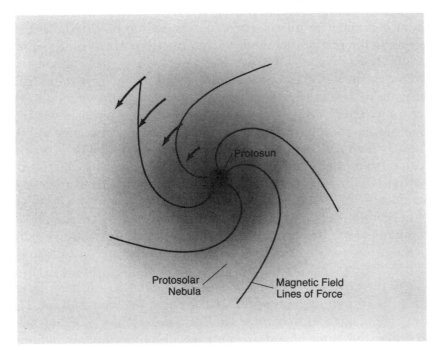

Figure 16.6 Magnetic Deceleration

As the protosolar nebula rotated, the magnetic field lines became increasingly twisted. We can think of them as exerting a tension, like elastic strings. Their effect was to decelerate the rotation rate of the protosun, so that most of the angular momentum remained in the outermost material of the protosolar nebula. This deceleration enabled the central condensation to form the sun, and it accounts for the high angular momentum of the planets.

matter remains in a ring or disk toward the periphery. This effect may help us to understand the distribution of angular momentum in the solar system. Our calculations appear to show that the sun should nevertheless be rotating much more rapidly than it is today. Direct observations of younger stars of similar mass confirm this result. We know now that continuing mass loss by the solar wind over the first billion years of the sun's lifetime was mostly responsible for the eventual loss of angular momentum.

The collapsing cloud rapidly developed a dense, slowly rotating, opaque core, which was destined to become the sun. It was surrounded by a rotating disk of gas, the protosolar nebula. The gas contained many dust particles and gas atoms; it was prevented from falling into the protosun by the centrifugal acceleration in the rotating nebula. The dust particles are known to be present everywhere in the interstellar medium, because they absorb and redden starlight.

During the early contraction of the protosun, the gas would have become so hot (above 2000 k) that preexisting grains would mostly have melted. Eventually, as the gas outside the protosun cooled, new grains condensed, much as snowflakes form. Metallic and refractory grains formed first; as the temperature fell, more-volatile, icy grains formed. Such a condensation sequence is strongly supported by studies of the composition and structure of meteorite samples.

In the protosolar nebula, the solid dust particles cooled and fell through the gas into an extremely thin disk in the equatorial plane of the protosun (Figure 16.7). The dust particles are much heavier than individual gas atoms, and the gas would offer little resistance as the dust collapsed into a disk. The rings of Saturn may be an example of part of such a disk that has not evolved further. The rings could not coalesce into a satellite because they were too close to either the large planet or to other satellites—the gravitational tidal forces would have ripped apart any such satellite. The outer regions of Saturn's disk eventually coalesced to form Saturn's satellites.

The thin disk of cold dust was gravitationally unstable, much like a cold gas cloud. Tiny fluctuations in density exerted a slightly greater gravitational pull on their surroundings. The dust grains would not have been impeded by pressure forces, and they fell

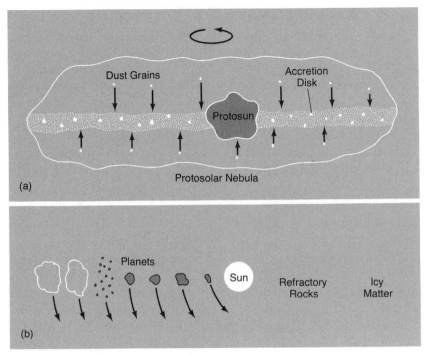

Figure 16.7 A Modern Theory of the Solar System

The protosun formed at the center of the protosolar nebula (a), which was supported by rotation as it collapsed. Dust grains fell through the nebula and accumulated into a thin disk of material. The disk broke up from gravitational instability into planetesimals; eventually, these formed the planets (b). Near the sun, only the refractory, rocklike materials could survive the high temperatures; in the outer region, the most abundant icy materials could condense. Thus, the inner planets are small and heavy, and the outer planets are large and light. As a consequence of the collapse, the planetary orbits all lie in the equatorial plane of the sun, and all the planets except Venus and Uranus rotate in the same direction as the sun.

toward the denser regions. As a result, the dust formed into small aggregations. The self-gravity of the dust overwhelmed its low pressure to make gravitationally bound clumps. Dynamical calculations indicate that these clumps would have the dimensions of asteroids. The term *planetesimals* has been invented to describe them. Asteroids and comet nuclei are likely to be the surviving vestiges of planetesimals that once filled the solar system.

Exactly how cold dust particles adhered to one another to form the clumps is somewhat of a mystery. One possibility is that because the dust was predominantly icy, the fluffy dust particles could have fused together with relative ease. The particles may have approached one another with insufficient velocity to cause shattering but with high enough impact velocity to adhere, just as snowflakes can be compressed into a snowball. The orbiting planetesimals would next collide with one another. Small rocks crashed into larger bodies and shattered. The debris accumulated and was compressed into solid rock as more collisions occurred.

At this stage, gravity aided the largest planetesimals in accreting matter, and they eventually grew to the size of planets. Each planet swept out the debris around its approximately circular orbit. Studies of cratering on the moon, Mercury, and Mars indicate that, over a period of several hundred million years (some 4.6 billion years ago), there was a thousandfold increase in the rate of cratering as compared with the present rate. The impacts were by planetesimals with diameters of 100 kilometers or more, typical of many asteroids. Most of the planetesimals were accumulated by planets and their satellites within 100 million years of the formation of the solar system. The remainder were mostly expended during the following few hundred million years by collisions with the larger bodies. The largest planetesimals naturally grew most rapidly, accumulating and trapping most of the debris from collisions with smaller fragments. Except for the asteroid belt, the rings of Saturn and Uranus, and a cloud of icy debris that surrounded the solar system, only dust remained behind in interplanetary space.

Meanwhile, the young sun was becoming luminous. As the solar rays penetrated the shroud of dust around the sun, the input of energy affected the properties of the forming planets. Near the sun, the temperature became very hot, and the volatile ices evaporated. Only refractory, rocklike, and metallic particles could remain. The inner planets thus formed predominantly from dense, rocky materials. They are of relatively low mass, and they did not retain very much hydrogen or helium. In the outer regions of the solar system, the temperatures were low enough to keep the ices from melting. More-massive planets could form there, and they could be of sufficient mass to retain hydrogen and helium. Thus, the giant, outer planets are massive but of relatively low density; they consist mostly of hydrogen and helium. Jupiter and Saturn have cores of liquid metallic hydrogen, in whose center the heavier elements form a rocky nucleus. The hydrogen is under such great pressure that electrons are stripped, and it behaves like a metal. As a consequence of their rapid rotations (10 hours in the case of Jupiter), the planets generate very strong magnetic fields. These manifest themselves by accelerating electrons in radiation belts around Jupiter and by provoking outbursts of radio emission. The satellites of the outer planets have largely retained the abundant light elements that constitute the ices, but they may have lost much of their hydrogen and helium.

According to this modern *accumulation theory*, most planets formed from the accumulation of many smaller bodies that orbited in a flattened disk around the protosun (Figure 16.8). Protoplanetary disks may have been detected through their infrared emission around nearby stars such as Vega, Beta Pictoris, and MWC 349. This theory provides a natural explanation of the direction of revolution and rotation of the planets. Uranus is an exception. It apparently formed from the coalescence of a few—perhaps only two—large bodies. This would result in a random orientation of its axis of rotation, and could account for its tilt of about 90 degrees

Figure 16.8   A Protoplanetary Disk

Observations in the infrared suggest that the star MWC 349 is very young and is surrounded by a protoplanetary disk of gas and dust from which planets could eventually condense. A disk has not actually been photographed, but observations at different wavelengths enable us to sketch a model of this system.

to the ecliptic. Only planets that formed from many smaller bodies, whose individual directions of spin and motion average out, would result in a planet whose spin axis was parallel to that of the sun.

The retrograde rotation of Venus suggests that a strong tidal deceleration of the planet has occurred. A similar effect has occurred in the earth-moon system, so that the moon always turns the same face to the earth. We understand this effect rather well: the moon rotates once on its own axis each orbital period. This locking of the rotation and orbital periods is due to the tidal bulge of the moon, which is slightly football-shaped. Gravitational forces make the largest axis point toward the earth at all times, even though the difference in length of the lunar axes amounts to only about 2 kilometers.

We can speculate that current observations of the asteroids provide a glimpse of the early solar system before any planets formed. Possibly because of the gravitational pull from Jupiter, the coagulation process was inhibited in the asteroid belt. The outer half of the asteroid belt contains predominantly black, carbon-containing asteroids. The inferred composition, similar to that of carbonaceous chondrites, indicates that condensation could not have occurred above about 400 degrees kelvin. One would expect the carbonaceous material to be more common at the cooler outer edge of the asteroid belt. A similar gradual transition is indicated between the condensation of refractory material in the inner solar system and icy volatile material in the outer solar system. The asteroids also overlap in size with many of the planetary satellites, which formed by similar processes of accumulation. Because the giant planets collected more debris than the smaller planets, they acquired relatively large numbers of satellites. For example, Jupiter and Saturn each have about 20 satellites; Uranus has 15. A number of these satellites were discovered during the Voyager flyby missions.

The more massive satellites resemble planets. Titan, the most massive satellite in the solar system, exceeds the mass of Mercury and has a methane atmosphere with pressure of about 1 earth atmosphere. Smog has been detected spectroscopically in the atmosphere of Titan. Io was observed to have volcanic eruptions during one of the Voyager passages. Other satellites such as Callisto and Prometheus have polar caps and other prominent ice features.

The accumulation theory predicts that a satellite system should resemble a solar system in miniature. The satellites should all lie within the equatorial plane of the parent planet, have circular orbits, and revolve and spin in the same direction. The few exceptions (such as the outer satellites of Jupiter) are often small bodies that could be recently captured asteroids. It is possible, too, that Saturn's rings are the primordial debris left over from the formation of the planetary satellites.

Pluto has the most eccentric orbit of all the planets; its orbit takes Pluto farther out of the ecliptic than any other planet. Pluto's orbit actually brings it within the orbit of Neptune. Unlike the other planets, Pluto is of very low mass and probably weighs less than the moon. Pluto's satellite, Charon, has about the same mass as Pluto. Astronomers have speculated that Pluto originally may have been a satellite of Neptune. The satellite Triton has an unusual retrograde orbit around Neptune and may be a vestige of a catastrophic encounter between two satellites that led to the ejection of Pluto into its present orbit.

Comets also may be primordial material that collected at the boundary of the solar system. A vast swarm of cometary nuclei is believed to orbit beyond Pluto. Orbital velocities are very low, and consequently chance gravitational deflections from passing stars may occasionally throw such a large chunk of icy debris toward the sun. In the inner solar system, the heat from the sun begins to vaporize the ices and create a cometary tail. The frequency of observed comets is consistent with this presumed cometary reservoir.

FORMATION OF THE EARTH AND LIFE

The many planetesimals that crashed together to form the earth must have melted on impact, forming a hot molten core. The heating of the inner core is maintained by radioactive decay of very heavy, unstable nuclei, and the earth still possesses a molten liquid core. The core density indicates that strong *chemical differentiation* (separation of heavier elements from lighter elements) has occurred. Most of the iron ended up in the core, with oxygen, silicon, and magnesium in the surrounding mantle. By studying the reverberations in the earth's crust produced by earthquakes, we have found that most of the mass of the earth consists of an iron-nickel core surrounded by a mantle of silicate-type rocks. One other important consequence of a molten core is that convective currents (much like atmospheric turbulence or winds) are driven as the hot material tends to rise. The silicate rocks, which have lower density, drift up to the surface as a result of the convection. The continental land masses may have formed in this manner. The currents in the outer mantle are responsible for continental drift. The present continents formed as the great land masses of the earth's crust drifted apart. The constant pressure from the hot interior results in occasional rifts and earthquakes and continued volcanic activity.

A number of lighter elements in small amounts form the atmosphere and oceans of the earth. The earth's atmosphere presently accounts for about 0.025 per cent of the mass of the earth. Outgassing of water vapor and carbon dioxide from the hot interior, such as is observed during volcanic eruptions, may have produced the terrestrial atmosphere. Elements such as hydrogen and helium would have been very much more abundant on the young earth. The primordial atmosphere largely consisted of hydrogen-rich gases, such as ammonia, methane, and water vapor. The atmospheres of Jupiter and Saturn are presently very similar to the atmosphere of the primitive earth, which contained virtually no free ozone or oxygen molecules. Oxygen was present almost entirely in the form of water vapor. Without ozone in the upper atmosphere to absorb ultraviolet radiation, the young earth was relatively unshielded against ultraviolet radiation from the sun. There is little doubt that full exposure to the solar ultraviolet radiation would be injurious or even fatal to life.

However, on the young earth, the ultraviolet rays provided a crucial catalyst in the synthesis of organic matter. In a remarkable series of laboratory experiments, scientists have studied the effects of flashes of ultraviolet radiation and electrical discharges on a gas mixture resembling the atmosphere of the primitive earth. In these experiments, complex molecules were actually created, including many different amino acids and other organic compounds that comprise the basic constituents of genetic material. It thus seems possible that lightning storms on the young earth may have been the original stimulus of the basic ingredients necessary for life (Figure 16.9).

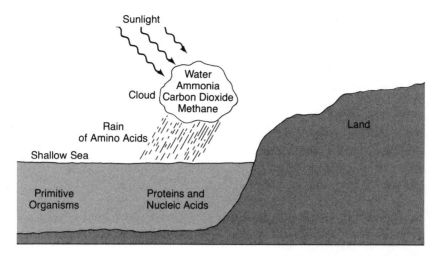

Figure 16.9 Origins of Life

The interaction of ultraviolet radiation from the young sun and electrical discharges or lightning storms with vapor clouds of methane, water, and other simple gases in the steamy atmosphere of the primordial earth led to the production of amino acids, which rained down on the earth. In the lagoons and shallow seas, the amino acids were sheltered from the intense radiation of the young sun and gradually developed more complex chains of molecules. Proteins and nucleic acids (the tracers of the genetic code) formed, and primitive single-celled organisms were eventually created. Many different chemical combinations must have remained inert, until the emergence of the first self-replicating systems signaled the origin of life.

Because amino acids are asymmetrical in structure, the mirror image of an amino acid molecule is different from the original molecule. We speak of left-handed and right-handed molecules, and these have intrinsically different structures. Terrestrial amino acids are predominantly left-handed molecules and are a unique signature of terrestrial life. It is possible that nonterrestrial amino acids, with a different proportion of right-handed and left-handed molecules, have been generated elsewhere in the universe. Finding such evidence would help to substantiate further our theories of chemical evolution. Extraterrestrial amino acids have actually been found in several meteorite samples.

The complex amino acid molecules must have collected in the oceans, where water shielded them from destruction by solar radiation. The watery environment enabled the molecules to interact with one another and gradually to synthesize even more complex molecules. The oceans provided a protective environment, a sort of gigantic primordial womb, enabling the long chains of molecules that form living cells to develop. Most biologists agree that it is overwhelmingly improbable for the key to life, such genetic tracers as DNA and RNA, to have emerged from this primordial broth, if we use our current understanding of chemical reaction rates. Nevertheless, somehow this happened. My view is that such complex systems are entirely unpredictable. Probability estimates are worthless. Even a simple pendulum can develop chaotic behavior. How then can we presently presume to know the outcome of this infinitely more complex environment? Only by looking around us, and realizing that, somehow, if we persist with scientific reasoning, molecular mutations must eventually have led to the development of primitive forms of plant life. Algae and similar plants originated first in the oceans; later, plant life developed on land.

Meanwhile, most of the hydrogen and helium, the lightest elements, escaped from the earth's atmosphere. Oxygen was produced, partly by ultraviolet dissociation of water, but mostly by photosynthesis. The oldest rocks found on earth, those older than about 2 billion years, contain evidence that the earth's atmosphere was highly oxygen-poor. However, younger rocks show evidence of rust, or oxidation. This finding confirms our idea that the development of an oxygen-rich atmosphere was a secondary phenomenon on earth. The oxygen, and particularly ozone, was crucial in shielding the earth against ultraviolet radiation. The presence of vast oceans played a further crucial role in aiding the development of life by providing a uniform temperature for the environment. The oceans retained heat and kept the earth warm. At the same time, water vapor evaporated from the oceans into the atmosphere. The water vapor acted like a greenhouse, trapping the infrared radiation from the ground, which was directly heated by sunlight. In this way, a safe environment was provided where animal life could eventually evolve.

CATASTROPHIC EVOLUTION

Life did not necessarily evolve smoothly. In fact, palaeontological studies of fossil extinctions suggest that evolution was punctuated by a series of often catastrophic changes. Over brief periods on a geological time scale—tens of thousands of years—many species underwent sudden extinction. One of the more dramatic of these periods occurred about 60 million years ago, when the dinosaurs died out.

Although theories about the death of the dinosaurs abound, one that has received much attention appeals to an astronomical event. An iridium layer found at the *Cretaceous/Tertiary* boundary layer led to the hypothesis that an asteroid or comet impacted the earth some 60 million years ago, showering our planet with debris. This event would account for the deposition of the iridium at a well-defined point in geological time: the iridium is more concentrated in meteoritic material than we would predict over widely separated regions on the earth. Some paleontologists speculate that the at-mospheric dust darkened the sky and resulted in a climatic change, possibly leading to an ice age, or "nuclear winter," lasting for sev-eral years. During this time, much of the tropical vegetation foraged by dinosaurs would have disappeared. A similar climatic cooling would, it is speculated, be the byproduct of nuclear war. There are many other speculations on the nature of the precise mechanism, including acid rain and ozone depletion, but the meteorite trigger is widely, if not unanimously, accepted. Data on fossil extinctions of many plant species are suggestive of a 27-million-year period-icity, and a similar periodicity has been reported in studies of im-pact cratering on the earth. The idea behind cratering studies is that just as we can distinguish fresh from old footprints in the snow, radioactive dating of rock ages, when combined with cratering fre-quency, can yield the time variation of cratering impacts on the earth. Although the evidence we have for periodicity is marginal, the theory has astounding implications. Every 27 million years, infall of asteroids, meteorites, comets, or similar objects is en-hanced on the earth. One hypothesis for the cause of this effect is that the sun has a low-mass stellar companion, christened Nemesis, with a highly eccentric orbit, that disturbs the cometary cloud out-side Pluto's orbit every 27 million years. This would lead to a pe-riodic occurrence of enhanced cometary encounters with the inner solar system, possibly accounting for the infrequent collison of a large extraterrestrial object with the earth some two periods ago. Needless to say, searches for Nemesis have failed to find a can-didate star, and the precise origin of what may be periodic mass extinctions and of iridium deposition remains uncertain. Indeed, most experts believe that purely stochastic events, perhaps associated with the onset of ice ages and changes in the sea level, can account for the fossil extinction data.

LIFE IN THE UNIVERSE

The evolution of the earth and of life finds a natural place in our cosmological scenario. Although we can only speculate, the conclusion seems plausible that planetary systems may accompany many stars in space. No system has been directly observed, but one of the best-studied candidates is the companion of Barnard's star. Studies of irregularities in the orbital motions of Barnard's star have revealed the possible presence of an unseen companion, whose mass is only 1.5 times that of Jupiter. Several other nearby stars have been carefully monitored and orbital motions sought that amount to only a fraction of a kilometer per second; Jupiter-mass planetary companions cause perturbations of the parent star's orbit that are, in principle, detectable. Although there are as yet no definitive detections, our theory of the protosolar nebula suggests that the formation of the solar system is unlikely to have been a unique event in the history of the galaxy.

Whether life is similarly inevitable is an open question. Biologists have agreed on three basic requirements for the development of life: a reasonably uniform temperature, a solvent such as water, and an atom such as carbon that is capable of forming complex chains of molecules. However, the probability of the resulting evolution of life cannot be estimated, given the complex, nonlinear nature of the chemical system. Self-organization is essential for a successful outcome. The complex patterns that are beginning to emerge in our tentative dabblings with nonlinear systems make me optimistic that progress in our understanding of the origin of life will eventually be forthcoming. It is conceivable that other forms of life could develop in oceans of ammonia or from silicon-based molecules. One is tempted to speculate that life does not seem to be unique to the terrestrial environment, particularly now that many complex molecules have been discovered in interstellar space. Water vapor is known to be a common constituent of interstellar gas in nearby galaxies. However, we as yet have no examples of life outside the earth. Perhaps earthlike conditions are necessary for life to occur.

Although there may be many other planetary systems and environments capable of creating life, we are still far from the assertion that intelligent civilizations exist elsewhere in the galaxy. Nevertheless, this possibility is taken seriously by astronomers, and active searches for extraterrestrial civilizations are being planned (Figure 16.10). So far, our search has focused on detection of electromagnetic radiation for radio and television communication. The largest radio telescopes are capable of detecting radio

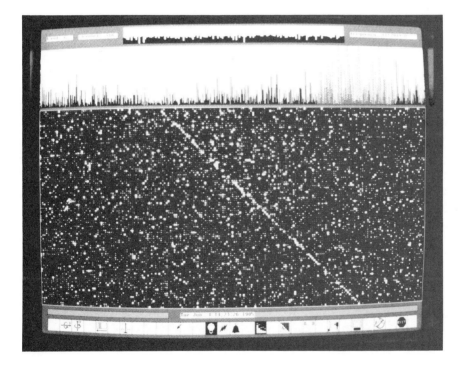

Figure 16.10 SETI

NASA's Search for Extraterrestrial Intelligence (SETI) program is investigating the most technologically efficient, cost-effective approaches to search for intelligent life elsewhere in the universe. A prototype SETI detector was used with the 26-meter Goldstone Observatory radio telescope to detect a 1-watt signal from outside the solar system, transmitted by the Pioneer 10 spacecraft at a distance of 3.2 billion miles from earth. The planned SETI radio spectrum analyzers will be at least 100 times more sensitive.

waves from a civilization that is less than 100 light-years away and that also has a level of electronic noise similar to that produced on earth. Of course, two-way communication over these distances is not possible in our lifetimes. However, we can hope to eavesdrop on any intelligent or not-so-intelligent neighbors in space who may be broadcasting.

· 17 ·

INTO THE INFINITE FUTURE

This is the way the world ends
Not with a bang but a whimper.
—*T. S. Eliot*

Some say the world will end in fire,
Some say in ice.
From what I've tasted of desire
I hold with those who favour fire.
—*Robert Frost*

Our survey of the evolution of the universe has brought us to the present, and we naturally are curious about the future. Will the universe continue to expand forever? The answer to this question is probably yes. Several approaches lead us to this same conclusion. Of course, as is often the case in astronomy, we lack the ability to perform precise experiments under carefully controlled conditions in the cosmic laboratory. Nor would any of the arguments stand up in a court of law. Nevertheless, it is possible to assert that the weight of evidence presently rests marginally in favor of an open, infinite universe that will expand indefinitely. We shall first assemble the case for an open universe; then we shall point out the various loopholes that could conceivably allow a closed universe.[21]

367

"THE BIG BANG? BELIEVE ME, IT WAS VERY, VERY, VERY, VERY, VERY, VERY BIG."

MASS DENSITY OF THE UNIVERSE

The most obvious technique for discriminating between an open and a closed universe is to measure the average density of matter. We saw in Chapter 5 how a simple equation (the Friedmann equation) describes the competition between the attractive gravitational force and the expansion of the universe. The gravitational attraction exerted at the center of an arbitrary sphere cut out of the universe is proportional to the average density of matter. The measured value of the Hubble constant (H) yields the kinetic energy of the expansion of the sphere. If the present density is below the critical value at which the expansion and gravitational attraction balance, gravity cannot halt the expansion, and the universe must be open. The critical density for closure of the universe is

$$d_{critical} = 3H^2/8\pi G = 5 \times 10^{-30} \text{ gram cm}^{-3}$$

where G is Newton's constant of gravitation. Another way to express this critical density is in the number density of atoms, which amounts to 3×10^{-6} atoms per cubic centimeter (cm^{-3}), or only 3 atoms per cubic meter.

When we search for the matter content of the universe, we actually appear to find rather less than the critical value for the density of matter in space. Of course, in the Milky Way, the average density of matter is high (about 10^{-23} gram cm^{-3}), but we are within the narrow confines of a galaxy. If we allow for the vast intergalactic spaces, we find that the average density of luminous material in the universe is quite low. The typical separation between galaxies like the Milky Way is about 10 million light-years. The Milky Way is about 100,000 light-years across the plane of the galaxy, and perhaps only one-tenth of this distance in a perpendicular direction. Only about 1 part in 100 million of the volume of space is therefore occupied by stars. This reasoning gives a mean density of some 10^{-31} gram cm^{-3}, or about 2 percent of the closure value.

This measurement of the stellar or luminous component of the mass density provides only a lower limit on the actual mean density of matter in the universe. We have already seen that rotation of spiral galaxies such as the Andromeda galaxy is measured out to distances as great as 100,000 light-years from the center of the galaxy and indicates that a considerable amount of mass must be in the outer regions, or halo, of the galaxy. The halos must contain dark matter, probably in very low mass stars, collapsed remnants of massive stars or in some species of weakly interacting particles.

We can estimate a more realistic value for the actual density of matter in the universe by computing the amount of dark mass. We know what the overall masses of the galaxies are from the dynamical measurements of rotation. Recall that the effective mass-luminosity ratio for a galaxy is about 30 solar masses per solar luminosity, which is considerably greater than the average value of about 5 solar masses per solar luminosity for the luminous regions of spiral or 8 for elliptical galaxies. Studies of pairs of galaxies and galaxy clusters suggest that still more dark matter may be present at even greater radii. However, even incorporating this dark mass leaves us short of closure by a factor of 10 in density.

We may conclude that the mass densities estimated for galaxies and even for clusters of galaxies do not suffice to close the universe. Of course, we cannot rule out the possible existence of mass that is uniformly distributed throughout space. Our density estimates do not yield any information on the contribution of such a component to the average mass density. Such uniformly distributed mass would have little dynamic effect on galaxies or clusters of galaxies. We can only speculate on its possible characteristics. The fact that in several rich clusters of galaxies the mass-luminosity ratio is about 200 solar masses per solar luminosity indicates that matter with a very high mass-luminosity ratio exists. But the clusters do not make a large contribution to the overall mass density of the universe. No rich clusters of galaxies have ever been found to have a mass-luminosity ratio as high as 1000.

However, the amount of matter required to yield a closed universe must have a mass-luminosity ratio of 2000 solar masses per solar luminosity. No known population of stars or galaxies can satisfy this ratio. The idea that the dark mass could take the form of massive planets (like Jupiter) or collapsed remnants of stars seems unlikely. The dark mass required to close the universe must be more uniformly distributed than the galaxies are, and both massive planets and collapsed stars would surely be associated with galaxies, perhaps forming early in their evolution. Of course, astrophysicists are sufficiently ingenious to overcome this objection: the uncertainty in galaxy-formation models is such that most of their gas once in halos may well have collapsed to form compact objects of low visibility. These might be white dwarfs, neutron stars, or black holes.

The most plausible form of the cosmological dark mass is gas left over from the era of galaxy formation. Astronomers have searched extensively for intergalactic gas. The detectability of the gas depends on how hot it is. If the gas is cold, it would be readily observable, and we observe very little cold intergalactic gas that can contribute to the dark mass. Only if the gas is very hot can large amounts of it be present in intergalactic space. In this case, the gas would emit copious amounts of ultraviolet or x-radiation. X-ray astronomers have found evidence for a uniform and isotropic background of cosmic diffuse x-radiation, and some have argued that these x-rays are being emitted by intergalactic gas at a temperature of 500 million degrees kelvin. Although this issue remains controversial, if this interpretation is correct the universe could contain about one-third of its critical density, in the form of hot gas. However, other explanations are offered for the diffuse x-radiation, most notably that it is emitted from a very large number of unresolved extragalactic x-ray sources. Somewhat cooler but ionized gas could also be present. It is certainly premature to dismiss the possibility that a substantial amount of mass in this form could be present; this amount conceivably could be enough to close the universe.

The difficulty with such options involving ordinary matter is that if the dark matter is baryonic, then the beautiful coincidence breaks down between the simple big bang prediction of light-element nucleosynthesis and the observed light-element abundances. For example, we would now have to resort to a contrived mechanism for forming deuterium, and perhaps resort even to destroying some of the helium, which tends to be overproduced in a dense baryonic universe.

The alternative route to dark matter is the particle zoo. Nonbaryonic particles, provided they are sufficiently weakly interacting, can account for the dark matter, but they do not affect primordial nucleosynthesis. As we noted earlier, no experiment has confirmed the existence of a single candidate particle, even though such particles remain theoretically plausible. If we adopt the assumption that what can exist does exist, we have some motivation for pursuing nonbaryonic dark matter as a means of closing the universe. Fortunately, we need not close our investigation here. We have other, indirect ways of determining the average mass density if we use measurements of the curvature of space.

CURVATURE OF SPACE

We can take an entirely distinct approach to the mass-density measurement to discriminate between the open and closed models of the universe. In principle, we can hope to measure the curvature of space directly. The key point to recall is that in nearby regions such as the solar system or even the galaxy, spatial curvature is entirely negligible, apart from in the immediate vicinity of any black holes that may be present. Only on the very largest scales of the observable universe does the curvature of space become significant. As we examine galaxies at greater and greater distances, the effects of space curvature play an increasing role in determining the actual distances.

Suppose that the faintest galaxies whose redshifts can be determined all have a similar intrinsic luminosity. Then, by measuring their apparent magnitude, we can hope to unveil the effects of curvature (Figure 17.1). Within a positively curved space, galaxies at a given recession velocity (or redshift) would have smaller apparent distances than in a flat space. For a loose analogy, imagine trying to flatten a shape drawn on a sphere: the edges of the shape will crinkle, become distorted, and draw closer together. In a negatively curved space, galaxies at a similar redshift would be more remote. The analogy in this case describes our attempts to flatten a shape drawn on a saddle-shaped surface: the result will be a stretching at the edges.

However, we can use another, more physical explanation. If the universe were expanding faster in the past (as would have occurred for a closed universe) galaxies with a given redshift would not be as far away as they would be if there were no deceleration (as in an open universe). It is primarily for this reason that galaxies appear brighter in a closed, as compared with an open, universe.

Thus, at a given redshift, the apparent magnitude of a galaxy would be least in a closed universe and greatest in an open universe. At a redshift of 1, close to the greatest redshift measured for a normal galaxy, the difference in apparent magnitude of a galaxy in either an open or a closed universe amounts to approximately 1 magnitude. This is about a factor of 2 in brightness. The galaxy appears brighter at a given redshift if the universe is closed. However, typical galaxies seen at this redshift are expected to be very faint. If such galaxies resemble nearby galaxies, they should barely be detectable on photographic plates taken with the world's largest telescopes.

GALAXIES AS COSMOLOGICAL PROBES

In fact, the few remote galaxies known whose redshifts are large are in some instances found to be rather brighter than the limit of detectability on a good photographic plate. Evidently, they are not well-balanced *standard candles*. Remote galaxies occasionally seem considerably more luminous than we would have expected them to be, even in a closed universe, if these galaxies are similar to nearby galaxies in their intrinsic brightness. The record redshift for a remote galaxy is 3.8 (in 1988); this galaxy is nearly 100 times more luminous than nearby bright galaxies and is forming stars at a prodigious rate.

We suspect that this phenomenon occurs because of evolution. The remote galaxies we observe are young galaxies, and galaxies should have been much brighter in their youth. The evolution of a galaxy is a complex process. At first, galaxies contain many young, massive, bright stars. As time passes, only the less massive and longer-lived stars survive from the early eras. Some young stars continue to form, but at a greatly reduced rate. The net effect seems to be that young galaxies are very luminous and gradually fade as they age. Evolution probably results in the dimming of galaxies with time.

Because the remote galaxies are brighter than we expect, we infer smaller apparent distances for them, and if we see them as less distant, we in turn reason that the galaxies have greater density. We therefore observe a universe that seems more likely to be closed than is really the case. Allowance for the effects of galactic

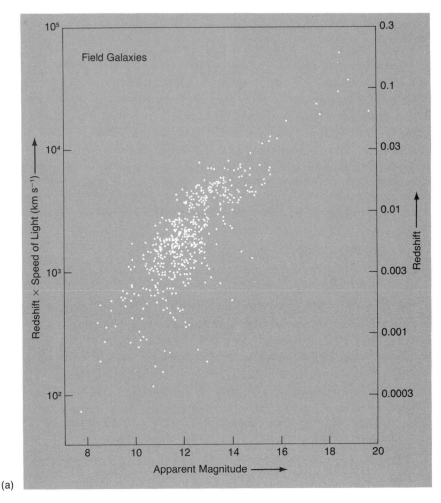

(a)

Figure 17.1   The Redshift–Magnitude Test

The redshift (*right vertical axis*), also interpreted as a velocity when multiplied by the speed of light (*left vertical axis*), can be plotted against the apparent magnitude (*horizontal axis*) (or brightness, which yields a measure of distance if the intrinsic brightness is known) for distant galaxies. For field galaxies, a proportionality between velocity and distance is barely discernible (*a*), because of the wide range of intrinsic luminosities at any given redshift. However, if only the brightest members of galaxy clusters are used (*b*), a linear relation between velocity, or redshift, and distance (Hubble's Law) is found. The box in the lower left indicates the range over which Hubble first formulated his recession law in 1929. Deviations from a straight line indicate that the universe may be noneuclidean and either open or closed. The predicted curves (*c*) for several different Friedmann-Lemaître cosmological models are labeled by the ratio of mean density to the critical closure density. The curve SS is that predicted for a steady state universe and is well outside the observational range unless strong evolutionary effects are invoked; this, however, would violate the principal tenet of steady state cosmology. Thus, if the series of dots (representing observed clusters) appears to curve upward and to the left, the universe is closed. If the dots curve to the right, the universe is open.

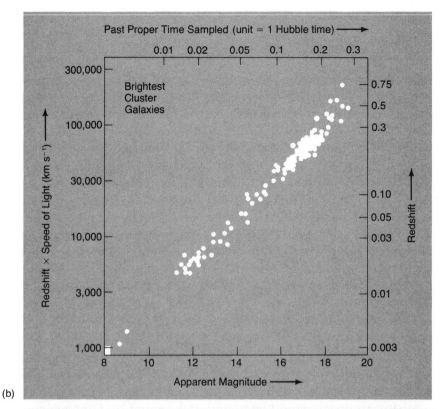

(b)

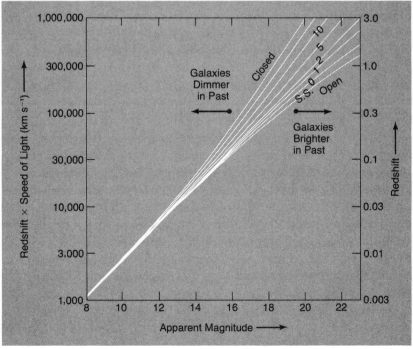

(c)

evolution lengthens the distance estimates, lowers the density estimates, and tends to make the universe more open than naive application of the *redshift-magnitude test* would indicate. Assuming that no evolutionary changes in the light of remote galaxies have occurred, the redshift-magnitude relation appears to show that the spatial curvature is negative. On this basis, we might feel justified in concluding that the universe is open. However, we are not at all confident in our knowledge of galactic evolution. Without an empirical measurement of evolution, cosmological arguments alone cannot carry a great deal of certainty. It is conceivable, for example, that the brightest galaxies were dimmer in the remote past, at least in the visible part of the spectrum. Such an effect has been predicted as a consequence of galaxy "gobbling." If galaxies swallowed their smaller neighbors in rich clusters, they would grow and brighten with time. We would have overestimated the distances to remote galaxies. The resulting correction would tend to favor a closed universe.

There is considerable hope that evolution and space curvature may eventually be measured simultaneously. To achieve this, we would combine measurements of the redshift-magnitude relation with other types of cosmological tests using distant galaxies. A promising approach would be to undertake counts of faint galaxies in a selected region of the sky (Figure 17.2). We find increasing numbers of galaxies if we lower the brightness threshold for counting them. A larger volume of space would be sampled as fainter galaxies are counted, and the number of galaxies is expected to increase proportionately with the volume of space being sampled. As we have seen previously, this simple proportionality applies only if the galaxies are distributed uniformly and space is flat, or euclidean. The cosmological expansion and the curvature of space introduce corrections to this prediction. The resulting *number-density-redshift test* allows, in principle at least, an accurate determination of the cosmological model, but it requires the painstaking acquisition of a complete sample of galaxies with hundreds or even thousands of measured redshifts. If space is not expanding and is flat, the number density of galaxies at a given redshift, or distance, is equal to its euclidian value. The effect of the expansion reduces the number counted in a static space by a factor that measures the

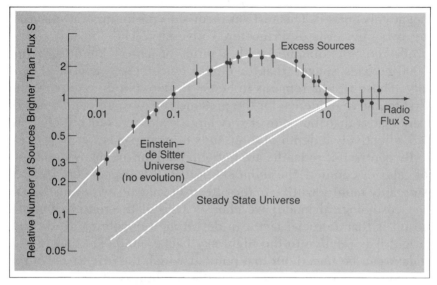

Figure 17.2 The Number–Count Test

The number of radio sources brighter than a certain flux (relative to that expected in an infinite, static universe) is plotted against their flux. A straight horizontal line would be expected if the universe were static. This horizontal line indicates that the number of sources increases with the volume of space sampled; the number of sources is expected to increase as the negative three-halves power of the flux. The effect of the redshift is to diminish appreciably the number of expected sources. Predictions are shown for the steady state universe and for an Einstein-de Sitter universe (with no evolutionary effects included). We in fact observe the expected result for nearby bright sources, followed by a dramatic increase for the number of fainter sources above that expected for a closed universe and, finally, a decay in the number of the very faintest sources. The excess number of sources can be attributed to evolution, particularly to a much higher birth rate for the remote sources at early eras in the expanding universe (corresponding to redshifts of 3 or 4). Steady state cosmology fails to explain the observations, since, by definition, the possibility of any such evolution is excluded. A similar test can be performed with optical galaxies, by counting the number of galaxies detectable on a photographic plate as a function of their brightness. This test should eventually yield information on the evolution of galaxies that are observed at optical wavelengths.

contraction of the proper-volume element at earlier epochs. Curvature acts to diminish this number still further if the universe is closed and positively curved by reducing the relative-volume element. Curvature increases this number if the universe is open and negatively curved. This test has received a preliminary application by two physicists at Princeton University. Their results, using crude redshifts obtained with a series of narrow band filters for 1000 galaxies, suggest that the universe is precisely at critical density. These results remains to be confirmed, but they provide the first tentative evidence for the presence of substantial amounts of dark matter distributed over very large scales to a redshift of about 0.5, effectively a depth of thousands of megaparsecs.

By contrast, if redshifts are not obtained for the many hundreds of galaxies counted, the *number-count test* is expected to be considerably more sensitive to the effects of galactic evolution than to the cosmological model we choose. One of the reasons for this result is that a crucial factor in detecting a remote galaxy is how bright it is relative to the night sky background. The image of a galaxy can be traced out to a point at which the brightness of the image is only a small fraction of the brightness of the night sky. Because the total light measured from a galaxy determines its brightness, measurement of the effective edge of the galaxy is essential. The surface brightness of the galaxy, relative to the sky background, determines the effective outer boundary of the galaxy. This effect is independent of the cosmology; it depends only on the redshift of the galaxy. Choice of a specific cosmological model does enter in determining the volume of space out to which the galaxies are being sampled. Nevertheless, the net effect is that the number-count test is relatively insensitive to cosmology.

Evolutionary effects can be crucial to these tests. For example, suppose that young galaxies had a burst of star formation in their halos. Indirect arguments about their early evolution have led us to suspect that this process might be common. The consequences of such an event would include an increase in the apparent angular size of the galaxies, an increase in their luminosity, and detection of greater numbers of galaxies at a given apparent magnitude. Galaxies that are more distant would be detected; in the absence of such a process, these galaxies would not have been visible.

The number-count test has indeed been found to constrain galactic evolution (Figure 17.3). By obtaining counts of galaxies down to limiting magnitudes in different regions of the spectrum, it should be possible to extract more detailed information about how galaxies evolve. For example, we might hope to determine whether the younger blue stars are more numerous than the older red stars at early eras; this information would give us a clue to the ages of the galaxies. Just as the number-count test provides a measurement of evolutionary effects for radio sources, it may enable us to determine the evolutionary history of ordinary galaxies.

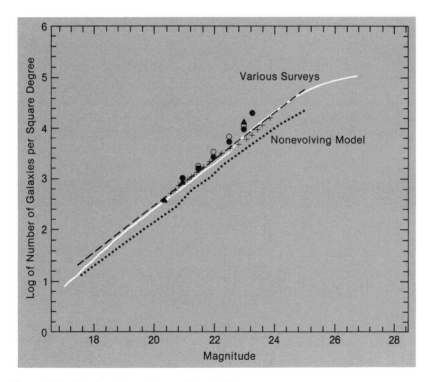

Figure 17.3   Number Count Versus Magnitude

Number counts plotted versus magnitude for several deep galaxy samples. Surveys are carried out using either photographic plates or CCD detectors. If evolutionary effects are neglected, the model predictions are seen to lie well below the data points.

QUASARS AS COSMOLOGICAL PROBES

Two astronomers, then at the Lick Observatory in California, have argued that quasars can be used to probe the universe. At first sight, this suggestion seems contradictory, for as we saw in Chapter 12, if galaxies evolve with time, then quasars must evolve far more strongly. Their frequency, and perhaps also their intrinsic brightness, increases greatly at earlier eras of the universe, and quasars exhibit very different properties from one another. Surely such effects would grossly distort any attempt to use quasars as standard candles.

Such was certainly the situation, until the Californian astronomers noted that there does appear to be a means of calibrating quasars and inferring their intrinsic luminosity. They found that the strengths of certain spectral features (or emission lines) correlated very well with the absolute brightness of the quasars. Thus, quasars could perhaps be used, much as galaxies are used, to probe the curvature of space. Because quasars are much more luminous than galaxies, we can observe them at greater redshifts, and they provide a potentially important probe. A preliminary application of this technique appears to favor a closed universe; however, the effects of quasar luminosity evolution on this result remain very uncertain.

In a more productive application of quasars to cosmology, quasar absorption lines have been studied as a probe of the intergalactic medium. Many absorption-line systems are believed to be produced by interstellar gas in intervening galaxies. These absorption lines include the heavy-element species, as well as hydrogen, that are prevalent in an ordinary galaxy. However, a considerable number of absorption lines are found to be produced in gas that is apparently free of heavy elements. The line is due to the resonant absorption of atomic hydrogen in its ground state, known as the Lyman alpha transition (Chapter 12). Indeed, an entire forest of Lyman alpha absorption lines is found, always blueshifted relative to the Lyman alpha emission line of the quasar. This relative blueshift means that the Lyman alpha lines are produced in clouds between us and the quasar and are therefore redshifted less than the quasar. The Lyman alpha absorbing clouds are as numerous as small galaxies and very weakly clustered in space, unlike most galaxies. It is believed that they are primitive intergalactic gas clouds, either about to form galaxies or somehow aborted from doing so. No constraints on cosmological parameters have yet been deduced from these clouds, but the Lyman alpha forest may provide a glimpse of a primitive stage of evolution of pregalactic structure before any stars had formed. The Lyman alpha clouds have been used to study the radiation field in the early universe; excessive radiation would destroy them, and there is evidence that they indeed are not present in the vicinity of quasars. Very strong Lyman alpha lines are associated with intervening gas-rich galaxies. From the frequency of occurrence of these lines, we learn that the intervening galaxies must be very large; they have been inferred to be protogalactic disks (galactic disks in the process of formation). About 20 percent of the sky is covered by such protogalaxies, at redshifts between 2 and 3. However, they are exceedingly faint, and there is, at present, no definitive image of any of these systems, which are only seen via their absorption against background quasars.

DEUTERIUM AND MASS DENSITY

We saw in Chapter 7 that the amount of deuterium produced in the big bang depends critically on the density of matter when the primordial nuclear reactions occurred. In a closed universe, the density would be relatively high when the universe was about one minute old. At this era, nucleosynthesis of deuterium would occur. The high density would result in burning of the deuterium, and almost no deuterium would have survived. If the density was sufficiently low at the nucleosynthetic era, as would occur in an open universe, enough primordial deuterium would have survived when helium was synthesized to account for the observed abundance of deuterium. It is difficult to conceive how deuterium could be synthesized in any astrophysical environment other than the early minutes of the big bang. Stars, for example, only burn and destroy their initial abundance of fragile deuterium. Failure to account for deuterium production by any means other than the big bang lends support to a theory of cosmological origin for deuterium. The implication is that the universe is open.

We should acknowledge, however, that although an open-universe model accounts for the production of sufficient deuterium to satisfy our theoretical requirements, it is not the only model that can do so. Simple closed models indeed fail to account for synthesis of enough deuterium. However, more complicated versions of the early evolution of the universe could account for an environment favorable to the synthesis and survival of deuterium.

One such mechanism involves generation of a local neutron excess during the quark-hadron phase transition, when the universe was at a temperature of about 200 million electron volts (MeV) (or 200 trillion degrees) (page 149). Both neutrons and protons are produced during this transition, which demarcates the epoch when ordinary hadronic and leptonic matter—protons, neutrons, electrons, muons—first came into existence. However, the phase transition may not be smooth but, just as boiling water bubbles away by generating steam, produces bubbles of the new phase. At first, there are bubbles of hadrons surrounded by quarks as the universe tries to cool but rests for a few picoseconds at 200 MeV. Finally, there remain a few pockets of quarks surrounded by the sea of hadrons. Now the net number of baryons is higher for the quarks than for the hadrons, since the phase transition produces entropy and dilutes baryon number. This means that at about one second, when neutrons freeze out, there are patches which are neutron-rich. The neutrons and protons diffuse at different rates: the result is that at the epoch of nucleosynthesis, when the temperature is about 0.1 million electron volts, we still have neutron-rich pockets of matter; nucleosynthesis there will consequently differ greatly from the process in neutron-poor regions. Such variation could lead to large variations in the light-element abundances predicted in the standard models. The principal difference is that substantial amounts of deuterium are synthesized even in a universe closed by baryons. This is an advantage; the major problem is that too much lithium is synthesized. This may be fatal to this model, and indeed, it would be somewhat remarkable if we ended up with the observed pregalactic abundances of the light elements that arise in a low-baryon-density big bang universe.

Alternatively, we can suppose that the very early universe, at eras of 1 minute or less after the big bang, was highly chaotic and turbulent. This turbulence could take the form of large-scale shearing motions. In this case, the rate of expansion would tend to be accelerated. One consequence is that the density would drop more rapidly by the nucleosynthetic era. Weak interactions would go out of equilibrium at a slightly higher temperature, corresponding to an earlier epoch, than in the standard model. There would be more neutrons left over, and these would synthesize more deuterium and more helium. The shearing motions might then decay rapidly as the universe continued to expand. (Such hypothetical motions are not, of course, observed at present.) We might therefore still argue that a closed universe could contain appreciable amounts of primordial deuterium. The plausibility of the assumptions required to arrive at this conclusion are questionable, however; the simplest model is that of the standard big bang, which requires a low baryonic density, but not necessarily an open universe, because dark matter could still be present. An acceptable model would have baryons present at one-tenth the critical density, and nonbaryonic dark matter contributing the remaining 90 percent.

THE HUBBLE EXPANSION

A number of other arguments have been proposed that are equally inconclusive about the curvature of the universe. One example is the study of the Hubble expansion of the galaxies in the vicinity of our local supercluster. This system is centered on the Virgo cluster of galaxies. A glance at the distribution of galaxies on the sky reveals an immense concentration of galaxies toward the Virgo region. The local inhomogeneity is perhaps 100 million light-years in extent. The Virgo cluster of galaxies is itself a central peak in the galaxy distribution that spans about 5 million light-years. One effect of this local excess of galaxies is to decelerate slightly the overall Hubble rate of expansion. There is excess mass and therefore excess gravity, and galaxies accordingly recede at a lower velocity than they would in the absence of the local supercluster.

The amount of this deceleration turns out to be sensitive to cosmology. Consider a universe of critical density. Any small localized excess of mass will clearly affect the delicate balance between gravity and expansion and will tend to lower the recession velocities in the local region. However, in an open universe, the expansion energy would greatly exceed the gravitational energy. In this case, small enhancements in the local density would slightly increase the gravitational energy but would not appreciably slow the rate of expansion of any galaxies. Because galaxies would be expanding freely, they simply would not be affected by any gravitational force. Only a very large local inhomogeneity would produce enough force to decelerate the galaxies in a local region of an open universe.

Unfortunately, our calculations of redshifts and distances to many remote galaxies are so uncertain that we cannot yet pursue this test unambiguously. Tentative indications are that the local expansion flow of the galaxies around us is smooth and undistorted, which favors an open-universe model. However, this uniformity, quite surprisingly, has been found to break down over large scales, in excess of 15 megaparsecs, or the distance to the Virgo cluster. One clue was found in an infrared selected sample of galaxies, which has yielded a somewhat contrary view. The advantage of an infrared catalog of galaxies is that we can undertake an all-sky survey: optical samples are restricted to high galactic latitudes away from the obscuring dust clouds in the Milky Way. Preliminary studies based on a catalog of galaxies selected by the IRAS satellite indicate the presence of a weak density inhomogeneity that extends to about 200 million light-years from us. This inhomogeneity is in the same direction as the motion of the Local Group of galaxies relative to the microwave background. If mass traces light on such large scales, then the magnitude of this motion can best be explained if the universe is at critical density. Without a complete set of redshifts, however, it is difficult to ascertain the precise depth of the survey, and therefore of the inhomogeneity.

Studies of a sample of some 400 elliptical galaxies with measured redshifts that extend to a similar distance have provided a more detailed view of the velocity field on such large scales. By using an apparently universal correlation between luminosity (the determination of which depends on distance) and intrinsic velocity dispersion of the galaxy (which is independent of distance), distance to groups and clusters could be obtained to an accuracy of a few percent. Comparison of these actual distances with those inferred from the redshifts, assuming Hubble's law, led a group of astronomers (nicknamed the Seven Samurai for their challenge to conventional wisdom) in California, Arizona, Massachusetts, and Cambridge, England, to infer that large-scale flows as large as 600 kilometers per second were occurring. In particular, our local group, the Virgo supercluster, and the more remote Centaurus cluster, all appear to be rushing toward a point in space that is obscured by the Milky Way but inferred to be the center of a great mass concentration (Figure 17.4). However, these results will be confirmed only when the studied samples of galaxies are shown to provide an unbiased probe of the large-scale matter distribution in the universe.

We should be reluctant to accept such far-reaching conclusions about gross deviations from the Hubble flow, with their corresponding implications for cosmology, without unimpeachable evidence. The bulk flows, if gravitationally driven, require considerable large-scale power in the primordial fluctuation spectrum. Yet the cosmic microwave background tells us from its impressive uniformity that on very large scales (1 billion light-years or more) the universe was and is exceedingly quiescent and expanding uniformly. Whether there are large deviations at intermediate scales of 100 million light-years is still a controversial issue. For the present, we have to conclude that the samples of space studied have not been deep enough to let us bring in a definite verdict. Whether the local Hubble rate is significantly decelerated by gravity—whether the universe is open or closed—remains uncertain.

Another argument depends on a precise determination of the age of the universe. If the age can be shown to equal H_0^{-1} rather than, say, $(2/3)H_0^{-1}$ or a smaller time scale, this evidence would confirm an open universe. The best determination of the age applies stellar-evolution theory to the *Hertzprung-Russell diagram* for stars in globular star clusters (Fig. 17.5). As in Figure 4.3, the age can be inferred from the main-sequence turn-off. The best value is about 14 billion years: presumably, if our galaxy took about 1 billion years to form, the age of the universe is about 15 billion years. Unfortunately, because we cannot refine the Hubble constant to better

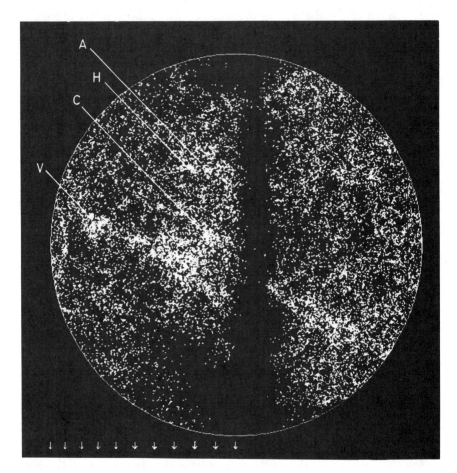

Figure 17.4 Computer plot of galaxies

This computer plot shows all galaxies in several catalogs of the hemisphere of the sky centered on the direction of the apparent large-scale streaming of the elliptical galaxies. The picture contains a hemisphere of the sky in an equal-area projection, with the zone of avoidance caused by our galaxy being the dark vertical band. The galactic center lies 37° up from the bottom. The Virgo, Centaurus, Hydra, and Antlia clusters are indicated. The great concentration of galaxies just below the Centaurus cluster, dubbed the "Great Attractor", may be responsible for the bulk of the streaming motion out to a disance of 100 million light-years.

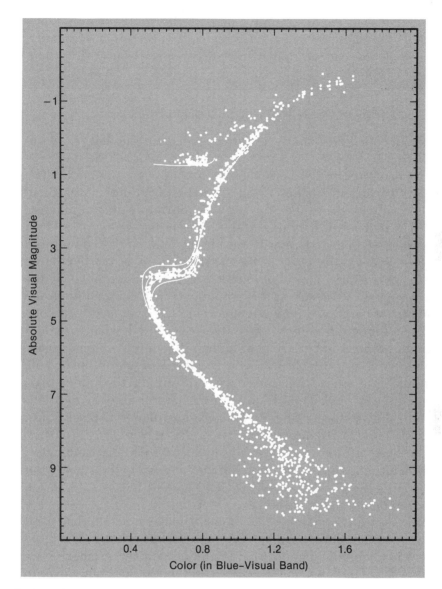

Figure 17.5   Absolute Magnitude Versus Color

A Hertzsprung-Russell diagram for the globular cluster 47 Tucanae. In this plot of absolute magnitude versus color, the hydrogen-burning main sequence is occupied by stars in the lower diagonal. The main sequence turn-off marks the location of stars that have just exhausted their hydrogen and have only about 100 million years or so of luminous life remaining. The evolutionary models (solid lines) are sensitive to this turn-off and give a satisfactory fit only if the cluster is about 14 billion years old.

than a factor of 2, we can only compare the stellar evolution age with H_0^{-1} equal to between 10 and 20 billion years. Consequently, no conclusion can be drawn from this comparison until a more reliable value of H_0 is forthcoming.

THE FUTURE OF OPEN AND CLOSED UNIVERSES

We may conclude from these various tests that the balance of observational evidence does point to an open model of the universe. As we have seen, there are weak points in every argument. The theoretical arguments described in Chapter 6 strongly favor a critical density for the universe. This was not simply due to inflation, but was more generally due to the fine tuning argument, that for the universe to deviate now toward being open or closed required it to be unnaturally close to the critical density at the beginning of the expansion, unless the universe always was at precisely the critical density. However, the proponents of a marginally closed universe seem forced into a tight corner. They are faced with one overwhelming difficulty. If the universe is closed, what form does the hidden mass take? Despite human ingenuity, which can surmount such an obstacle in theory, in the absence of observable evidence, any suggestions must lack credibility. If the hidden mass is completely unobservable, anything is permissible. Suggestions for the dark mass have ranged from rocks and planets to black holes, even to snowflakes and back issues of the *Astrophysical Journal*. Then there are the exotic weakly interacting particle candidates, of which there must exist a number about equal to the square root of the number of particle physicists. Some of these, at least, are experimentally detectable.

Although an open universe seems the favored alternative at present, an open universe has two profound drawbacks. First, the future is distinctly unappealing: in an open universe, galaxies are destined to run down, and stars are destined to burn out, never to be reborn. Gravity can no longer counteract the expansion, and gravitational forces become insignificant on the very largest scales. Space will become blacker and blacker. The voids between the galaxy clusters will deepen immeasurably. As nuclear energy supplies dwindle, matter no longer will be able to support itself against gravity in gravitationally bound systems. Galaxies, and ultimately even the great clusters, will collapse to form gigantic black holes. Eventually, all matter will become utterly cold, attaining a temperature of absolute zero. All forces will fade and disappear, until a state is reached where nothing will ever change again. Space is infinite, and a cold, black, immutable future is inevitably destined to be attained throughout space. This fate will not occur for untold billions of years, but it is nevertheless inevitable in an ever-expanding universe.

Second, there is an experimental test. Because gravity is relatively unimportant at present, in an open universe significantly larger primordial density fluctuations are required in order to form galaxies through the mechanism of gravitational instability, which we described in Chapter 10. These fluctuations leave an imprint on the microwave background radiation, and most theories predict that the microwave radiation should vary from point to point in the sky by as much as 1 part in 10,000. In a closed universe, the fluctuations are about 10 to 30 times smaller than in an open universe. Detection of such fluctuations would be an important confirmation of the gravitational-instability theory of structure formation from tiny density fluctuations. At present, however, only upper limits on the smoothness of the radiation are available (Color Plate 12). These limits are sufficiently strong, however, to rule out some models of an open universe containing only adiabatic density fluctuations at the outset. Needless to say, one can always design an open universe with a suitably chosen spectrum of fluctuations to meet this challenge. The point is that the simplest, most natural models of an open universe are testable.

A closed universe has, it must be admitted, a decidedly brighter fate than has an open universe. Galaxies may run down, but as long as a supply of intergalactic gas is available, new galaxies can form. Gravitational forces always remain important. In any region of space, the gravitational self-attraction of matter exceeds and eventually overwhelms the expansion. Radiation from the galaxies will maintain a dim flicker of light and keep the universe warm. After attaining a finite size, the universe will eventually begin to recollapse. As the universe recollapses, everything within it becomes compressed. As radiation is compressed, it is heated. Eventually, galaxies will start to collide and disrupt one another. Stars will also collide. As the collapse unremittingly continues, in what has been aptly labeled the big squeeze, all structure will be destroyed. The universe will collapse into a dense, hot soup of compressed matter. This state of affairs will continue, for all we can tell, without limit, until the universe is compressed into an infinitely dense, infinitely small region. If the universe is closed, the big squeeze will occur within about 20 billion years.

What happens afterward is beyond the scope of our present knowledge of physics. General considerations about the structure of matter have been applied to this model, and they indicate that a future singularity—a return to the infinitely dense state of the big bang—awaits us. In a closed universe, there is both a past and a future singularity, when the density becomes arbitrarily large. Since all our concepts of physics become invalid under such extreme conditions, we can only speculate on the possible outcome.

An attractive conjecture is that a closed universe will bounce and reexpand. If this happens, the future of the universe will be to repeat the entire cycle of the big bang and again form galaxies and stars. There is one addition: the radiation produced by stars in the previous cycle of the big bang would still be present. We can postulate a whole series of successive cycles of expansion and collapse, each producing more and more radiation as a consequence of galaxy formation and star formation. In the high-density phase of the collapse, this radiation is absorbed and reemitted many times, and it loses all trace of any characteristic wavelengths or spectral features. The radiation ends up as blackbody radiation. Because we measure a specific amount of cosmic blackbody radiation in the background radiation, we infer that a closed universe can have undergone only a finite number of repeated bounces. Otherwise, too much radiation would have been produced during earlier cycles of the expansion.

The number of previous bounces that are permitted by this constraint is not large, being just the ratio of the blackbody radiation intensity to the average flux of starlight from remote galaxies. Compared with light from the Milky Way, the light from distant galaxies makes only a feeble contribution to the night sky brightness. There are a vast number of distant galaxies, however. The emission from distant galaxies actually covers the entire sky, but it amounts to only about 1 percent of the Milky Way brightness. This amount is not enough to make any detectable light, even on the darkest of nights. It happens that the intensity of the cosmic blackbody radiation is approximately equal to that of the light from the Milky Way. The background radiation is, of course, concentrated at radio wavelengths and is invisible to the human eye. If we assume that each previous cycle of the expansion results in the generation of as much light as there is in the present universe, we infer that there may have been about 100 previous expansion and collapse cycles of the universe. A much larger number (for example, an infinite sequence) can be excluded on the basis of this argument. However, the skeptic may feel justified in challenging the logic of this conclusion, resting as it does on the assumption that a bounce can actually occur and, moreover, will conserve the entropy generated in the previous cycle.

It would seem that even a bouncing closed universe probably cannot have an infinite lifetime, which removes one of the principal aesthetic appeals of a closed universe. The beginning of time is unavoidable. Many cosmologists also philosophically prefer a closed universe as an alternative to infinite space. However, such highly subjective issues should play no role in science. The case for an open universe ultimately rests on observational data alone.

Astronomy in the 1990s should take us much farther in our study of observational cosmology. The greatest potential advance comes from a new generation of telescopes. There will be a major astronomical observatory in space, centered on the space telescope (Figure 17.6). This telescope is not exceptionally large by terrestrial standards (its aperture will be 2.4 meters) but it should be much more sensitive than the largest telescopes we now rely on, because it will be observing in a dark sky with no blurring atmosphere. Study of many remote galaxies at a redshift of 1 or 2 will be possible, and spectroscopy with this telescope should tell us much about the evolution of young galaxies. Once we understand (or can at least model) galactic evolution, there will no longer be any obstacle to using distant galaxies to measure the curvature of space directly.

Once distant galaxies can be routinely detected, a new type of curvature measurement becomes feasible that bypasses many of the uncertainties in galactic evolution which make galaxies such poor standard candles. Suppose a supernova is detected in a distant galaxy; near maximum luminosity, the galaxy doubles in brightness, so this detection should be feasible. Measurement of the su-

Figure 17.6   The Space Telescope
A 2.4-meter telescope, due to be launched by the space shuttle in 1989, will become a more or less permanent observatory in space.

pernova spectrum yields the velocity of the expanding shell; we then infer the radius of the supernova shell from the observed rate of decay in light. But the measured flux of radiation can be used to infer a radius if the distance is known, since at a given temperature, the larger the radius is, the greater is the luminosity of a shell of glowing gas. Thus, we can infer the actual distance and therefore the curvature of the universe. Although this test is sensitive to uncertainties in supernova atmosphere models, such problems are not insuperable, and a direct curvature measurment should eventually be possible.

Astronomers are also planning to construct much larger telescopes, with apertures of 10 meters or more, on the earth. An optical telescope of this size requires innovative engineering technology; one such project, the Keck 10-meter telescope being built by the University of California and the California Institute of Technology, will not be a single structure but will consist of 36 contiguous hexagonal mirror segmental elements that are intricately aligned by sophisticated electronic gadgetry. Sensors continuously monitor the shape of the overall mirror, and actuators on the back of each segment control its orientation, so as to be equivalent to a 10-meter–diameter primary mirror. The light-collecting power of a 10-meter telescope should enable us to study galaxies that are truly at the edge of the observable universe. The morphology of these galaxies will tell us much about how elliptical and spiral galaxies acquire their characteristic shapes. Another notable space project will be a major new x-ray telescope in space. This telescope should help to resolve the origin of the diffuse x-ray background, which is of crucial importance to the issue of the mean density of matter in the universe. Still another planned project is a large infrared telescope in space. Astronomers hope to use it to detect distant galaxies in the process of formation, when protogalaxies are still shrouded by a dense cloud of dust. We can confidently hope that the evidence will accumulate to allow a definite verdict within the next decade. For now, the choice of an open universe remains a gamble with marginally favorable odds.

· 18 ·

ALTERNATIVES TO
THE BIG BANG

We shall not cease from exploration
And the end of all our exploring
Will be to arrive where we started
And know the place for the first time.
—*T. S. Eliot*

Big bang cosmology lays no claim to being the only possible description of the universe. It does, however, provide a satisfactory framework for explaining much of what astronomers observe. It also has provided the basis for predictions that have subsequently been borne out by experiment and observation. Most cosmologists today accept big bang cosmology as a description of the accessible universe to era as far back as nucleosynthesis, when the universe was about one minute old.

Before that moment, however, big bang cosmology becomes removed from the realm of conventional physics. Moreover, it cannot be said, even by enthusiastic proponents, that big bang cosmology explains all we need to know about the universe. The big bang theory has not yet resolved three fundamental issues: what happened prior to the initial instant, the nature of the singularity itself, and the origin of the galaxies. St. Augustine indicated one approach to the first of these issues with his response that, prior to creation, there existed a hell for those who raised such awkward questions.

A modern speculation might be that the present expansion is one cycle of many that a closed universe has undergone. If this is so, one must then face the issue of how one collapse phase becomes a subsequent expansion phase. An even more modern conjecture is that the observed universe originated as a quantum fluctuation out of nothing, and came into existence at the Planck instant. This neatly sidesteps two of our fundamental issues, but raises others. Why is the universe so isotropic, and was it always so in the past? And how did the initial fluctuations arise from which galaxies formed? Or have such fluctuations always been present? To avoid a confrontation with these awesome questions, which are unsolvable in terms of our present physics but should become feasible once quantum cosmology is fully developed, alternative descriptions of the universe have been sought. In this chapter, we shall examine some alternative theories to the big bang and explore ways of resolving the remaining fundamental issues.

TIRED-LIGHT COSMOLOGIES

The expansion of the universe, as both an observational and a theoretical concept, has proved to be the most important contribution of the twentieth century to cosmology. This concept has been the basis for many other important ideas and discoveries. Yet the nagging thought lingers: Could we be completely wrong? Is it possible that we are living in a static universe?

In principle, light could be redshifted by effects other than the Doppler shift of light from the recession velocity of distant galaxies. One such effect could be *tired light*—quanta of light could lose energy during their traversal of space from remote galaxies. This decrease in photon energy would result in an increase in wavelength, or reddening, that would be proportional to the distance traversed. Scattering by intergalactic dust particles could also cause a reddening of the light from remote galaxies. However, we now measure redshifts at wavelengths that range from the optical (5000 Ångstroms) to the radio (21 centimeters). Precise agreement is found when the redshift of a galaxy is observed at both optical and radio wavelengths. Dust particles are generally transparent to radio waves, unless they are about the same size as the wavelength. Space could hardly contain enough boulders to scatter radio waves from a distant galaxy. Dust particles alone could not produce the required reddening effect in intergalactic space. Also, scattering would broaden the spectral lines, which is contrary to what is observed.

Although we have no evidence whatsoever that light can lose energy in traversing space, it could be argued that over the vastness of intergalactic space, our terrestrial laws of physics may be wholly inapplicable. The only resource seems to be to test the tired-light-cosmology, which makes decidedly different predictions for the results of the various cosmological tests that we discussed in the last chapter. In particular, the predicted energy flux from a distant galaxy would decrease only by an amount proportional to the redshift factor. The rate at which photons arrive would be diminished by this amount because a tired-light universe is static, and the rate at which the energy is being radiated would be the same as the rate at which it is observed. In the standard big bang models, there is an additional decrease of the energy flux by a second redshift factor, because the expansion produces a time-dilation effect: time is measured at a slower rate in our frame relative to that of the distant galaxy, resulting in a reduction in the measured energy flux. In principle, this predicted difference leads to an observable test, although such tests are not yet completely conclusive.

However, tired-light cosmology can be ruled out without recourse to such a test, to most cosmologists' satisfaction. We cannot account for the origin of the cosmic microwave background or for the light-element abundances in a cosmology that dispenses with a hot big bang. In addition, it seems desirable to dismiss the tired-light cosmology, if only because of its ad hoc nature. Nevertheless, it is probably premature to assume that the cosmological tests discussed in Chapter 17, or any other astronomical observations, enable us at the present time to eliminate definitively the tired light cosmology as an alternative to the big bang.

ARPIAN OBJECTS

Tired-light cosmologies are unsatisfactory because they invoke a new law of physics, but other attempts to dispense with the apparent expansion of the universe are based on observations. The American astronomer Halton Arp, who has specialized in taking superb photographs of unusual galaxies and quasars with the 200-inch telescope on Mount Palomar, has proposed a phenomenological origin for large cosmological redshifts. He has discovered a number of systems linked by exceedingly faint jets of nebulosity to objects with a totally different redshift. Arp presents these objects as evidence that the redshift, or at least a considerable fraction of it, is to be associated with an origin intrinsic to the objects being studied rather than with distance.

Arp's peculiar objects and associations have failed to convince most astronomers, because the statistical basis of his studies has never been clearly presented. How carefully must we search before finding something unusual? Even in the apparently anomalous examples, we cannot tell whether the faint nebulosity is a widespread phenomenon around one of the objects and is simply viewed in projection against the other object or whether it is a unique physical link in space connecting objects of widely different redshift. Some of Arp's bridges are almost certainly due to chance projection in the sky of objects at disparate distances. Cursory inspection of photographs of remote galaxies often reveals very bright stars in our own Milky Way that are viewed in projection against a background galaxy. Even if we view such a star at the tip of a spiral arm in the distant galaxy, no one would argue that the star was recently ejected from the distant galaxy. We know such projections occasionally occur by chance. For analogous reasons, the majority of astronomers take a skeptical view of Arp's interpretations of his photographs. With the launch of a large space telescope, it may be possible to obtain spectra that will finally decide the issue. If a continuous range in redshift can be measured along a jet of gas linking a low redshift galaxy to a high redshift quasar, Arp will doubtless be vindicated.

STEADY STATE COSMOLOGY

We have previously examined the principal motivation, avoidance of a possible time-scale difficulty, for the steady state universe. The infinite duration of time was its most appealing feature. Even the notion of creation was used to advantage: the failure to detect a large flux of gamma rays (expected when newly created particles occasionally annihilated) led to the proposal that creation preferentially occurred only in dense galactic nuclei and quasars. This proposal initially met with some enthusiasm, for it appeared to solve two problems: continuous creation was revived, and an ample energy source was provided for the most energetic objects in the universe.

However, three shortcomings eventually led to the demise of steady state cosmology. First, astronomers have discovered a large increase in the number of faint radio sources as the limiting flux level is systematically decreased. The number they estimate is well above that expected for a uniform source distribution in euclidean space, which suggests that strong evolutionary effects are occurring at great distances. The notion of an excess number of distant faint sources did not find universal acceptance. For several years, Hoyle was able to argue that an equally plausible hypothesis was a deficit of nearby bright sources. Now, however, it seems clear with measured redshifts that the highly redshifted sources, most notably the radio galaxies and the quasars, reveal strong evolutionary effects. Equal volumes of space contain progressively more quasars and powerful radio galaxies at greater distances. Only by disputing the interpretation of quasar redshifts as a cosmological distance indicator can this conclusion be avoided. With the continuing discoveries of absorption redshifts in quasars that are associated with intervening galaxies, this position seems increasingly difficult to maintain. Radio galaxies are essentially stellar systems, and any challenge to the cosmological nature of their redshifts is even more dubious. Second, the redshift-magnitude test and other cosmological tests have probably eliminated steady state cosmology as a viable model strictly on the issue of the predicted curvature of the universe. Finally, the discovery of the cosmic microwave background radiation and confirmation of the blackbody nature of its spectrum have presented overwhelming evidence for an initial hot and dense phase of the universe. Cosmologies that do not include the big bang have not produced any plausible alternative interpretation of the background radiation.

GALAXIES AND ANTIGALAXIES

Another attempt to dispense with the big bang was made by the Swedish physicists Hannes Alfvén and Oskar Klein. Their basic motivation was the idea that a symmetry between matter and antimatter should be a fundamental assumption in any cosmology. In their model, the universe began as a giant spherical metagalaxy of diffuse, slowly contracting gas, which contained equal amounts of matter and antimatter. When the density became sufficiently high, matter and antimatter annihilated (Figure 18.1). Enormous amounts of radiation resulted from the annihilation. This radiation halted and eventually reversed the collapse of the remaining matter. The universe began to reexpand, and galaxies eventually condensed. In this theory, comparable numbers of galaxies and antigalaxies are expected to exist in the observable universe.

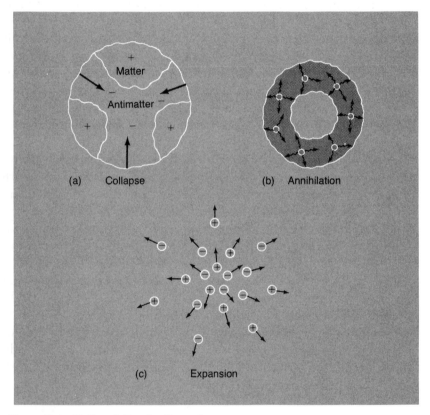

Figure 18.1   Matter–Antimatter Cosmology

A huge, spherical metagalaxy, consisting of equal amounts of matter and antimatter, is initially very diffuse and collapses (a). The density reaches a point at which annihilation can commence (b); the pressure of the annihilation radiation is assumed to reverse the collapse and initiate a general expansion (c). At the same time, clouds of matter and antimatter collect; eventually, these form galaxies and antigalaxies.

Many objections have been raised to the *Alfvén-Klein cosmology*. First, if the Alfvén-Klein cosmology is correct, some mechanism must separate regions of matter and antimatter. Because even intergalactic space contains some matter, galaxies could not be completely separated from antigalaxies. Alfvén has proposed a possible way of separating regions of matter and antimatter, but most other astrophysicists remain skeptical about his theory. Second, there must be many regions where annihilation of matter and antimatter would occur. One result of annihilation is to produce copious amounts of high-energy gamma rays, which should be observable by space experiments. Astronomers have indeed measured a diffuse cosmic background of gamma rays, but the flux is relatively low. The implication is that very little annihilation can be occurring at present.

Perhaps the most significant objection to this cosmology stems from the presence of the cosmic background radiation. Like the steady state theory, the Alfvén-Klein cosmology must attribute a noncosmological origin to this radiation. Moreover, the high degree of isotropy of the radiation would imply that the Milky Way galaxy must be located very close to the center of the metagalaxy. We could not be off center by more than 0.1 percent, or an excessive asymmetry in the background radiation would be detected. This strong conflict with the copernican cosmological principle, that our location in the universe should not be preferred in any sense, makes the Alfvén-Klein version of matter-antimatter cosmology seem extremely dubious.

It is also possible, of course, to develop a big bang cosmology in which the notion of symmetry of matter and antimatter is incorporated at the outset. If the symmetry is unbroken, there are similar difficulties in avoiding detectable gamma-ray emission by annihilation of matter and antimatter. Perhaps the key problem here is avoiding almost total annihilation of the entire universe during the high density phase of the expansion. Accomplishment of this goal remains controversial.

VARIABLE GRAVITATION

Gravitation is the weakest force in the universe. The ratio of the gravitational force to the electric force between an electron and a proton has the infinitesimal value of approximately 10^{-40}. Why should such a small number appear in the laws of physics relating these two fundamental forces? By a most intriguing coincidence, we can construct a number of similar magnitude from the ratio of two time scales. One number is the unit of time for light to cross a distance equal to the classical radius of the proton. The other number is the Hubble time scale. This dimensionless ratio is also approximately 10^{-40}. Thus, the ratio of two fundamental forces and the ratio of atomic time scales to cosmic time scales are of the same infinitesimal order of magnitude. Can this be merely a coincidence?

The physicist Paul Dirac, one of the founders of modern quantum theory, hypothesized in 1937 that such a coincidence is so remarkable that it should be accepted as one of the laws of nature. Dirac argued, however, that because the Hubble constant varies with the particular era of the universe, the gravitational constant also had to vary with time to maintain the ratio. Newton's constant G should therefore be very large at early eras of the universe. A secular decrease in G over billions of years would have interesting implications for the evolution of the solar system; for example, the sun would have been much more luminous when the earth was formed. The best modern experimental limit on the rate of change of G comes from interplanetary radar ranging, which suffices to rule out Dirac's original theory.

Others have since reformulated Dirac's hypothesis and have constructed new theories of gravitation, in which G varies with time, to replace Einstein's theory of general relativity. Carl Brans and Robert Dicke made the most notable of these attempts in 1961. The *Brans-Dicke theory* makes a number of interesting predictions, some of which are testable. The most dramatic effect has come from observing a pulsar in orbit around a companion star. This so-called binary pulsar offers a unique laboratory for testing gravitational theory, for it is in effect a highly accurate clock: the periodic radio pulses from the pulsar are produced by a freely falling clock in the strong gravitational field of its companion. The most recent observations of this system appear to vindicate Einstein's theory and to eliminate the Brans-Dicke theory from serious consideration.

A SHRINKING UNIVERSE

It is perhaps appropriate that we devote a section to a cosmology reminiscent of the adventures of Alice in the universe created by Lewis Carroll. Fred Hoyle and Jayant Narlikar have argued that the expansion of the universe is indistinguishable from the alternative hypothesis that the atomic dimensions are actually decreasing with time. According to this view, space is unchanging. Galaxies are not flying apart from one another, but everything within the galaxies, including ourselves, is shrinking. This shrinkage can be understood in basic physical terms if the masses of all the elementary particles are assumed to increase over a cosmological time scale. The mass of an atom increases, but the electric charges do not change. Consequently the electrons must orbit more and more tightly around the atomic nucleus. This situation would correspond to one in which electrons attain a higher energy state, requiring more energy to dislodge them; conversely, more energy would be released when an electron is captured in an inner orbit. Radiation emitted by such an atom would be more energetic and have a shorter wavelength than radiation from a less tightly bound atom. In a remote galaxy, the atoms that emitted the light would have been larger than the galaxies' atoms would be now. The wavelength of this light would be longer, or redder, than light produced by these atoms in a terrestrial laboratory. Thus, the cosmological redshift is explained in terms of the shrinking of atoms and consequent reddening of light.

The extreme ingenuity of this theory makes it susceptible to the accusation of being merely a clever mathematical model that bears little relation to reality. The hypothesis it rests on is superficially simple and elegant but lacks any basis in physical experiment. Predictions are the ultimate test of a new theory, and the *variable-mass theory* fails to surmount this hurdle. However, the Hoyle-Narlikar theory does possess one merit that deserves serious attention. It illustrates a possible way around one of the fundamental issues of the big bang theory—the mystery of the initial singularity. The proposed solution is actually a general property of the big bang models. To understand its significance, we must probe more deeply into the nature of the initial singularity.

AVOIDING THE SINGULARITY

The pioneers of the big bang theory were not greatly concerned about the singularity in space-time that was apparently required by the Friedmann equation. The open models possess a singularity in the finite past, and the closed model has both a past and a future singularity (Figure 18.2). To such cosmologists as Richard Chase Tolman, who was one of the principal developers of the hot big bang model, the predicted singularities were a manifestation of the inadequacy of the relevant equations and inapplicability of the simplest physics. In a more complex and subtle physical theory, the singularities would vanish. One common analogy was that of a bouncing ball: the equation of motion of the ball must be modified at the moment of impact by consideration of the surface irregularities of the ball and the floor. Tolman argued that we also might expect a realistic universe to bounce: no singularities would occur, and there could be a sequence of alternate cycles of compression and expansion. We have seen previously that production of radiation by galaxies and stars in each cycle imposes a limit to the number of possible cycles; nevertheless, the vision of a closed universe that has undergone many cycles of expansion and contraction has seemed attractive to many.

Unfortunately, we now know that this idea is ill conceived. It is premature to state flatly that we cannot inhabit a cyclical universe, but we can say definitely that subject to certain reasonable assumptions, a past singularity is unavoidable. This realization has come from an important theorem proved by the British relativity theorists Stephen Hawking and Roger Penrose.

In fact, the past singularity does have an interesting property in the standard big bang model. In principle, one can pass through it, despite the fact that the density becomes infinite. The singularity can be transformed away, as is apparent from the space-time diagram (Figure 18.2). However, this particular property of the singularity only applies to the idealized Friedmann universe. The presence of any inhomogeneities guarantees that the singularity cannot be transformed away. This result is fatal to the variable-mass cosmological model. Moreover, even if the universe initially passed through a singularity, a subsequent collapse phase would be highly inhomogeneous. We reluctantly conclude that a future

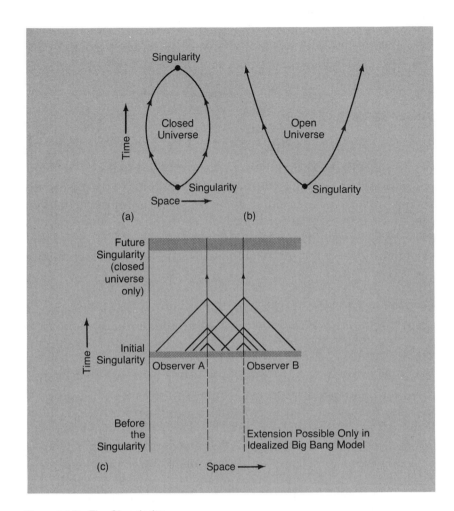

Figure 18.2 The Singularity

The closed universe (*a*) has singularities in the past and in the future; the open universe (*b*) only has a past singularity. Only if the universe was initially perfectly regular (that is, highly homogeneous and isotropic), then by defining new length scales, the singularity can be stretched out as in (*c*), and hypothetical observers can pass through time zero. In general, if the universe is not so highly idealized, the world-lines of all observers begin abruptly at the past singularity and, in the closed universe, end abruptly at the future singularity. The existence of a singularity that abruptly terminates the world-lines of observers (either in the past or in the future) is a general property of the universe and is not specific to the big bang theory.

singularity is inevitable in a closed universe: hypothetical observers cannot pass through it, and so the universe probably cannot be cyclical, unless we introduce new physical laws that would significantly modify our theory of gravitation.

CHAOS VERSUS ORDER

Why is the universe isotropic? More specifically, why does the cosmic blackbody radiation temperature equal 3 degrees kelvin when observed in opposite directions? If these regions initially were at different temperatures, it is necessary for radiation to travel from one region to another in order to equalize the temperatures, as is now observed. However, these regions cannot have had time to communicate over the age of the universe, because the radiation is just reaching us now.

As we saw earlier, the postulate of primordial chaos might alleviate this difficulty by allowing different expansion rates in different directions and thereby opening up the horizons of the early universe. Unfortunately, we cannot describe chaotic initial conditions well enough to evaluate this hypothesis conclusively (Figure 18.3). Another approach to the question of initial chaos comes from consideration of *entropy*, which is a measure of disorder. A fundamental law of thermodynamics states that entropy can never decrease. Work by Stephen Hawking and others has shown that entropy can also be associated with the evaporation of black holes. A chaotic early universe would form black holes prolifically, the smallest of which would evaporate and produce prodigious amounts of entropy. A direct measure of the entropy of the universe is obtained from the background radiation, which incorporates all products of dissipation and black-hole evaporation in the early universe. The fact that this radiation is finite directly constrains the allowable degree of primordial chaos: the universe cannot have been highly chaotic at very early eras; otherwise the measured entropy, which is the ratio of the cosmic blackbody-radiation photon number to the number of protons, would exceed the observed value, based on any reasonable scheme for baryon genesis (as we saw in Chapter 6). We already saw how the success of primordial nucleosynthesis probes the universe at an age of only a few minutes

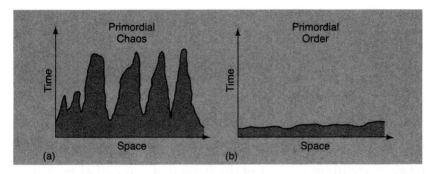

Figure 18.3 Order or Chaos?
It is attractive to speculate that the early universe was chaotic (*a*) and might subsequently have developed into the ordered state that we observe at large scales. Primordial chaos could conceivably lead to sufficient mixing to account for the uniformity of the cosmic blackbody radiation. Alternatively, we can argue that the early universe was very (but not completely) regular and ordered (*b*). This hypothesis also accounts for the uniformity of the cosmic blackbody radiation and is more in accord with the thermodynamic concept of increasing entropy (or degree of disorder). Since the chaos hypothesis is unproven (and specifically in the mixmaster universe is known not to work), many cosmologists prefer an orderly beginning to the universe.

and tells us that the expansion then must have been quite uniform. Indeed, the universe could not have been even mildly chaotic at epochs any later than 1 year; otherwise, the cosmic microwave radiation would reveal gross distortions from a blackbody spectrum. Production of blackbody radiation is efficient only during the first year or so of the big bang.

These measures make a strong argument against a chaotic early universe. The requirement that entropy should increase also suggests that the early universe was highly regular. This strong limitation on the possible early state of the universe also helps us to understand the isotropy of the cosmic blackbody radiation, although its precise degree of isotropy seems indeterminate from any such argument.

The isotropy of the universe is explained by inflationary cosmology. Prior to inflation, excessive chaos would be a disaster, for the universe could recollapse back to its singular state. If one considers quantum cosmology, there may be restrictions on what could emerge prior to the subsequent initiation of inflation. Superstrings remain to be compactified to low-dimensional space-time, and there is no guarantee that a successful cosmology will emerge.

Nevertheless inflation, in its several guises (chaotic, quantum, GUT), is the great hope for resolving the fundamental issues unresolved in the big bang theory.

A COLD UNIVERSE

If entropy is readily produced in a moderately chaotic universe, then the very early universe might have been cold. The blackbody radiation could be a result of dissipation of primordial irregularities, or it could even have formed at the later era of galaxy formation. One important advantage of a cold universe is that it can probably decide one of the fundamental issues of the standard big bang model—the origin of the galaxies. Cold matter is inherently unstable. At an early era in a cold universe, hydrogen is solid, and it may undergo a phase transition that results in spontaneous generation of asymmetrical fissures as the universe expands. Some cosmologists have speculated that the seed fluctuations for galaxy formation could be produced in this way, even with an initially homogeneous universe.

The principal objection to a cold universe is its failure to account for the cosmic blackbody radiation. It has been argued that radiation from an early generation of stars could be absorbed by dust grains and ultimately reradiated as blackbody radiation. However, it is difficult to reconcile such an origin with both the high degree of isotropy and the blackbody spectrum at 3 degrees kelvin.

FUTURE TESTS

None of the alternative cosmologies we have discussed seem very promising, and we conclude that the big bang theory is the best cosmology we can provide. There is a crucial test, however, of at least one of the major assumptions that is yet to be performed. We have seen how density fluctuations with amplitudes of roughly 1 percent over galactic or galaxy-cluster mass scales were present at the decoupling era. These density fluctuations have formed galaxies by the present era. The imprint of these fluctuations must be left on the cosmic blackbody radiation at decoupling. By searching for small-scale angular variations, radio astronomers can test the theory of galaxy formation. The techniques are very difficult, however, and a sensitivity of 0.001 percent or better must be attained. Such measurements should be forthcoming within the next few years.

The birth of galaxies must have been a very energetic event. Because the redshift hides newly born galaxies from us, we may ultimately expect to detect galaxies at birth at redshifts of 10 or more by extending our observations to the infrared. In an open universe, galaxies must have formed at high redshift, because gravity is relatively ineffective today at driving fluctuation growth, whereas in a closed universe, galaxy formation proceeds unimpeded up to the present era. The search for young galaxies is accordingly fundamental to developing an improved model of the big bang.

At present, observations of galaxies at a redshift below 1 or 2 have yielded tantalizing indications that some galaxies may be evolving. Nearby galaxies are mature and settled in their characteristics, but some distant galaxies seem to be excessively blue, which is usually a sign of active star formation. Evolution becomes significant at a redshift of about 1, and many galaxies in this redshift range should become accessible with the new generation of very large ground-based telescopes that are being constructed or planned. Galaxies are the prime building blocks of big bang theory. Having painstakingly studied mature galaxies and had tantalizing glimpses into their youth, observational astronomers are about to probe the birth of galaxies and even their conception. These achievements should help to decide the remaining issues concerning galaxy formation and conceivably even resolve the issue of whether the universe is open or closed.

A NEW PHYSICS

We can also reasonably hope that a fundamental new theory of physics will eventually enable us to probe the third fundamental issue of the big bang, the conditions at the initial singularity. This new physics has been given various names, including quantum gravity, supergravity, and superstring theory. All these theories aim to unify gravity into the framework of the other fundamental interactions, nuclear and electromagnetic. Recent developments in elementary particle physics indicate that we may be on the threshold of such an advance. One of the great successes of elementary particle physics has been the unification of electromagnetic and weak nuclear interactions into a single theoretical framework. As we saw in chapter 6, recent attempts have also been made to incorporate the strong interactions (which hold atomic nuclei together) into a unified theory. One implication of this grand unified theory is that during the first 10^{-43} second of the universe, the temperature was so high that supermassive particles (with mass $\sim 10^{-6}$ gram) existed during this phase, in which the strong, weak, and electromagnetic interactions played an equal role. The universe was symmetric initially, when these particles were in equilibrium with the radiation and consisted equally of matter and antimatter. As the universe expanded, the temperature declined, and shortly after the singularity, the supermassive particles decayed. Recall that according to particle theory, the decays occurred asymmetrically, producing a slight excess of matter over antimatter. Although we have seen that the precise amount of the excess is not well determined by current theory, the fascinating idea emerged that the net asymmetry of the early universe—by 1 part in 10^8 of matter over antimatter—can perhaps eventually be understood by a more sophisticated theory of elementary particle physics.

Other theoretical developments have led to some understanding of how matter behaves in the immensely strong gravitational field surrounding a black hole, where spontaneous creation of particles and antiparticles can occur. At the high energies attained in such an extreme environment, one has some prospect of being able to experimentally test some of the predictions of the grand unification theory. Of course, this requires discovery of a miniblack hole of the appropriate mass to just be evaporating today. Such objects may exist; if one adopts the philosophy that whatever can exist, does exist somewhere in the universe, then one has reason for optimism. Of course, others would argue that the degree of fine-tuning required to result in the requisite abundance of miniblack holes is so extreme as to make their surival highly unlikely.

The ultimate goal of particle-physics cosmology is to include, in a unified fashion, all four fundamental forces: gravity, electromagnetism, and the weak and strong nuclear interactions. We have seen in Chapter 6 how the bizarre objects called superstrings, in either 10 or 26 dimensions of space-time, promise some prospect of achieving this goal at the Planck time, 10^{-43} second after the big bang. Much work remains to be done, however, to relate these objects to our four-dimensional universe at energies well below the Planck scale. The era of creation is where quantum physics and gravitation merge. Quantum gravity comes of age with superstrings, which potentially provide a theory of "everything." We have not yet accomplished coupling the high-energy world of the superstring, where the Planck scale of 10^{19} billion electron volts is the only characteristic energy, to the low-energy universe that we inhabit. The big bang may well provide this elusive connection. We hope that it will arise naturally and uniquely, perhaps driven by a spontaneous fluctuation at the quantum gravity era.

Such developments provide substantial grounds for hope that we may eventually be able to understand the era of creation near the singularity, where similar physical conditions are attained. For now, however, we must reluctantly admit that big bang theory is not complete: it lacks a beginning, and we cannot yet confidently predict its ending. If a better theory of the universe is forthcoming, there seems little doubt that it will incorporate the big bang theory as an appropriate description of the observable universe. Perhaps a new theory will encompass the big bang in the same way that Einstein's theory of gravitation encompassed and generalized the concepts of newtonian gravitation. Although the ultimate theory of the universe is still beyond our vision, we can feel fairly confident that we have at least seen its form emerging.

MATHEMATICAL NOTES

CHAPTER 3

1. Parallax and Parsecs

Small sectors of a circle can be approximated by straight lines. Angles are measured not in degrees but in radians. A circle of radius r has a circumference equal to $2\pi r$ and is said to subtend an angle of 2π radians all around the circle. Thus, 2π radians equals 360 degrees. Moreover, if D is the sector length of part of a circle, r is the radius, and p is the angle (in radians) subtended at the center of the circle by the sector, then

$$D = rp$$

This relation is applied to the measurement of stellar parallaxes by taking D equal to the mean radius of the earth's orbit around the sun. This distance, 1 *astronomical unit*, is equal to 1.5×10^{13} cm. Since D is known and p is measured for each star, we can then determine the star's distance by

$$r = D/p$$

In practice, measurements are performed six months apart, so that the baseline D is equal to the diameter of the earth's orbit, and the parallax is defined to be equal to half the corresponding angular shift of the star between the two observations. A parallax of 1 second of arc corresponds to a distance of 1 parsec. One radian $= 2.1 \times 10^5$ arc-seconds, and so 1 parsec $= (2.1 \times 10^5) \times (1.5 \times 10^{13}) = 3.1 \times 10^{18}$ cm.

2. The Doppler Shift

Consider a star emitting light waves at a rate of N waves per second. We say that N is the frequency of the radiation. If the star is moving away from us at a velocity v, then in 1 second the star will have moved a distance v.

The interval between two successive waves emitted by the star is $1/N$ second, and, during this time, the star has moved a distance v/N centimeters. Therefore, the waves have farther to go, and the observer, who is stationary, measures a longer time interval.

$$N^{-1} + \left(\frac{v}{c}\right) N^{-1} \text{ seconds}$$

where c is the velocity of light (3×10^5 km s^{-1}) between the two successive wave crests. In other words, the observer finds a lower frequency for the light than the frequency of emission at the star. The ratio of the frequency measured by the observer to that of the star is

$$\frac{N_{observed}}{N_{emitted}} = \frac{1}{1 + v/c}$$

We can equally well use wavelength (λ), the separation of successive wave crests, given by

$$\lambda = c/N.$$

The corresponding ratio is

$$\frac{\lambda_{observed}}{\lambda_{emitted}} = 1 + \frac{v}{c}$$

This formula indicates that recession results in an increase of wavelength, or a redshift. If the star is moving toward us, similar formulas apply, provided that v is everywhere replaced by $-v$. Approach of the star to us therefore results in an increase in frequency, which is equivalent to a reduction of wavelength, or a blueshift of the light from the star.

We define redshift z by

$$z = (\lambda_{observed} - \lambda_{emitted})/\lambda_{emitted} = v/c$$

This is true only for velocities that are small relative to the speed of light. When v approaches c, we must make use of a special relativistic formula,

$$1 + z = \frac{1 + v/c}{(1 - v^2/c^2)^{1/2}}$$

This formula shows that the redshift can become arbitrarily large as v approaches c.

3. Magnitude

The unit of *magnitude* is a logarithmic measure of stellar brightness. A reduction of 1 magnitude corresponds to an increase in brightness by a factor of 2.5. The brightness of a star in the sky is its *apparent magnitude, m.* Therefore,

$$m_1 - m_2 = 2.5 \log_{10}(\text{brightness of star 2} - \text{brightness of star 1}).$$

The brightest stars are about first magnitude. We can see stars down to about sixth magnitude with the naked eye. The faintest galaxies detectable on a deep CCD image are about 27th magnitude in the red with a 4-meter telescope.

Absolute magnitude (*M*) is defined as the magnitude a star would have at a distance of 10 parsecs from the earth. Since the same distance is always used, absolute magnitude is a measure of intrinsic luminosity. The absolute magnitude of the sun is +4.8.

Magnitudes can be positive or negative—the more negative the magnitude, the brighter is the object. The absolute magnitude of the Milky Way galaxy is −20; the absolute magnitude of a giant elliptical galaxy is −22; and the absolute magnitude of the most luminous quasar is about −27. Since the light received from a star decreases in intensity with the square of its distance, we can replace $\log_{10}(\text{brightness})$ in the formula for apparent magnitude by −2 $\log_{10}(\text{distance})$:

$$m_1 - m_2 = 5[\log(\text{distance to star 1}) - \log_{10}(\text{distance to star 2})]$$

The apparent magnitude (*m*) and absolute magnitude (*M*) of a given star are related by

$$m - M = 5 \log_{10} r - 5$$

where the distance to the star is expressed in parsecs. This formula reduces to $m = M$, if $r = 10$ parsecs (since $\log_{10} 10 = 1$).

CHAPTER 4

4. Blackbody Radiation

The spectrum, or wavelength distribution, of blackbody radiation per unit of volume is given by the Planck distribution, which expresses the energy density of radiation:

$$u_\lambda = \frac{8\pi hc}{\lambda^5} [e^{(hc/kT\lambda)} - 1]^{-1}$$

between a narrow range of wavelengths centered on λ. In this formula, T is the radiation temperature in degrees kelvin (a scale equivalent to the centigrade scale but based on absolute zero at -273 degrees centigrade), c is the speed of light, h is Planck's constant (equal to 6.625×10^{-27} erg sec), k is Boltzmann's constant (1.38×10^{-16} erg per degree kelvin), and e is a numerical constant equal to 2.718. At wavelengths long compared with hc/kT, the Planck distribution reduces to a simpler expression,

$$u_\lambda = \frac{8\pi kT}{\lambda^4}$$

which is known as the *Rayleigh-Jeans limit*. It is customary to introduce the frequency (ν) of the radiation, defined by $\nu = c\lambda^{-1}$. The Rayleigh-Jeans law is now expressed in terms of the energy density in a narrow frequency interval centered on ν and is proportional to $kT\nu^2$. This is usually a very good approximation at radio wavelengths.

This expression breaks down toward wavelengths shorter than hc/kT. In fact, quite generally, the Planck distribution reaches a maximum at a wavelength $\lambda = 0.2(hc/kT)$. At shorter wavelengths, it declines rapidly. The total energy density of black-body radiation is obtained by adding the contributions to u_λ from all wavelengths, both large and small. This summation (actually an integration over infinitesimal wavelength intervals, each centered on λ, with λ varying from 0 to ∞) yields

$$u = \frac{8\pi^5 (kT)^4}{15(hc)^3}$$

This is written for convenience as

$$u = aT^4$$

where T is in degrees Kelvin and the radiation density constant

$$a = 7.56 \times 10^{-15} \text{ erg cm}^{-3}(\text{degree})^{-4}$$

The 2.7-degree radiation has an energy density amounting to

$$u = a(2.7)^4, \text{ or } 4.0 \times 10^{-13} \text{ erg cm}^{-3}$$

To express this in more useful units, we note that 1 electron volt $= 1.6 \times 10^{-12}$ ergs; hence, the radiation density is 0.25 electron volts cm^{-3}. This is roughly equal to the density of starlight in the Milky Way. The peak of the 2.7-degree radiation occurs at a wavelength of 0.2 $(hc/2.7k)$, or 1 millimeter.

CHAPTER 5

5. The Friedmann Equation

Consider an arbitrary spherical shell of matter expanding with the universe. The density in the interior of the shell is uniform and equal to the mean cosmological mass density. The kinetic energy of expansion of the shell, with expansion velocity v and present radius r, is $\frac{1}{2}v^2$ per unit mass of the shell. If the shell encloses a mass M, its gravitational potential energy is $-GM/r$ per unit mass. The mass of the sphere contained within the shell is $M = 4\pi dr^3/3$, where d is the average density of matter. Introducing the Hubble law $v = Hr$ required for isotropy, where in general $H(t)$ denotes that H is a function of time, we must have, by applying the law of energy conservation to the motion of the shell,

$$\frac{1}{2}\,H^2 r^2 - \frac{4\pi}{3}\,G\,dr^2 = \text{a constant}$$

But, by hypothesis, our shell was an arbitrary shell. It follows that this equation must apply to all possible shells and therefore to all particles in the universe. More generally, suppose that the spherical shell at some instant has radius r_0. Then

$$r = R(t)r_0$$

where $R(t)$ is the scale factor that expresses the amount by which the sphere (containing a fixed amount of matter) has expanded. We usually define $R(t)$ so that it equals 1 at present. We define the time-independent curvature constant for the particular sphere we are considering as $-\frac{1}{2}kr_0^2$. This leads to the Friedmann equation,

$$H^2 - \frac{8\pi}{3}\,Gd = -kR^{-2}$$

The distance between successive wavecrests (or the wavelength) of radiation changes in precisely the same way as the radius of a sphere containing a fixed amount of matter. We can think of both waves (or photons) and particles as being conserved quantities. (This assumes that we neglect absorption or emission processes in the intervening space.) Thus, the redshift z is given by the ratio of the difference between observed and emitted wavelengths to emitted wavelength, or

$$z = \frac{R(t_{\text{observed}})r_0 - R(t_{\text{emitted}})r_0}{R(t_{\text{emitted}})r_0} = \frac{1}{R(t_{\text{emitted}})} - 1$$

We refer to the epoch t_{emitted} as t, an arbitrary time after the big bang, and so

$$z = \frac{1}{R(t)} - 1$$

expresses the relation between z, R and t. All three provide equivalent measures of time.

6. The Einstein-de Sitter Universe

If $k = 0$, the Friedmann equation reduces to the Einstein-de Sitter universe. In this universe, the density is

$$d = \frac{3H_0{}^2}{8\pi G}$$

Since H_0 is a measured quantity (at present equal to 20 kilometers per second per million light-years), we can infer that there is a critical value of the density (noting that $G = 6.67 \times 10^{-8}$ cm^{3}g^{-1}s^{-2} and 1 light-year $= 9.5 \times 10^{17}$ cm) given by

$$d_{\text{crit}} = \frac{3}{8\pi} \left(\frac{20 \times 10^5}{1 \times 10^{24}}\right)^2 \frac{1}{(6.7 \times 10^{-8})} = 8 \times 10^{-30} \text{ g cm}^{-3}$$

In terms of atoms (the mass of a hydrogen atom equals 1.66×10^{-24} g), we can define a critical particle density as

$$n_{\text{crit}} = 5.3 \times 10^{-6} \text{ cm}^{-3}$$

This corresponds to 5 atoms per cubic meter.

The Friedmann equation, with k set equal to zero, is particularly easy to solve exactly. This limit can be shown to be always an accurate description of the big bang at early times in the matter-dominated era. Any effects of pressure are neglected. By conservation of particle number, the product of d with the volume of a sphere (proportional to R^3) is constant, or,

$$dR^3 = d_{\text{crit}}$$

where d_{crit} is the present particle density. We choose the scale factor to be unity at present. In terms of redshift z, we have

$$R = (1 + z)^{-1}$$

Since d increases as R^{-3}, but the k-term increases only as R^{-2}, we see that, at early times, the k-term must become negligible.

At the era when the dominant contribution to the density is from matter (rather than radiation),

$$H^2 = \frac{8\pi}{3} G d_{\text{crit}} R^{-3}$$

where $H(t) = v/r = $ (rate of change of R)/R. This is a simple differential equation for R, which can be written as

$$\frac{\text{rate of change of } R}{R} = \left(\frac{8\pi}{3} G d_{\text{crit}}\right)^{1/2} R^{-3/2}$$

To solve this, we shall quote an elementary result of differential calculus: If the rate of change of $R = AR^{-n}$, then the time t elapsed for $R(t)$ to acquire its present value (if we assume that $R = 0$ at $t = 0$) is

$$t = \frac{A^{-1}}{n + 1} R^{n+1}$$

In the present example, $n = \frac{1}{2}$, so

$$t = \frac{2}{3}\left(\frac{8\pi G d_{\text{crit}}}{3}\right)^{-1/2} R^{3/2}$$

or

$$R(t) = (6\pi G d_{\text{crit}})^{1/3} t^{2/3}$$

and

$$d(t) = \frac{1}{6\pi G t^2}$$

It follows also that the age of the Einstein-de Sitter universe

$$t = \frac{2}{3} \cdot \frac{1}{H}$$

Adopting $H = 20$ kilometers per second per million light-years yields an age of

$$\frac{2}{3}\,(20 \times 10^5/10^{24})^{-1}\ \text{s}$$

or roughly 10 billion years.

The possible uncertainty in H is consistent with a range from 15 to 30 kilometers per second per million light-years; thus, the age of an Einstein-de Sitter universe could be as short as 7 billion years or as long as 14 billion years.

It is useful to write the expression for d and t in terms of redshift:

$$d_{crit} = 8 \times 10^{-30}(1 + z)^3 \text{ g cm}^{-3}$$

and

$$t = 3 \times 10^{17}(1 + z)^{-3/2} \text{ s}$$

7. The Friedmann-Lemaître Models

Solving the general Friedmann equation exactly is beyond the scope of these notes. However, a simple trick allows us to study the behavior of the solution. We shall first consider a closed universe, with k negative. The Friedmann equation can be written as

$$H^2 = \frac{8\pi}{3} Gd - kR^{-2}$$

Recall that $d \propto R^{-3}$. Evidently, as R increases, there comes a time when H^2 is first equal to zero. Subsequently, the kR^{-2} dominates (if R were to increase any further); however, this would make H^2, a positive quantity, become negative. This is impossible, and we have arrived at a mathematical contradiction that is resolved only if R reaches its *maximum* value at the first zero of H^2. In other words, the scale factor reaches a maximum size, when the expansion rate slows down to zero, and the universe subsequently recollapses.

To examine the nature of the solution for the scale factor, we consider the following two equations:

$$R = \frac{4\pi Gd_0}{3k} (1 - \cos y)$$

and

$$t = \frac{4\pi Gd_0}{3k^{3/2}} (y - \sin y)$$

where y is a parameter that we leave unspecified for the moment. The density at the present era is denoted by d_0. As y is varied, R and t acquire a definite value for each value of y. It can be shown that the equations for R and t actually satisfy the Friedmann equation. We say that they provide a *parametric representation* of the solutions to this equation. In particular, as y varies from 0 to π, R increases from 0 to a maximum value and subsequently declines, vanishing at $y = 2\pi$. Note that $\cos 0 = 1$, $\cos \pi = -1$, and $\cos 2\pi = 0$. The time t steadily increases as y goes from 0 to 2π. Note that $\sin 0 = 0$, $\sin \pi = 0$, and $\sin 2\pi = 0$. If we consider R to be a function of time, we see that this representation provides us with the periodic behavior of $R(t)$ in a closed Friedmann universe.

Now let us consider an open universe. We write the Friedmann equation as

$$H^2 = \frac{8\pi G}{3} d + kR^{-2}$$

Note that now H^2 is the *sum* of two terms that are positive. Consequently, however large R becomes, H^2 will never vanish (as long as R is not infinite). There will always be a solution with a nonzero expansion rate at arbitrarily large (but not infinite) values of R. In this case, the universe expands forever. The parametric solution in the Friedmann equation now has a very different form:

$$R = \left(\frac{4\pi G d_0}{3k}\right)(\cosh y - 1)$$

and

$$t = \left(\frac{4\pi G d_0}{3k^{3/2}}\right)(\sinh y - y)$$

Superficially, it looks remarkably similar to the closed solution, except that we have replaced the cosine and sine by cosh and sinh, symbols that denote hyperbolic cosine and sine functions. These functions are not periodic (or cyclical), like the trigonometric functions, but are *monotonic*, continuing to increase without limit. We actually have

$$\cosh y = \tfrac{1}{2}(e^y + e^{-y})$$

and

$$\sinh y = \tfrac{1}{2}(e^y - e^{-y})$$

The solution for $R(t)$ is accordingly now very different from the previous result—it increases without limit as time increases, which is a characteristic property of an open universe.

From the Friedmann equation (Note 5),

$$H^2 - \frac{8\pi G}{3}\frac{d_0}{R^3} = \frac{-k}{R^2}$$

it is apparent that, when R is large and k is negative (for an open universe), the equation can be approximated by

$$H^2 = \frac{(-k)}{R^2}$$

Let us evaluate this for the present era, when $H = H_0$, (the observed Hubble constant), and $R = 1$. We obviously have $(-k) = H_0{}^2$. This neglects the density term entirely. A more accurate statement is

$$k = -H_0{}^2(1 - \Omega_0)$$

where $\Omega_0 = d_0/d_{crit}$, d_0 is the present matter density, and $d_{crit} = 3H_0{}^2/(8\pi G)$ is the matter density in an Einstein-de Sitter universe. We shall write

$$R = (1 + z)^{-1}$$

where z is the redshift, and we find that the Friedmann equation for an open universe (substituting for k) becomes

$$H^2 - \frac{8\pi G}{3} d_0(1 + z)^3 = H_0{}^2(1 + z)^2 (1 - \Omega_0)$$

In the Einstein-de Sitter universe, we would have

$$H^2 - 8\pi \frac{G}{3} d_0(1 + z)^3 = 0$$

Thus, we can ask, at what value of z is it appropriate to approximate an open universe by an Einstein-de Sitter universe? The answer is obtained by equating the terms $8\pi G/3d_0(1 + z)^3$ and $H_0{}^2(1 + z)^2(1 - \Omega_0)$; this yields

$$1 + z = \frac{3H_0{}^2(1 - \Omega_0)}{8\pi G d_0}$$

Hence, we find that

$$1 + z = d_{crit}/d_0(1 - \Omega_0) = \Omega_0{}^{-1} - 1$$

Recall that Ω_0 is the ratio of the actual density of matter at present to the critical value. We see that, at a redshift much larger than $1/\Omega_0$, the Einstein-de Sitter model provides an excellent approximation to an open universe.

CHAPTER 6

8. The Radiation Era

At very early times, the radiation density made a larger contribution to the total density than does the matter. We found that the radiation density is

$$u = aT^4 \text{ erg cm}^{-3}$$

The corresponding mass density of radiation, according to Einstein's prescription $E = mc^2$, is

$$d_{\text{radiation}} = u/c^2 = aT^4/c^2$$

Now we will show that the blackbody radiation temperature decreases with the expansion as R^{-1}. The energy of each blackbody photon of wavelength λ is equal to hc/λ (this represents the quantum of mass-energy that we associate with each photon, according to quantum theory). We denote the Planck distribution energy density per unit volume per unit wavelength interval by u_λ. The number density of photons is obtained by dividing u_λ by the energy of a photon of wavelength λ and summing over all wavelengths. The result, equal to the total number of photons per unit volume, is

$$n = 60.4(kT/hc)^3 = 20.3T^3 \text{ photons cm}^{-3}$$

where T is in degrees kelvin. There are about 400 photons per cubic centimeter in the 2.7-degree blackbody radiation. The average energy of a blackbody photon is equal to 2.7 kT, or about 1.0×10^{-15} erg, or 6×10^{-4} electron volt.

The effect of the cosmic expansion leaves the spectrum of the blackbody radiation unchanged. If the wavelength is redshifted by a factor R, the energy density is decreased by a factor R^4 (one factor of R due to the energy loss of each photon and three factors of R due to the increase in volume). However, the formula for the blackbody distribution u_λ, when we make these changes, can be reexpressed in terms of the new wavelength, equal to λR. The new expression looks exactly like the old formula, provided that we replace T with a lower temperature equal to T/R. In other words, blackbody radiation maintains its spectral distribution and remains blackbody radiation at lower temperatures as the universe expands.

With $T(t_0)$, the radiation temperature at present, equal to 2.7 degrees kelvin, the present mass density of radiation is only

$$d_{\text{radiation}}(t_0) = 4.5 \times 10^{-34} \text{ g cm}^{-3}$$

However,

$$d_{\text{radiation}}(t) = d_{\text{radiation}}(t_0)R^{-4}$$

whereas

$$d_{\text{matter}}(t) = d_{\text{matter}}(t_0)R^{-3}$$

so we infer that, at early eras, radiation becomes dominant over matter in its gravitational influence. (We write d_0 as $d_{\text{matter}}(t_0)$ for clarity.) This matter dominance first occurred at a time when the scale factor was equal to $d_{\text{radiation}}/d_{\text{matter}}$, or at a redshift

$$z_{\text{eq}} = d_{\text{matter}}/d_{\text{radiation}} = (8 \times 10^{-30})(4.5 \times 10^{-34})^{-1} = 2000$$

if the present density is given by its critical value for closure.

At earlier eras, the Friedmann equation is well approximated by

$$H^2 = \frac{8\pi G}{3} d_{\text{radiation}}(t_0)R^{-4}$$

This equation can be rewritten as

$$\text{rate of change of } R = \left[\frac{8\pi G}{3} d_{\text{radiation}}(t_0)\right]^{1/2} R^{-1}$$

The solution to this differential equation is given by the simple general formula of Note 6, with n set equal to 1. The result is

$$R(t) = \left[\frac{32\pi G}{3} d_{\text{radiation}}(t_0)\right]^{1/4} t^{1/2}$$

and

$$d_{\text{radiation}}(t) = \frac{3}{32\pi G t^2}$$

In terms of the radiation temperature,

$$T = \left(\frac{3c^2}{32\pi Ga}\right)^{1/4} \frac{1}{t^{1/2}} = \frac{10^{10}}{t^{1/2}} \text{ degrees kelvin}$$

where t is measured in seconds from the big bang.

At $t = 1$ second, the radiation temperature is 10^{10} degrees kelvin. The average energy of a blackbody radiation photon is equal to 2.7 kT, or 2.3 ×

10^6 electron volts at this era. This exceeds the rest mass energy of an electron, which has mass equal to 9×10^{-28} g or equivalent energy $9 \times 10^{-28}\, c^2$ erg, or 5×10^5 electron volts. Somewhat later (for example, at $t = 30$ seconds), the photon energy has dropped below the rest mass energy of an electron. Consequently, the creation of particle pairs occurred only at eras earlier than about 1 second.

At later eras, we have the general result that the radiation temperature is given by

$$T = 2.7(1 + z) \text{ degrees kelvin}$$

One implication of these results is that the number density of photons and the number density of particles vary in a similar way (both vary as T^3). Their ratio is a constant of the expansion. This statement is valid only after annihilation of electron-positron pairs has been completed, at times later than 1 second. Subsequently there are approximately 10^8 photons per hydrogen nucleus. Prior to this era, a number of particle pairs comparable with the number of photons was present. This is the basis of the statement that the early universe was asymmetric (containing an excess of matter over antimatter) by only 1 part in 10^8. The only remaining trace of all this antimatter is in the blackbody radiation.

CHAPTER 7

9. Nucleosynthesis

a. Neutron production. When electron–positron pairs are in thermal equilibrium with the radiation field, at temperatures greater than 10^{10} degrees, some of the reactions that occur are:

$$\text{proton} + \text{electron} \rightleftarrows \text{neutron} + \text{neutrino}$$

and

$$\text{proton} + \text{antineutrino} \rightleftarrows \text{positron} + \text{neutron}$$

The double arrows signify that the reactions can proceed in either direction. The electron-positron pairs annihilate rather abruptly at 10^{10} degrees kelvin, when the average energy of a blackbody photon drops below the creation threshold. One consequence is that these reactions stop, and neutrons are no longer created or destroyed. The number of neutrons becomes frozen relative to the number of protons and electrons. While thermal equilibrium is maintained, the neutron abundance can be expressed as

$$\frac{n_{\text{neutron}}}{n_{\text{proton}}} = e^{-(m_n - m_p)c^2/kT}$$

where m_n and m_p are the masses of the neutron and proton. (Here, e denotes the number $2.718\ldots$.) An expression like this is always characteristic of a thermal equilibrium process. The mass difference, equal to an equivalent energy of 1.3 million electron volts, determines the relative proportion of neutrons. The residual neutron abundance can be obtained to a rather good approximation by choosing $T = 10^{10}$ degrees (equivalent to 9×10^5 electron volts) in the above expression, since the reactions cease and the neutron abundance freezes at this temperature. The result is that one neutron remains for every five protons.

b. *Helium synthesis.* The subsequent fate of the neutron might be to decay into a proton and an electron. However, this decay takes approximately 11 minutes for a free neutron. Another fate awaits neutrons long before the universe is 11 minutes old. Nuclear reactions can occur, and neutrons tend to react with protons to make deuterium. However, at first, the temperature is too high for the deuterium to survive. Once deuterium is made, more complex nuclei are synthesized. In fact, there is a narrow range of temperatures over which reactions are possible, around 10^9 degrees kelvin. A high temperature is needed for nuclei to get close enough for the nuclear forces to take over and form more complex nuclei. At the same time, if the particles are moving too fast, they will not stay near each other long enough for the nuclear forces to work at all.

Once the temperature drops to 10^9 degrees, when the universe is about one and one-half minutes old, the reactions begin. First we have

$$\text{neutron} + \text{proton} \rightarrow \text{deuterium} + \text{photon}$$

producing a deuterium (or heavy hydrogen) nucleus. This is rapidly followed by a sequence of reactions:

$$\text{deuterium} + \text{deuterium} \rightarrow \text{helium}^3 + \text{neutron}$$
$$\text{deuterium} + \text{deuterium} \rightarrow \text{tritium} + \text{proton}$$
$$\text{deuterium} + \text{tritium} \rightarrow \text{helium}^4 + \text{neutron}$$

Helium^3 is an isotope of helium, consisting of two protons and one neutron; helium^4 is the common helium nucleus, consisting of two neutrons and two protons; tritium is unstable ultraheavy hydrogen, consisting of one proton and two neutrons. The chemical nature of a nucleus is determined by the number of charges it contains; different isotopes of the same element have similar numbers of protons but different numbers of neutrons in the nucleus.

It so happens that there are no stable nuclei of masses 5 and 8. Consequently the nuclear reactions rapidly diminish after making helium. Small amounts of other light elements, notably lithium, beryllium, and boron, are made. But it is not possible to make significant amounts of heavier nuclei such as carbon (mass 12). The higher densities in stellar interiors are required to allow three helium nuclei to fuse into a carbon nucleus.

Practically all the neutrons end up forming helium nuclei. It is not difficult to understand why this happens. The probability that a neutron will collide and react with a proton is roughly given by its geometrical size (or cross-section), πr^2, where r is the nuclear radius. If there are n protons per cubic centimeter moving at velocity v, the number of collisions that a neutron will undergo per second is simply $(\pi r^2)nv$. We require each neutron to undergo at least one collision with a proton within the age (t) of the universe at 10^9 degrees. This condition requires that the quantity

$$Q = (\pi r^2)nvt$$

must be equal to unity for most of the neutrons to react with protons. But we can easily evaluate Q. We know that t is approximately 100 seconds; v is the velocity of a neutron at 10^9 degrees, equal to $(kT/m_n)^{1/2}$ or 3×10^8 cm s^{-1}; the nuclear radius r is 10^{-13} cm; and $n = n_{crit}(T/3$ degrees$)^3$ is the particle density in a universe of critical density. Consequently we find that

$$Q = (3 \times 10^{-26})(3 \times 10^{-6})(3 \times 10^8)^3(3 \times 10^8)(10^2) = 7 \times 10^4$$

A more accurate rate than the geometrical rate $(\pi r^2 v)$ for the process

$$\text{neutron} + \text{proton} \rightarrow \text{deuterium} + \text{photon}$$

is 4.6×10^{-20} cm^3 s^{-1}. This leads to $Q \approx 10^3$, still a large number. Evidently there are overwhelming odds (10^3 to 1!) that a neutron will react with a proton.

The remaining nuclear reactions can be shown to follow rapidly. We conclude that, since each helium nucleus contains two neutrons and we begin with one neutron for every five protons, the Big Bang produces one helium nucleus for every eight protons. Moreover, because Q came out to be so large, this conclusion is essentially independent of the highly uncertain value of the present density (had we overestimated n by even 100, our result would not change).

c. Deuterium. Deuterium is an important by-product of the nuclear reactions that synthesize helium. A few neutrons can remain locked up in deuterium nuclei, as many as 1 per 10,000 protons in an open universe. In a closed universe, however, essentially no deuterium survives. The reason is that, if we evaluate the likely fate of a deuterium nucleus by estimating Q for it to react, a far smaller value is obtained than for the previous case. Detailed computations of the nuclear reactions show that the deuterium abundance attains a peak value of about 1 percent of the hydrogen abundance and subsequently decays. The decay is primarily a result of burning to form heavier nuclei (notably helium[3] and tritium, and ultimately helium[4]). Very roughly,

428 · MATHEMATICAL NOTES

we see that, when the deuterium density has dropped to one-thousandth of the proton density, if the particle density is at one-tenth of its critical value for closure, Q becomes less than unity, and the decay of deuterium can be halted. However, if n is close to its critical value, Q is larger than 10 when the reactions occur, and almost all the deuterium is destroyed. In an open universe, a significant fraction of the big bang deuterium would be preserved.

CHAPTER 8

10. Decoupling

A critical factor in determining the state of ionization of matter in the early universe is the ionization of newly formed atoms by blackbody radiation photons. The photons most effective at ionizing are the high energy photons in the short wavelength limit of the blackbody photon distribution described in Note 4, at wavelengths small compared with hc/kT. The number of these ionizing photons (relative to the particle density) can be approximately written as

$$\frac{n_i}{n} = \frac{16\pi}{n}\left(\frac{kT}{hc}\right)^3 e^{-I/kT} = \frac{3 \times 10^7}{\Omega} e^{-157,000/T}$$

Here I, equivalent to a temperature of 157,000 degrees Kelvin, denotes the energy required to ionize an unexcited hydrogen atom. Since n_i/n drops rapidly once T falls below 10^5 degrees, we may expect hydrogen atom formation and decoupling to occur as soon as there are insufficient ionizing photons to keep the atoms ionized. Since there are a large number of blackbody photons (about 10^8) for every atom, the temperature has to drop well below 10^5 degrees for the exponential factor $(e^{-157,000/T})$ to become sufficiently small. In fact, n_i/n becomes less than unity only when T has fallen below about 8000 degrees kelvin.

The equilibrium between ionized and neutral phases involves a delicate balance between ionizations by the remaining high energy blackbody photons and electron recombinations. An equation that describes this equilibrium, known as the *Saha equation*, can be written in terms of the ionized fraction x as

$$\frac{x^2}{1-x} = \frac{(2\pi m_e kT)^{3/2}}{nh^3} e^{-I/kT}$$

where m_e is the electron mass, k is Boltzmann's constant, and h is Planck's constant. Solution of this equation as T drops below several thousand degrees reveals a very abrupt change in the state of the matter from being ionized to being neutral at about 4000 degrees kelvin, as is shown in the following table.

Values for x

T (kelvin)	Ratio of present density of universe to closure value	
	1.	0.01
5400	0.99	1
4800	0.73	0.995
4200	0.11	0.664
3600	0.004	0.034
3000	4×10^{-5}	3×10^{-4}

This procedure slightly overestimates the decoupling temperature for the following reason. The atoms tend to be excited by absorption of trapped Lyman alpha photons. These photons are produced when hydrogen forms; they correspond to an electron jumping from the first excited state of hydrogen to the ground state. They are absorbed by other hydrogen atoms, which subsequently deexcite, producing a further Lyman alpha photon for each one absorbed. An appreciable fraction of hydrogen atoms may therefore be in the excited state during the decoupling era. However, excited atoms, being more loosely bound, are more easily ionized than unexcited atoms. If we allow for the presence of excited hydrogen atoms, the temperature at which decoupling is effective is reduced to about 3000 degrees kelvin.

CHAPTER 9

11. Interaction of Radiation with Density Fluctuations

If we consider density fluctuations prior to the decoupling era on scales where gravitational effects are important, it is simplest to discuss separately the isothermal, or constant-temperature, fluctuations (in which the radiation field is unperturbed) and adiabatic fluctuations, in which both matter and radiation are compressed.

    a. Isothermal fluctuations. Let us imagine an electron incorporated in the density fluctuation that is moving at a small velocity v relative to the isotropic background radiation. An observer in the rest frame of the electron would see a net flux f of radiation, amounting to $\frac{4}{3} v a T^4$, in the direction of motion. The fact that there is a flux of radiation means that a drag force is exerted on the electron, since the photons will scatter off the electron predominantly in the direction of its motion. If the mean distance between scatterings (or *mean free path*) of a photon is λ, then the force exerted on the electron is due to the excess radiation pressure over a distance λ. The excess pressure is just

f/c, where c is the velocity of light. The drag force is equal to $f/\lambda c$, or

$$F_{\text{drag}} = -\tfrac{4}{3}\,(v/c)(aT^4/\lambda)$$

We have written the drag force as a negative quantity, because it is always in the opposite direction to that of the motion of the electrons. The photon mean free path is inversely proportional to the electron density and is written

$$\lambda = (\sigma n)^{-1}$$

where $\sigma = 0.7 \times 10^{-24}$ cm^2 is the electron scattering cross-section. We can think of σ as the effective target area of an electron that is seen by an impinging photon. It is considerably larger than the geometrical size of the electron, because the charge of the electron determines the degree of scattering of the photon. We finally have

$$F_{\text{drag}} = -\tfrac{4}{3}\,(v/c)\sigma(At^4)n$$

The gravitational force is

$$F_{\text{grav}} = \frac{GM}{L^2}\,d = \frac{Gd}{L^2}\,(\tfrac{4}{3}\,\pi dL^3)$$

We assume that the mass of the fluctuation is approximately equal to the amount of matter at the cosmological density d within a sphere of radius L.

Taking the ratio of these two forces (the electron and proton must always move together because of the electrostatic forces that result if they separate), we have

$$\frac{F_{\text{drag}}}{F_{\text{grav}}} = \frac{\sigma(aT^4)v}{L\pi Gcm_p\,d}$$

$$= 10^{-8}(1+z)^{5/2}$$

In obtaining the numerical value, we have assumed an Einstein-de Sitter cosmology (Notes 7 and 9), with $T = 3(1+z)$ degrees kelvin, $d = 6 \times 10^{-30}$ $(1+z)^3$ g cm^{-3}, and $\tfrac{2}{3}\,L/v = t = 3 \times 10^{17}(1+z)^{-3/2}$ sec. It is clear that, prior to decoupling at $z = 1000$, the radiation drag force will always appreciably exceed the gravitational force.

The radiation drag force therefore has the effect of inhibiting any motion of isothermal fluctuations. It also helps us to understand why the matter and radiation temperatures are almost identical prior to decoupling, even though processes that are ordinarily characteristic of thermal equilibrium (such as absorption and emission of radiation) may not be occurring. For, if we consider v to be due to thermal motions of individual electrons rather than to mass motion of the fluctuation, a similar argument applies. Any excess energy of the electrons (if the electron temperature exceeds the radiation temperature) is lost. The process also works in reverse, if the electron temperature is lower than the radiation temperature. In this case, the electrons will heat up until the matter and radiation temperatures are equal.

b. *Adiabatic fluctuations.* A dynamic process is important for adiabatic fluctuations, because the radiation tends to resist compression. The fact that radiation has a finite mean free path means that it can eventually escape by diffusion from fluctuations of any given size. For sufficiently small-scale fluctuations, this diffusion time is rapid and will be less than the age of the universe. These small-scale fluctuations are doomed; the radiation will diffuse away and smooth (or damp) the fluctuations. Because of its short mean free path, the diffusing radiation homogenizes the fluctuation as it escapes.

We can easily estimate a critical size for fluctuations to be damped by this process of radiative diffusion. If the fluctuation size is L, we can imagine that photons must perform a random walk with step length λ (the scattering distance) to escape and diffuse away. The solution to this problem is that it takes N^2 scatterings to diffuse a distance equal to N scattering lengths. A similar problem confronts a drunkard who has taken N^2 steps to get to the next bar, when, if sober, he could do the journey in N steps. The time taken is therefore the square of the number of scatterings $(L/\lambda)^2$ multiplied by the time per scattering (λ/c), or $L^2(\lambda c)^{-1}$. This diffusion time is less than the age of the universe, if

$$L^2(\lambda c)^{-1} < t \text{ or } L < (ct\cdot\lambda)^{1/2}$$

Now, ct is essentially the radius of the observable universe (or the distance light can travel in time t). L is the geometric mean of the photon mean free path λ and the horizon size.

Diffusion becomes ineffective, once scattering becomes unimportant, after atomic hydrogen has formed. The maximum scale of fluctuations that have been damped is best expressed in terms of their mass,

$$M_d = \frac{4\pi}{3} dL^3 = \frac{4\pi}{3} n^{-1/2} m_p (ct/\sigma)^{3/2}$$

where m_p is the proton mass. Adopting values for n and t appropriate to decoupling ($n = 3000 \, \Omega_0 \text{ cm}^{-3}$ and $t = 10^{13} \, \Omega_0^{-1/2}$ sec), this expression yields $M_d = 10^{12} \, \Omega_0^{-5/4}$ solar masses. We have used $m_p = 1.7 \times 10^{-24}$ g and 1 solar mass $= 2 \times 10^{33}$ g in deriving this result.

This actually underestimates M_d for the following reason. During the process of decoupling, the proton scattering distance is rapidly increasing. Initially it is very small compared with the size of any fluctuations of interest. However, it ends up being comparable with the size of the observable universe, because the decoupling process decouples radiation from matter by removing almost all the free electrons that are responsible for the scattering.

Once the photon mean free path is very large, radiative diffusion becomes irrelevant—the radiation has already left. During the brief instant when the photon mean free path is comparable with the size of the fluctuation, considerable additional damping can occur. The net effect is to increase M_d to about 10^{13} solar masses in an Einstein–de Sitter universe. If the universe is open, with present density equal to only $0.03\, d_{crit}$, M_d will be increased by about a factor of 100.

12. Gravitational Instability

Fluctuations that survive decoupling are subject to gravitational instability, if they are of sufficiently large scale. The criterion for gravitational instability, originally formulated by Sir James Jeans, states that fluctuations of scale greater than a critical length (the *Jeans length*, which we denote by R_{Jeans}) will be unstable and grow larger. In the absence of expansion, the instability is strong. It is basically due to the competition between pressure and gravitational forces.

To examine the Jeans criterion, we consider a medium of density d, a sound speed of v_s, and a fluctuation of diameter R. We compare the typical kinetic energy (due to thermal or random motions) $\sim(dR^3)v_s^2$ with the gravitational potential energy due to the fluctuation $\sim G(dR^3)/R$ per gram. Evidently the thermal energy dominates for small R, but gravity takes over for large R. The thermal and gravitational energies (and implicitly the pressure gradient and gravitational forces) are in balance at $R = R_{Jeans}$, where

$$dv_s^2 = Gd^2R_{Jeans}^2, \text{ or } R_{Jeans} = v_s(Gd)^{-1/2}$$

A more rigorous derivation yields

$$R_{Jeans} = v_s(\pi/Gd)^{1/2}$$

Study of the behavior of fluctuations in an expanding universe leads to the conclusion that, on scales less than R_{Jeans}, the fluctuations oscillate like pressure waves. On larger scales, the density fluctuations are weakly unstable. The instability manifests itself as a slow growth in amplitude of the initially weak large-scale density fluctuations. The mass contained within a sphere of diameter equal to the Jeans length is called the *Jeans mass* and is given by

$$M_{Jeans} = \pi d[v_s(\pi/Gd)^{1/2}]^3$$

Prior to decoupling, the mixture of matter and radiation maintains a high sound speed, close to that of light. In fact, while the universe is radiation dominated,

$$v_s = c/(3)^{1/2}$$

This means that the Jeans mass encompasses practically all the mass within the observable horizon. Immediately before decoupling, this amounts to a very large mass indeed, exceeding that of the largest clusters of galaxies.

After decoupling, the sound velocity drops abruptly to a value characteristic of hydrogen at 2000 degrees kelvin, given by $(kT/m_p)^{1/2}$, or approximately 4 kilometers per second. The Jeans mass is now reduced to about 10^5 solar masses. To compute this, set

$$v_s = 4 \times 10^5 \text{ cm s}^{-1}$$

and

$$d = 6 \times 10^{-30}(1 + z)^3 \Omega_0 \text{ g cm}^{-3}$$

in the Jeans mass formula, which yields

$$M_{\text{Jeans}} = 10^9(1 + z)^{-3/2}\Omega_0^{-1/2} \text{ solar masses}$$

The change in Jeans length means that many fluctuations present prior to recombination suddenly become gravitationally unstable. Their growth and eventual collapse must lead to the formation of the galaxies.

It is not difficult to estimate the rate at which fluctuations larger than the Jeans length can grow. We have previously seen in Note 7 how the evolution of the Friedmann background can be described by equations of the form

$$R = A(1 - \cos y) \text{ and } t = B(y - \sin y),$$

where A and B are constants, and y is a parameter that is allowed to take any arbitrary value. Let us imagine a region that is slightly denser than the background. It will tend to lag behind the expansion of the rest of the universe. Let us associate a deviation δR with respect to the scale factor R of the fluctuation and a corresponding density enhancement δd with respect to the background density d. At early times we can treat y as a small quantity, and we can write

$$\cos y = 1 - \tfrac{1}{2} y^2 + (\tfrac{1}{24}) y^4$$

and

$$\sin y = y - (\tfrac{1}{6}) y^3$$

We now obtain

$$R + \delta R \ A \left(\tfrac{1}{2} y^2 - (\tfrac{1}{24}) y^4\right) \text{ and } t = \left(\frac{B}{6}\right) y^3$$

It follows that since we are considering small perturbations, we must have $R \propto y^2 \propto t^{2/3}$ and $\delta R \propto y^4 \propto t^{4/3}$. Since $(d + \delta d)(R + \delta R)^3$ is equal to a constant by virtue of the conservation of mass in a comoving volume, the fractional change in density $\delta d/d$ must be equal to $3\delta R/R$. It follows that $\delta d/d \propto y^2 \propto t^{2/3} \propto R$, or the density contrast of fluctuations relative to the uniform background increases proportionately to the expansion scale factor.

For example, if fluctuations were present with an amplitude of 1 percent at decoupling, they would have become large by a redshift of 10, when the universe had expanded 100-fold. This result is only true if the curvature term is still negligible at a redshift of 10. Recall that we can adequately approximate the early universe by the Einstein-de Sitter cosmology at a redshift larger than Ω_0^{-1}. Since Ω_0 probably lies in the range from 0.1 to 1, and since other arguments suggest that many galaxies may have formed at large redshifts (greater than 10), neglect of curvature effects on the growth of galactic perturbations may be a reasonable assumption.

In a closed universe, the rate of growth of density perturbations tends to be enhanced over the rate in an Einstein-de Sitter universe. In an open universe, however, their growth is inhibited, because a low density universe, where the curvature term is important, is essentially coasting along on the kinetic energy of the expansion. The gravitational energy term in the Friedmann equation is relatively small. This means that the self-gravitational attraction of fluctuations becomes unimportant. Unless density fluctuations are large enough to recollapse, they cease to grow any further, once the curvature term becomes significant. This phenomenon first occurs at a redshift roughly equal to $1/\Omega_0$, if Ω_0 is small. Such an effect may be important on very large scales, corresponding to superclusters of galaxies, where large inhomogeneities in the galaxy distribution do not appear to be accompanied by similar deviations from the Hubble expansion, as would be expected if the superclusters were gravitationally bound configurations. This result remains to be confirmed; if correct, it would provide support for an open universe.

CHAPTER 10

13. Collapse of Density Fluctuations; Galaxy Formation

We have described in Chapter 5 how the arguments that justify Newtonian cosmology in a homogeneous universe are equivalent to the argument that we can neglect the gravitational effects of the rest of the universe outside an arbitrarily large but finite sphere. This result is actually a special case of a theorem that applies to any spherically symmetric matter distribution: only the matter contained within a spherical shell contributes to the gravitational forces acting on that shell. One consequence is that the equations we have used to describe the Friedmann universe can easily be generalized to follow the evolution of a spherically symmetric density fluctuation.

Let us compose the fluctuation into a series of spherical shells. Suppose that r is the radius of a shell. Then the equations

$$r = A(1 - \cos y) \quad \text{and} \quad t = B(y - \sin y)$$

describe the evolution of this shell. We must still take A and B as constants, which, however, may be allowed to vary from shell to shell (they will depend on the mass within any given shell). We can fix A and B by specifying the mass M within a shell and the total energy ($\frac{1}{2} E$) per unit mass of the shell:

$$\tfrac{1}{2} E = -\tfrac{1}{2} V^2 + GM/r$$

The resulting relations (which we shall not derive here) are $A = GM/E$ and $B = GM/E^{3/2}$. Note the similarity with the equations of Note 7, if we identify E with kr^2; now, however, k is no longer a constant but varies from shell to shell.

There is an interesting consequence of these equations for r and t. Time increases with increasing values of y, but r reaches a maximum value (at $y = \pi$) and then decreases. The shell collapses to zero radius at $y = 2\pi$. In other words, at a time t_*, where $t_* = B(\pi - \sin \pi) = \pi B$ (since $\sin \pi = 0$), the shell has expanded to the greatest radius it will attain. As the shell turns around near time t_*, its velocity lags more and more behind the Hubble expansion, dropping to zero at this time. The shell subsequently collapses. By the time $2t_*$ (equal to $2\pi B$), it completely recollapses. At the turn-around time t_*, the density of matter within the shell is not very much greater than the density of the unperturbed background. At $2t_*$, the matter density has formally become infinite (since the shell radius has shrunk to zero). Of course, what really happens is that nonradial motions will develop, and pressure forces will intervene to halt the collapse. This would lead to galaxy formation at a finite (nonzero) value of the radius.

14. Characteristic Masses and Dimensions of Galaxies

Let us consider the fate of a density fluctuation that begins to collapse. Will it indeed become a galaxy? We have said that pressure forces must eventually decelerate the collapse. The balance between pressure and gravity is determined by the Jeans criterion. If the cloud is much more massive than the Jeans mass appropriate to its density and temperature, it can continue to collapse. During the initial collapse, the density is low, and cooling is unimportant. The cloud heats up as it collapses, and the Jeans mass will increase. The collapse stops when the pressure and gravitational forces are in balance, or when the Jeans mass first becomes comparable with the cloud mass. Realistically, one might expect other sources of pressure, such as turbulence and nonradial mass motions, to play a role; in general, such motions tend to dissipate rapidly, if they are highly supersonic. Thus, the simple concepts of temperature and density provide an adequate description of the forces that can counterbalance gravity.

The systematic rise of temperature is crucial to this scenario. As the density rises, the gas may be able to begin to radiate effectively. Cooling may at first prevent the temperature from increasing and delay the time when equilibrium must prevail. If the gas cloud can radiate away the compressional energy acquired as it collapses, it will continue to collapse. Whenever the Jeans mass is less than the cloud mass, fragmentation can occur. Fluctuations will develop and collapse on scales comparable to the Jeans mass. If cooling prevents the temperature from rising, the Jeans mass (which is proportional to $T^{3/2}d^{-1/2}$) decreases, since the density increases with the collapse. The result is that the cloud fragments into smaller and smaller subfragments. These eventually are stars.

We can derive a simple condition for a hydrogen gas cloud to be able to cool, by requiring that the cooling time be comparable with the characteristic collapse time scale. These two time scales can be expressed in terms of the average temperature and density of the cloud.

At temperatures below about 3×10^5 degrees kelvin, cooling is predominantly due to combination of electrons with protons. The cooling time is very approximately given by the expression

$$t_{cool} \doteq 3 \times 10^5 T^{3/2} n^{-1} \text{ s}$$

for a gas at temperature T and density n atoms per cubic centimeter.

To obtain the gravitational collapse time, let us consider a gas cloud of mass M and radius R that is allowed to suddenly cool. It will collapse at a velocity $V = (GM/R)^{1/2}$. The time for it to collapse appreciably is R/v or $R^{3/2}/(GM)^{1/2}$. If $M = 4/3 \; mdR^3$, the gravitational collapse time

$$t_{grav} = (Gd)^{-1/2} = 3 \times 10^{15} \; n^{-1/2}$$

(We have made use of $d = nm_p$ to convert mass density into particle density.) The ratio of these time scales is therefore

$$\frac{t_{cool}}{t_{grav}} = 10^{-10} T^{3/2} n^{-1/2}$$

However,

$$M_{Jeans} = 10^{35} T^{3/2} n^{-1/2} \text{ g}$$

and we can therefore derive

$$\frac{t_{cool}}{t_{grav}} = \left(\frac{M_{Jeans}}{10^{45} \text{ g}}\right) = \left(\frac{M_{Jeans}}{5 \times 10^{11} \text{ solar masses}}\right)$$

This result suggests that only when the Jeans mass has decreased to about 5×10^{11} solar masses will a collapsing cloud be able to cool, efficiently subfragment, and make stars. Early collapse will not be associated with star-formation, since the gas remains hot.

This argument must be modified, if the initial temperature exceeds 3×10^5 degrees kelvin, where the gas is too hot to decouple and radiates mostly by a process known as *free-free radiation*, from electrons passing near and being accelerated by other electrons and protons. The brief acceleration causes the electrons to radiate. The cooling time scale for free-free radiation is approximately given by

$$t_{\text{cool}} = 3 \times 10^{11} T^{1/2} n^{-1} \text{ s}$$

and we now obtain

$$\frac{t_{\text{cool}}}{t_{\text{grav}}} = 10^{-4} \frac{T^{1/2}}{n^{1/2}} = \left(\frac{R_{\text{Jeans}}}{3 \times 10^{23} \text{ cm}} \right) = \left(\frac{R_{\text{Jeans}}}{3 \times 10^5 \text{ light-years}} \right)$$

This expression indicates that star-formation in a hot collapsing cloud will only occur once the Jeans length has fallen to about 3×10^5 light-years.

It is probably no mere coincidence that these characteristic cloud masses and radii for galaxy formation are similar to scales appropriate for a massive galaxy. Although a more detailed theory is lacking at present, it seems entirely plausible that the elementary physical processes of radiation and collapse in a primordial gas cloud should lead to formation of galaxies similar to those we observe.

15. When Did Galaxies Form?

If we know both the age of our galaxy and the age of the universe, we could infer the redshift at which our galaxy had formed. Unfortunately, the uncertainty in both of these age determinations leaves the era of galaxy formation indeterminate—most galaxies (including our own galaxy) could have formed either at a large redshift ($z = 50$, say) or relatively recently ($z = 3 - 4$).

One approach to inferring the era when our galaxy formed relies on the nature of the orbits of the oldest stars. These orbits are often highly eccentric or elongated about the center of the galaxy. These stars apparently formed at the maximum distance from the center reached on their present orbits, more than 150,000 light-years. Subsequently, the stars fell in toward the galaxy. Their kinetic energy of motion, acquired at infall, is sufficient to bring them out again to the radius at which they formed. These stars will retain their highly elongated orbits for a very long time. We infer that the galaxy as a whole must have collapsed from a region extending beyond the orbits of the oldest stars.

We can estimate the density our galaxy would have if it were to fill such a volume of radius 200,000 light-years, or 2×10^{23} cm. Adopting a mass of 10^{12} solar masses for the galaxy yields a mean initial density of 10^{-25} grams per cubic centimeter. The density of the universe at any redshift z is

$$6 \times 10^{-30}\Omega_0(1 + z)^3 \text{ g cm}^{-3}$$

where $\Omega_0 = 1$ for a universe of critical density for closure and is less than 1 for an open universe. Consequently, if we seek the redshift at which the initial galactic density was just equal to the cosmological density, we obtain

$$10^{-25} = 6 \times 10^{-30}\Omega_0(1 + z)^3$$

or

$$1 + z = 25\Omega_0^{-1/2}$$

At larger redshifts, the universe was denser than our galaxy could have been initially. Thus, our galaxy could not have formed by collapse at an earlier era than given by this condition. At a redshift of 24 (if $\Omega_0 = 1$), the density fluctuation that was to become our galaxy was therefore twice as dense as the background medium.

If our galaxy was initially much larger, then its density would have been rather low, and it could have formed relatively recently. Unfortunately, even the present extent of our galaxy is poorly known. Globular clusters are seen to distances of more than 300,000 light-years from the galactic center. The halos of other spiral galaxies may extend to comparable distances, where there is little evidence of luminous material, although a considerable amount of mass may be present. Galaxy formation could occur at a redshift as low as 3 or 4, and it may even be an ongoing process.

The best estimate of the era of galaxy formation will come when we directly observe young galaxies that have formed recently. The characteristics of a newly born system should be very apparent in its preponderance of young massive stars and considerable gas and dust content.

CHAPTER 11

16. Formation of Galaxy Clusters

A galaxy cluster may have initially been a density fluctuation that was rather small in amplitude relative to the background matter. The fluctuation grew by gravitational instability. Eventually galaxies formed and fell toward one another. The collapse of a cluster proceeds in the same way as the evolution of a gaseous density fluctuation from which galaxies developed, except that the concept of pressure has no meaning, when describing the evolution of cluster galaxies. The galaxies do not frequently collide with one another, as do atoms in a gas (or, if collisions do occur, they are relatively infrequent). This makes it rather easy to estimate the final radius of the cluster, since no energy is lost from the system.

When the galaxies just begin to turn around and fall into the forming cluster under their mutual self-gravity, the available energy is entirely in the form of gravitational potential energy. It amounts to

$$E = -\frac{GM}{R_*}$$

per unit mass, where R_* is the radius and M the mass of the system. Now let us follow the galaxies through recollapse. Because their energy is conserved, one might at first think that they would emerge to the same radius from which they began falling (just as a perfectly elastic rubber ball would bounce back to the height from which it was dropped). At first, the motions of the galaxies are probably inward (or radial). After the galaxies have passed by and been deflected by other galaxies, they rapidly acquire nonradial motions. Since energy is conserved, the mean radial velocity is lowered. Consequently the galaxies will bounce back to a smaller radius than that from which they fell.

If the final radius of the cluster is R_0, then the energy per unit mass

$$E = \frac{-GM}{R_0} + \tfrac{1}{2} v^2$$

where v is the average velocity dispersion of the cluster galaxies. If the cluster is now in equilibrium, we can also make use of the *virial theorem*, which requires that the sum of twice the kinetic energy of the galaxies and the gravitational potential energy must vanish:

$$2 \cdot \tfrac{1}{2} v^2 - GM/R_0 = 0$$

or

$$v^2 = GM/R_0$$

Consequently, we find that

$$E = -\tfrac{1}{2} GM/R_0$$

But the energy of the cluster is unchanged; hence, we must have

$$R_0 = \tfrac{1}{2} R_*$$

The cluster therefore collapses by a factor of two in radius, when the galaxy motions randomize and equilibrate in the gravitational field of the cluster.

A similar argument may also apply to the formation of an elliptical galaxy, if fragmentation into stars occurs before the protogalaxy has recollapsed. Some astronomers have argued that such a dissipationless collapse is required to explain the observed roundness of many elliptical galaxies.

CHAPTER 13

17. Star Formation

Fragmentation of a collapsing cloud continues until star formation occurs. This happens when fragments form that are sufficiently opaque to inhibit radiation from escaping, thereby diminishing the role of cooling. Once the fragments heat up, pressure forces can balance gravity, and a protostar will form.

The Jeans mass gives us the minimum size of a self-gravitating fragment at any stage. Larger fragments can also exist, but the smallest scale on which gravity is important is given by the Jeans criterion. We can also estimate the Jeans mass for a fragment when opacity first becomes significant. Consider the first opaque fragment to form that satisfies the Jeans criterion. The rate at which a particle in this fragment (at temperature T) will cool is, on the average,

$$\sigma T^4 / R_{\text{Jeans}} \text{ erg cm}^{-3} \text{ s}^{-1}$$

where σ is the Stefan-Boltzmann constant that determines the flux of blackbody radiation at a specified temperature and equals 4.6×10^{-5} erg cm^{-2} K^{-4}. This follows since we can very crudely regard the fragment as radiating like a blackbody. Radiation escapes from a depth in the fragment roughly equal to the Jeans length. The cooling time scale is the ratio of the thermal energy in a volume divided by the rate at which cooling occurs by radiation, or

$$t_{\text{cool}} \approx \frac{n \, (\tfrac{3}{2} \, kT)}{(\sigma T^4 / R_{\text{Jeans}})}$$

where n is the particle density, k is Boltzmann's constant $= 1.38 \times 10^{-16}$ erg/degree kelvin, and $(3/2)kT$ is the average energy of a particle. The numerical factor is appropriate only for a gas of pure atomic hydrogen; if the gas contains heavy elements and is mostly molecular, as will generally be the case, a small correction factor must be included.

Let us equate t_{cool} with the gravitational collapse time,

$$t_{\text{grav}} \approx (Gd)^{-1/2}$$

If $d = nm$, where m is the average mass of a particle (equal to about 1.3 times the mass of a proton, if we allow for the presence of helium), we obtain

$$(Gd)^{-1/2} = \frac{3}{2} \frac{nkT}{\sigma T^4} \left(\frac{kT/m}{Gd}\right)^{1/2}$$

or

$$n = \frac{2}{3} \sigma T^{5/2} m^{1/2} k^{-3/2}$$

The fragment mass

$$M_{\text{Jeans}} = \pi \frac{d}{6} R_{\text{Jeans}}^3 = \frac{\pi}{6} d \left(\frac{kT}{m} \frac{1}{Gd}\right)^{3/2}$$

Inserting the above expression for n, and appropriate values for the various constants, we find that the critical fragment mass is equal to $(kT/mc^2)^{1/4}$ solar masses. The factor $(kT/mc^2)^{1/4}$ varies between 0.001 and 0.01 for T spanning the range from 10 to 10^4 degrees kelvin. Clouds will cool (in the absence of any extraneous heat source) and will adjust to a temperature within this range. In fact, T is about 10 degrees kelvin in a cold molecular cloud. Because this expression for M_{Jeans} is fairly insensitive to T, we have demonstrated that the minimum fragment size will be rather less than that of the smallest known stars, with masses of about 0.01 solar mass.

Processes other than fragmentation, such as accretion and the effects of magnetic fields, undoubtedly play an important role in determining the masses of the observed stars. However, the result that minimum fragment masses turn out to be comparable with the masses of giant planets or very small stars suggests that we may be on the right track in studying fragmentation. It is quite possible that most of the mass in the universe is in the form of low mass stars. The stars that we observe, which are mostly greater than a few tenths of a solar mass, could be formed as the result of coalescence, involving protostars and the gas cloud fragments from which they form.

Some idea of the actual mass range of stars can be inferred by following the evolution of the protostellar fragments to a later stage. Once they are opaque, a slow collapse continues, with the rate of collapse being determined by how opaque the fragment has become. For the radiation pressure to slow the collapse appreciably, the protostar mass must be below a certain critical size, which depends on the opacity of the material. For ordinary material, this critical mass is about 0.2 solar mass. Smaller protostars may have undergone appreciable coalescence, and larger protostars should form more easily. This result suggests that 0.2 solar mass should be a characteristic stellar mass. Indeed, most of the stars in the galaxy are between 0.1 and 1 solar mass.

The greatly lowered opacity of primordial material (in which dust grains and elements heavier than helium are absent) increases this critical mass to about 20 solar masses. This result provides evidence for the argument that the first generation of stars, formed when no heavy elements were present, may have contained a preponderance of rather massive stars. Such a hypothesis appears to be necessary to account for the generation of the heavy elements in the evolved stellar cores. These first stars subsequently explode as supernovae and eject enriched material into the surrounding interstellar medium. New generations of less massive stars can subsequently condense from the enriched material through cooling and fragmentation.

18. Lifetimes of Stars of Differing Mass

Massive stars have high luminosities and are profligate users of their nuclear fuel supply. Consequently they are short-lived. Low mass stars radiate in a relatively feeble manner; they use nuclear fuel thriftily and are long-lived.

To verify these statements, let us consider a star of mass M, radius R, and central temperature T in *hydrostatic equilibrium;* that is, we assume the star is static, neither collapsing nor exploding. The balance between pressure and gravity can be described by equating the thermal and gravitational energy, or

$$\frac{kT}{m_p} = \frac{GM}{R}$$

The luminosity of a star can be expressed approximately as

$$L \propto 4\pi R^2 \frac{\sigma T^4}{(dR)}$$

where $4\pi R^2 \sigma T^4$ is the amount of energy radiated by a blackbody of radius R. The factor dR in the denominator, where d is the average density, is the total mass of material in a column of unit area through the star. Radiation must overcome the opacity of this material to escape. The higher the column density, the lower will be the emergent flux. It is important to realize that opacity is modifying the escape of radiation from the center of the star: T is the *central* temperature. (At the surface of the star, the luminosity can also be expressed in terms of the much lower surface temperature T_s by $L = 4\pi R^2 \sigma T_s^4$).

More precisely, the opacity of the stellar material is proportional to its density. The emergent flux is attenuated by a factor equal to the total opacity of a column of material, which is therefore approximately proportional to dR. We accordingly have

$$L \propto \frac{T^4 R}{d} \propto \frac{T^4}{M} R^4$$

where $d = 3M(4\pi R^3)^{-1}$. By virtue of the equilibrium relation $T \propto M/R$, we infer that

$$L \propto M^3$$

The luminosity of a star therefore increases roughly as the third power of its mass. A star of 10 solar masses is expected to be 1000 times more luminous than the sun. We can write this relation as

$$L = L_\odot (M/M_\odot)^3$$

where the symbol $_\odot$ denotes solar units (solar luminosity and solar mass).

The nuclear fuel supply of a star is simply proportional to its mass, according to Einstein's formula $E = Mc^2$. In fact, when hydrogen is fused into helium, only 0.7 percent of the original rest mass is liberated as energy. We can now infer the lifetime of a star. Since its total nuclear energy supply is available in its core (which constitutes roughly 10 percent of the star's mass), we have

$$E_{\text{nuclear}} = 0.007 \times 0.1 \times Mc^2 = 7 \times 10^{-4} Mc^2 = 10^{51}(M/M_\odot) \text{ erg}$$

Therefore, the lifetime on the hydrogen-burning main sequence is given by

$$t = \frac{E_{\text{nuclear}}}{L}$$
$$= \frac{10^{51}(M/M_\odot)}{3.8 \times 10^{33}(M/M_\odot)^3}$$
$$= 10^{10}(M/M_\odot)^{-2} \text{ years}$$

Here we have used $L_\odot = 3.8 \times 10^{33}$ erg/sec, $M_\odot = 2 \times 10^{33}$ g, and 1 year = 3.2×10^7 sec in deriving the final expression. A star of 30 solar masses is rather short-lived (actually lasting about 2×10^6 years, according to more accurate calculations). The sun, which has only been burning hydrogen as a nuclear fuel for about 4.5 billion years, has some 5 billion years more to live before its fuel is exhausted.

Once the core supply of hydrogen is exhausted, the successive stages of nuclear burning of helium and heavier elements follow rapidly, until iron is synthesized. The star ends its life either quietly as a white dwarf or explosively as a neutron star or black hole. Whether a star is destined to have a peaceful old age or suffer a violent death depends on whether its initial mass is less or greater than about 6 $M_\odot$.

CHAPTER 15

19. Mass of a White Dwarf

To understand why a critical mass for stellar stability exists, we must consider what happens as a star evolves. Its core collapses to higher and higher density and temperature as heavier and more tightly bound nuclei are fused. The end comes with fusion of iron; any further collapse will not yield energy but in fact must absorb energy. Since the star is continuously losing energy by radiation, it loses pressure support and collapses to increasingly higher densities.

Collapse will be halted for stars below a certain mass when a critical density is reached, where the atoms have become so tightly packed that a new form of pressure arises. This degeneracy pressure can be understood, in the context of quantum mechanics, as being due to a fundamental characteristic of electrons—no two electrons can be in the same place at the same time. We can evaluate degeneracy pressure by using the Heisenberg uncertainty principle, which states that the velocity and location of an electron cannot be specified simultaneously. Expressed mathematically, the uncertainty principle is

$$(\Delta p)(\Delta x) = h$$

where Δp is the uncertainty in momentum of the electron, and Δx is the uncertainty in its position. The constant h is Planck's constant. Since we can imagine squeezing the electrons together until they are a distance Δx apart, the electron density is $(\Delta x)^{-3}$, and we have

$$\Delta p = h n^{1/3}$$

The pressure P exerted by the electrons is just equal to $n m_e (\Delta v)^2$ or $n(\Delta p)^2 / m$ (since $\Delta p = m_e \Delta v$, where m_e is the electron mass and Δv its uncertainty in velocity). We consequently find that

$$P = n^{5/3}(h^2/m_e)$$

for the pressure of a degenerate gas. What is unusual about this pressure is that it does not depend on the temperature of the gas (or the random thermal motions of the electrons).

For a star to be in hydrostatic equilibrium, gravity must balance the pressure force. For a star of radius R, we therefore infer that

$$(n m_p) \left(\frac{GM}{R^2} \right) = \frac{P}{R}$$

We found that $P \propto n^{5/3}$; hence,

$$\frac{M}{R} \propto \frac{P}{n} \propto n^{2/3} \propto (M/R^3)^{2/3}$$

It follows that the radius $R \propto M^{-1/3}$. The more massive the star, the smaller it must become. This suggests that there will indeed be a limiting mass, if only by the time R has shrunk to zero!

In fact, another physical effect intervenes to modify the expression we de-

rived for electron pressure. The electrons become so tightly squeezed together that their random motions approach the velocity of light. In this limit,
the pressure due to the electrons is

$$P = nc\Delta p$$

so that now $P = hcn^{4/3}$. Instead of increasing as the 5/3 power of density, the
pressure of a relativistic degenerate gas increases as the 4/3 power.

The argument about equilibrium must be modified with the new expression
for the pressure. In this case, we find that

$$(nm_p)\left(\frac{GM}{R^2}\right) = \frac{P}{R} = \frac{hcn^{4/3}}{R}$$

whence

$$\frac{M}{R} = \frac{hc}{Gm_p}n^{1/3} = \left(\frac{hc}{Gm_p}\right)\left(\frac{M}{\frac{4}{3}\pi R^3 m_p}\right)^{1/3}$$

Rather surprisingly, the stellar radius cancels out of this relation, which
simplifies to give a critical mass

$$M = \left(\frac{hc}{Gm_p{}^2}\right)^{3/2} m_p$$

This is the limiting mass (known as the *Chandrasekhar mass*) of a white dwarf
and is equal to about 1.4 $M_\odot$. Less massive stars, larger and less dense, are
supported by nonrelativistic degeneracy pressure. More massive stars cannot
be supported by electron degeneracy pressure and must continue to collapse,
until the atomic nuclei themselves become tightly squeezed together. Here,
protons and electrons are squeezed together to form neutrons, which now in
turn become degenerate. Even neutron degeneracy will not halt the collapse
of very massive stars, which will form black holes.

The quantity $hc/Gm_p{}^2$ in the expression for the Chandrasekhar mass is a
pure (or dimensionless) number equal to 10^{39}. It involves two dimensionless
numbers. One expresses the ratio of electrostatic forces to gravitational forces
between a pair of elementary particles. There is also a dimensionless factor
$2\pi e^2/hc$ (equal to 1/137), known as the *fine structure constant*, by which this
ratio of forces must be divided to obtain $hc/Gm_p{}^2$. The ratio of electric to
gravitational forces is so great, because gravity is an exceedingly weak force
at atomic scales. Because only the values of the fundamental constants (h, c,
G, m_p) comprise this expression, the Chandrasekhar mass is a very fundamental unit of stellar mass. All stars must have masses not very different

(within a factor of 10 or so) from that of the sun. Less massive stars acquire such dense cores that they become degenerate and never heat up sufficiently to ignite their nuclear fuel. Stars more massive than about 50 $M_\odot$ develop so much radiation pressure that they blow themselves apart.

CHAPTER 16

20. The Titius-Bode Law

Numerology seems to have played a role in the discovery of the planet Neptune in 1846. Peculiarities in the motion of Uranus indicated another planet might be responsible. However, the new planet could not be precisely located without knowing its approximate distance. The estimated distance came from the Titius-Bode law, a prescription for calculating planetary distances. The method is to calculate the distance D to the $(n + 2)$th planet by taking 0.4 and adding $0.3(2)^n$:

$$D_n = 0.4 + (0.3)2^n \text{ astronomical units}$$

For Mercury, we are supposed to take 0.4. This method gives 0.7 for Venus ($n = 0$), 1.0 for earth ($n = 1$), and 1.6 for Mars ($n = 2$). It predicts a planet for $n = 3$ at 2.8 astronomical units, where none is found; instead, we find the asteroid belt at roughly this distance from the sun. Proceeding outward, we obtain 5.2 for Jupiter ($n = 4$), 10.0 for Saturn ($n = 5$), 19.6 for Uranus ($n = 6$), and 39 for Neptune ($n = 7$). The method gives accurate distances for the inner planets but begins to break down for Saturn (actually at 9.6 astronomical units), Uranus (at 19.2 astronomical units), and Neptune (at 30.1 astronomical units). The Titius-Bode law fails to predict Pluto's orbit.

Many astronomers feel that, rather than be ascribed to chance, the coincidences of this law must be explained by a correct theory for the origin of the solar system. According to this viewpoint, Pluto, with its elliptical orbit that occasionally takes it within the orbit of Neptune, is often ascribed a different origin from that of the other planets. Perhaps Pluto began life as a satellite of Neptune that was lost during a chance encounter with a nearby star.

CHAPTER 17

21. Cosmological Tests

a. Density. We can evaluate the Friedmann equation at the present era, where $H = H_0$, and $d = d_0$. The result is an equation for the curvature constant:

$$k = \frac{8\pi G}{3}(d_0 - d_{\text{crit}}) = H_0^2(\Omega_0 - 1)$$

where $d_{crit} = 3H_0^2/8\pi G$ is the value of the density appropriate to the Einstein-de Sitter ($k = 0$) cosmology, and $\Omega_0 = d_0/d_{crit}$. It is immediately clear that, if $\Omega_0 > 1$, or, if d_0 exceeds d_{crit}, k is positive, and we have a closed universe, whereas, if $\Omega_0 < 1$, or, if d_0 is less than d_{crit}, k is negative, and the universe is open. Measurement of d_0 would therefore provide a conclusive cosmological test. Unfortunately, we have only a lower limit on the value of d_0 because of the many possible forms of nonluminous matter. This lower limit is approximately equivalent to $\Omega_0 = 0.1$.

b. Redshift-Magnitude Test. Consider a distant galaxy of luminosity L, whose distance, as measured in our coordinate system, is r. We require a more convenient way of calculating the distances of galaxies than by using coordinates that we cannot easily measure. One solution is to introduce the concept of *luminosity distance.* The flux of radiation received at earth from the galaxy is spread over a sphere (area $= 4\pi r^2$) and is diminished by the redshift. The energy of each photon is decreased by $1 + z$, and the rate at which the photons arrive is diminished by the same factor. The net result is that we receive from the galaxy

$$\text{flux} = \frac{L}{4\pi r^2(1 + z)^2} \text{ erg cm}^{-2} \text{ s}^{-1}$$

This provides us with an effective luminosity distance (by analogy with the concept of flux $= L/4\pi r^2$ from a nearby source at distance r),

$$r_{lum} = r(1 + z)$$

We can compute r_{lum} by measuring the flux, if we know L; it is therefore an experimentally determined distance.

To proceed further, we evaluate r along an actual light ray. This procedure requires use of general relativity; however, the result can be expressed fairly simply. For an Einstein-de Sitter universe,

$$r = \frac{2c}{H_0}\left\{1 - \frac{1}{(1 + z)^{1/2}}\right\}$$

We can now use the definition of magnitude,

$$m = M + 5(\log r - 1)$$

where r is the luminosity distance expressed in parsecs, to derive

$$m = M + 5 \log \frac{2c}{H_0}\{1 + z - (1 + z)^{1/2}\} - 5$$

This is the relation between magnitude and redshift for a flat universe. Curved Friedmann models yield different functions of redshift. All models, however, look alike at small redshifts, and the relation can be written as

$$m = M + 5(\log cz - \log H_0) + 1.086\ (1 - \tfrac{1}{2}\ \Omega_0)z - 5$$

This explicitly shows that the relation between apparent magnitude m and $\log z$ is linear at small redshifts.

Thus, a graph of m against $\log cz$, where we identify cz with the recession velocity of a galaxy if z is small, would result in a straight line whose intercept with either axis yields the value of H_0 (assuming that we know the absolute magnitude M). Comparison with the definition formula for m shows that, at small redshifts, z is a direct measure of distance, since we have $cz = H_0 r$. For more remote objects, a nonlinear term, $1.086(1 - \tfrac{1}{2}\ \Omega_0)z$, introduces a curvature into the $m - \log z$ relation that depends on whether the universe is open or closed. If the universe is open, a galaxy at a given redshift has a greater value of m and therefore appears to be fainter than it would be in a flat universe. Intuitively, we can think of space in an open universe as being stretched—distances in curved space increase relative to those in flat space. If galaxies were brighter in the past because of evolutionary effects, we would have to correct for this effect by reducing Ω_0 or making the universe more open than it appears to be.

  c. *Angular Diameters.* Consider a galaxy of actual diameter d and apparent angular diameter A at coordinate distance r. The relation between d and A can be shown to be

$$d = rA(1 + z)^{-1}$$

Again, although this result is taken from a more sophisticated analysis, its meaning is quite clear. We must take the product of A and the distance from the source and divide by $1 + z$ to yield the diameter of the source (as locally determined). At small redshifts, the formula reduces to that obtained in a local region of space, where expansion is unimportant, $d = rA$. The angular size is obtained by substituting the formula given previously for r in terms of z, valid for an Einstein-de Sitter universe. This yields

$$A = (1 + z)d/r = dH_0(2c)^{-1}(1 + z)[1 - (1 + z)^{1/2}]^{-1}$$

This formula has an interesting property. When z is small, A varies as z^{-1} (or as the inverse of distance, just as expected for nearby sources), whereas, at large z, A must increase as $1 + z$ (since $[1 + z]^{-1/2}$ becomes negligible compared with 1). In other words, the angular diameter goes through a minimum. This actually occurs in an Einstein-de Sitter universe at a redshift of 1.25. At larger redshifts, galaxies appear to grow bigger! This completely unexpected result is a consequence of the focusing of light by the gravitational field of the universe, which acts as a kind of gigantic lens.

This effect occurs even in the open Friedmann model, although at a larger redshift because of the reduced pull of gravity. In principle, it offers a way of distinguishing between different cosmological models by measuring the angular diameters of very remote galaxies. In practice, it has been very difficult to perform this test, because the images of galaxies are exceedingly fuzzy and the edges are not well defined. The angular diameter of a galaxy at redshift 1 is typically only about 3 arc-seconds, equivalent to the angle subtended by a penny at a distance of 1 kilometer. The angular diameter-redshift relation offers greater promise as a cosmological test when applied to clusters of galaxies, the sizes of which seem to be relatively uniform. Evolutionary corrections are very uncertain, however.

Having found the angular diameter of a galaxy, it is easy to obtain its surface brightness, which determines how detectable a galaxy is relative to the sky background. To obtain the surface brightness, we simply divide the flux from the galaxy by the apparent area it subtends as measured on a sphere of unit radius. (This angular area is called a *solid angle*.) This leads to

$$\text{surface brightness} = \left(\frac{\text{flux}}{\pi \dfrac{A^2}{4}} \right) = \frac{L}{4\pi r^2 (1 + z)^2} \left[\frac{\pi}{4} (1 + z)^2 \frac{d^2}{r^2} \right]^{-1}$$

$$= \frac{L}{\pi^2 d^2 (1 + z)^4} \ \text{erg/cm}^2/\text{s}$$

The factor $L/\pi^2 d^2$ is the surface brightness as determined by a nearby observer, and the redshift factor is due to the cosmological expansion. This result has not required any assumption about a cosmological model—it is generally true that surface brightness decreases as the fourth power of redshift. An earlier illustration of this result (in Note 8) was implicit in the variation of the background radiation energy density u_λ as $(1 + z)^4$(equivalent to the surface brightness divided by c).

d. *Number counts.* Let us first consider a static universe with a uniform distribution of sources of luminosity L. If f is the flux from a source at distance r, then

$$f = \frac{L}{4\pi r^2}$$

Hence, we can see sources brighter than f out to a distance r, where

$$r = \left(\frac{L}{4\pi f} \right)^{1/2}$$

The total number N of sources brighter than f is given by

$$N = \frac{4\pi}{3} r^3 n_0$$

where n_0 is the local number density of sources. In other words,

$$N = \frac{4\pi}{3} n_0 \left(\frac{L}{4\pi f}\right)^{3/2} \propto 1/f^{3/2}$$

Thus, the number of sources increases as the inverse three-halves power of the flux.

It is straightforward to adapt this result to the case of an Einstein-de Sitter universe. We must simply replace r in this derivation by the luminosity distance $r_{lum} = r(1 + z)$. We now obtain

$$N = \frac{4\pi}{3} n_0 \left(\frac{L}{4\pi f}\right)^{3/2} (1 + z)^{-3}$$

In other words, the number brighter than any given flux is reduced by a factor $(1 + z)^3$ relative to that in the nonexpanding case.

A similar but more complicated reduction is found when open or closed Friedmann models are considered, which give a shallower rise in the predicted number of sources than would be given by the inverse three-halves power of the flux. We can graph $\log N$ against $\log f$, and the slope of the resulting curve should therefore be flatter than $\frac{3}{2}$ because of the effects of the cosmological redshift, fewer sources being found at higher redshifts than expected in a nonexpanding universe.

In the case of the radio source counts, a steeper slope of approximately 1.8 was actually found. This finding can be explained most easily as resulting from evolution; either or both of the following effects must occur. Sources must be brighter (L must increase) or be more frequent (n_0 must increase) so as to overcompensate for the $(1 + z)^{-3}$ factor and give an increase at low flux levels over that expected in a nonexpanding universe. In this way, a slope steeper than $\frac{3}{2}$ will be obtained.

FOR FURTHER READING

For the reader who wishes to pursue in more detail some of the topics described in this book, there is considerable choice. Historical surveys of cosmology may be found in:

Barrow, J., and Tipler, F. *The Cosmological Anthropic Principle*. Oxford: Oxford University Press, 1986.
Charon, J. *Cosmology*. New York: McGraw-Hill, 1970.
Ferris, T. *The Red Limit*. New York: William Morrow, 1977.
Gribbin, J. *In Search of the Big Bang*. New York, Bantam, 1986
Munitz, M. K. *Theories of the Universe*. New York: Free Press, 1957.
North, J. D. *The Measure of the Universe*. Oxford: Clarendon Press, 1950.

Elementary discussions of Big Bang cosmology may be found in:

Gamow, G. *The Creation of the Universe*. New York: Viking, 1961.
Bondi, H. *Cosmology*. Cambridge: Cambridge University Press, 1961.
Harrison, E. *Cosmology*. New York, Cambridge University Press, 1981.
Wagoner, R., and D. Goldsmith. *Cosmic Horizons*. Stanford, Cal.: Stanford University Press, 1982.
Layzer, D. *Constructing the Universe*. New York: Scientific American Library, 1984.

The very early universe is discussed in:

Weinberg, S. *The First Three Minutes*. New York: Basic Books, 1977.
Barrow, J., and J. Silk. *The Left Hand of Creation*. New York, Basic Books, 1983.
Pagels, H. *Perfect Symmetry*. New York: Simon and Schuster, 1982.
Davies, P. *The Accidental Universe*. Cambridge: Cambridge University Press, 1982.
Reeves, H. *Atoms of Silence*. Cambridge, Mass.: MIT Press, 1981.

The solar system and its formation are described in:

Dermott, S. F., ed. *The Origin of the Solar System*. New York: Wiley, 1978.

Field, G. B., G. L. Verschuur, and C. Ponnamperuma. *Cosmic Evolution.* Boston: Houghton Mifflin, 1978.

Baugher, J.F. *The Space-Age Solar System.* New York: Wiley, 1988.

Wood, J. *The Solar System.* Englewood Cliffs, N. J.: Prentice-Hall, 1979.

Short books devoted to alternative cosmologies include:

Alvén, H. *Worlds–Antiworlds.* San Francisco: Freeman, 1966.

Lemaître, G. *The Primeval Atom.* New York: Van Nostrand, 1950.

At a somewhat more technical level are:

Berry, M. *Principles of Cosmology and Gravitation.* Cambridge: Cambridge University Press, 1976.

Rowan-Robinson, M. *Cosmology.* 2d ed. Oxford: Clarendon Press, 1981.

Sciama, D. *Modern Cosmology.* Cambridge: Cambridge University Press, 1975.

At a still more technical level, the many available books include:

Landau, L., and E. Lifshitz. *Classical Theory of Fields.* Oxford: Pergamon Press, 1962.

McVittie, G. C. *General Relativity and Cosmology.* 2d ed. London: Chapman and Hall, 1965.

Misner, C., K. Thorne, and J. Wheeler. *Gravitation.* San Francisco: Freeman, 1973.

Peebles, P. J. E. *Physical Cosmology.* Princeton: Princeton University Press, 1971.

———*The Large-Scale Structure of the Universe.* Princeton: Princeton University Press, 1980.

Tolman, R. C. *Relativity, Thermodynamics, and Cosmology.* Oxford: Clarendon Press, 1934.

Weinberg, S. *Gravitation and Cosmology.* New York: Wiley, 1972.

Zel'dovich, Ya. B., and I. D. Novikov. *The Structure and Evolution of the Universe.* Chicago: University of Chicago Press, 1983.

A number of review articles are devoted to more specific topics. These are often more detailed and up to date than any of the preceding books. A recent collection is:

Kolb, E., M. Turner, D. Lindley, K. Olive, and D. Seckel. *Inner Space, Outer Space.* Chicago: University of Chicago Press, 1986.

A useful compendium of review articles that cover a range of topics, including properties of galaxies, extragalactic radio sources, and observational cosmology, is:

Sandage, A., M. Sandage, and J. Kristian, eds. *Galaxies and the Universe.* Chicago: University of Chicago Press, 1975.

GLOSSARY

absolute brightness   The total luminosity radiated by an object.

absolute magnitude   The apparent magnitude that a star would have at the standard reference distance of 10 parsecs. The absolute magnitude of the sun is 4.85.

absorption   The process of transition of an electron from an inner orbit to an outer orbit around the nucleus is caused by incidence of light at a wavelength characteristic of the gain in energy by the electron. One quantum of light at this characteristic wavelength is expended for each electron that makes the transition.

accumulation theory   The theory by which planetesimals are assumed to collide with one another and coalesce, eventually sweeping up enough material to form the planets.

adiabatic fluctuations   Fluctuations in both the matter and radiation density, as though a volume of the universe were slightly squeezed but allowing no radiation to escape. Prior to the decoupling era, adiabatic fluctuations behaved like waves, on scales smaller than the horizon size. After decoupling, gravitational instability sets in on scales above about $10^{13} M_{\odot}$, smaller adiabatic fluctuations having been damped at earlier eras.

age of the universe   The time elapsed since the singularity predicted by the Big Bang theory, estimated to be between 7 and 20 billion years.

Alfvén-Klein cosmology   A cosmological model in which the early universe is depicted as a giant collapsing spherical cloud of matter and antimatter. When a critical density is reached, the matter and antimatter begin to annihilate, the resulting release of radiation and energy causing the universe to expand. There are many difficulties with this model of the expanding universe, which is largely discredited on observational grounds.

Ångstrom   A measure of wavelength. 1 Ångstrom (Å) = 10^{-8} centimeter.

angular momentum A measure of the momentum associated with rotation, computed about a specified origin by dividing the system into a number of elements or particles, and taking the sum of the vector products of the radius joining each particle to the origin and the momentum of each particle. The angular momentum of an isolated system is conserved.

anisotropy A lack of isotropy; that is, a dependence on direction.

anthropic cosmological principle Only a restricted range of possible universes is favorable to the evolution and existence of life.

antineutrino The antiparticle of a neutrino.

antiparticle An elementary particle of opposite charge but otherwise identical to its partner. Most of the observable universe consists of particles and matter, as opposed to antiparticles and antimatter.

antiproton The antiparticle of a proton, identical in mass and spin but of opposite (negative) charge.

apparent magnitude A measure on a logarithmic scale of the observed brightness of a celestial object. The brightest stars are of first magnitude. The fainter a star, the greater is its apparent magnitude; the faintest stars visible to the naked eye are sixth magnitude.

asteroid belt A region of space between Mars (at 1.5 astronomical units) and Jupiter (at 5.2 astronomical units) where most asteroids are located.

asteroids Small planetlike bodies of the solar system. More than 1800 have been discovered, and millions of smaller asteroids probably exist. The largest asteroid, Ceres, has a diameter of 1000 km and a mass equal to 2 $\times$ 10^{-4} that of the earth. The total mass of all the asteroids probably amounts to less than 10 times that of Ceres.

astronomical unit The mean earth—sun distance, equal to 1.496×10^{13} cm or 8.31 light-minutes.

big bang theory A model of the universe in which space-time began with an initial singularity and subsequently expands.

binary star A system of two stars orbiting around a common center. Visual binaries are those whose stellar components can be resolved telescopically; spectroscopic binaries can also be detected because of the variable radial velocities of the two stars.

blackbody An idealized body that absorbs radiation of all wavelengths incident on it. It is a perfect absorber and therefore a perfect emitter. The radiation emitted by a blackbody depends only on the temperature of the blackbody.

blackbody radiation Radiation whose intensity distribution with respect to wavelength is that of a blackbody.

black hole A gravitationally collapsed mass from which no light, matter, or any kind of signal can escape. A black hole is formed when the gravitational field has become so strong that the escape velocity of a body approaches the velocity of light.

blueshift The shift of spectral lines toward shorter wavelengths in the spectrum of an approaching source of radiation.

Bok globule   A compact quasi-spherical dark interstellar cloud. Characteristic masses are about 20 solar masses, and characteristic radii about 1 light-year. Bok globules could be precursors to protostar formation.

boson   A class of elementary particles with integer units of the basic unit of spin $h/2\pi$. Examples are the photon, X, W, and Z bosons. More than one boson can exist with identical quantum numbers. They do not obey the Pauli exclusion principle.

bottom-up theory   A scenario for large-scale structure formation in which the smallest objects (dwarf galaxies) form first and structure develops hierarchically unitl the largest superclusters of galaxies are formed.

Brans-Dicke theory   A modification of Einstein's theory of general relativity, in which gravitation is described both by a tensor field (as in Einstein's theory) and a scalar field.

carbonaceous chondrites   Stony meteorites characterized by the presence of carbon compounds. They are the most primitive samples of matter in the solar system.

Cepheid variable star   One of a group of very luminous supergiant pulsating stars named for the prototype δ Cephei. Classical Cepheids (Type I) have characteristic periods of 5 to 10 days and are found in relatively young galactic star clusters. Type II Cepheids have characteristic periods of 10 to 30 days and are primarily found in globular clusters. The luminosities of all Cepheids are proportional to their periods, but a different period-luminosity relation applies to each type. Polaris is the nearest Cepheid.

Chandrasekhar mass   A limiting mass for white dwarf stars, equal to 1.4 solar masses for a nonrotating white dwarf. Above this critical mass, a white dwarf cannot be supported by electron degeneracy pressure against gravitational collapse.

chemical differentiation   The separation of different elements, often heavier elements from lighter elements, as a consequence of different chemical reactions.

chondrules   Smaller spherical grains, usually composed of iron, magnesium, or aluminum silicates, with a glassy appearance, found as inclusions in primitive stony meteorites. The size of an inclusion varies from microscopic dimensions up to the size of a pea.

closed space   A space of finite volume but without any boundary (in the cosmological context).

closed universe   A Friedmann-Lemaître cosmological model in which space is finite and of positive curvature and which will eventually recollapse.

cluster of galaxies   An aggregate of galaxies. Clusters may range in richness from loose groups, such as the Local Group, with 10 to 100 members, to great clusters of over 1000 galaxies.

cold dark matter   A generic name for any species of weakly interacting particles that constitutes the dark matter of the universe and is cold, with negligible velocity dispersion or temperature, at the era when the universe first becomes matter-dominated. This allows preexisting fluctuations to grow without hindrance on all scales.

comet   A diffuse body of gas and solid particles that orbits the sun in a highly eccentric trajectory, ranging from within the earth's orbit to more than 10^4 astronomical units. Comets are unstable bodies with masses about 10^{-10} earth masses. They do not survive, on the average, more than 100 passages by the sun. Only about 4 percent of known comets recur periodically.

comoving sphere   A hypothetical and arbitrary spherical surface (about any point) that is expanding along with the rest of the universe. Relative to the comoving sphere, the particles on it are at rest.

composite particle theory   A class of elementary-particle theories, according to which there are increasing numbers of elementary-particle states of higher and higher mass.

Compton wavelength   The dual nature of particles and waves results in the association of a fundamental wavelength with each elementary particle. It is equal to $\hbar/mc$, where $\hbar$ is Planck's constant, c is the speed of light, and m is the particle mass.

conservation of angular momentum   The total angular momentum of an isolated dynamical system does not change during the course of its evolution.

conservation of energy   The total energy of a system (including kinetic energy and gravitational energy) is conserved and does not vary. Thus, kinetic energy can increase only at the expense of gravitational potential energy. Modern physics has modified the law of conservation of energy, because matter can be created or annihilated; a more general law is the conservation of mass and energy.

copernican cosmological principle   The point from which we view the universe should not be any preferred location in space. This principle leads to the hypothesis that the universe is approximately homogeneous and isotropic on the largest scales. It follows that at a given cosmic time, all observers everywhere in space would view approximately the same large-scale distribution of matter in the universe.

cosmic microwave background radiation   Diffuse isotropic radiation whose spectrum is that of a blackbody at 3 degrees kelvin and consequently is most intense in the microwave region of the spectrum.

cosmic strings   The tubelike configurations of energy that can arise in the early moments of the universe. Typically, a string would have a thickness of only 10^{-27} centimeter, but if extended across the entire observable universe would have a mass of about 10^{17} suns.

cosmogony   The study of the origin of celestial systems, ranging from the solar system to stars, galaxies, and clusters of galaxies.

cosmological constant   A term introduced by Einstein into his field equations of gravitation to permit a static model of the universe. It corresponded, as introduced originally, to a cosmic repulsion force that could withstand the attractive tendency of gravity.

cosmology   The study of the large-scale structure and evolution of the universe.

Coulomb barrier Reactions between atomic nuclei are inhibited by the need for the nuclei to overcome the repulsive force that acts between any pair of similarly charged particles. At high temperatures, nuclei move sufficiently fast to be able to overcome the Coulomb barrier. The greater the nuclear charge, the higher is the temperature that is required for nuclear reactions to occur.

Cretaceous-Tertiary layer A geological layer, deposited about 65 million years ago, that demarcates the age when the dinosaurs and many species of planktonic animals became extinct.

damping The gradual decay of an oscillation or wave resulting from viscous drag forces.

dark mass Matter whose presence is inferred from dynamical measurements but which has no optical counterpart. The luminous regions of galaxies have mass-luminosity ratios of about 10. However, the mass-luminosity ratio in the outer halos of many spiral galaxies is 100 or more; one sees the brightness fall off with distance from the center of the galaxy, but considerable mass is present. A similar situation prevails in galaxy clusters, where nonluminous matter must provide most of the self-gravitational attraction that holds the clusters together. The dark mass is not really missing; it is present but invisible (at least to current detectors). It is generally believed to consist either of the remnants of massive stars or of planetary-sized objects comparable in mass to Jupiter, or of weakly, interacting relic particles from the big bang.

decoupling The rapid transition from an ionized state at a redshift of 1000, when the blackbody radiation is scattered by the free electons, to an unionized state, when the matter is predominantly in the form of hydrogen atoms that do not scatter the radiation appreciably. Radiation subsequently does not interact with matter unless the matter becomes reionized at a later epoch by radiation from quasars or forming galaxies.

decoupling era The era some 3×10^5 years after the big bang when the cosmic blackbody radiation was last scattered by the matter. At this era, at a redshift of about 1000 and a temperature of about 3000 degrees kelvin, the protons and electrons combined to form hydrogen atoms, which are effectively transparent to the radiation.

degeneracy pressure Pressure in a degenerate electron or neutron gas.

degenerate matter A state of matter found in white dwarfs and other extremely dense objects, in which strong deviations from classical laws of physics occur. As the density increases at a given temperature, the pressure rises more and more rapidly, until it becomes independent of temperature and dependent on density alone. At this point, the gas is said to be degenerate.

density fluctuations Localized enhancements in the density of either matter alone or matter and radiation in the early universe. The amplitude need initially have been very small (less than 1 percent) to account for subsequent formation of galaxies.

deuterium The first heavy isotope of hydrogen, its nucleus consisting of one proton and one neutron. Like hydrogen, the deuterium atom has one electron, and it has similar chemical properties to hydrogen, forming, for example, heavy water (HDO). The abundance of deuterium in interstellar space is about 1.4×10^{-5} that of hydrogen.

Doppler effect Displacement of spectral lines in the radiation received from a source due to its relative motion along the line of sight. A motion of approach results in a blueshift; a motion of recession results in a redshift.

Eddington-Lemaître universe A cosmological model in which the cosmological constant plays a crucial role by allowing an initial phase that is identical to the Einstein static universe. After an arbitrarily long time, the universe begins to expand. The difficulty with this model is that the initiation of galaxy formation may actually cause a collapse rather than initiate an expansion of the universe.

Einstein-de Sitter universe The Friedmann-Lemaître expanding model in which space is Euclidean was advocated by Einstein and de Sitter in 1932.

Einstein static universe A cosmological model in which a static (neither expanding nor collapsing) universe is maintained by introducing a cosmological repulsion force (in the form of the cosmological constant) to counterbalance the gravitational force.

electron volt The energy acquired by an electron when accelerated through a potential difference of 1 volt; 1 electron volt = 1.6×10^{-12} ergs.

elliptical galaxy A galaxy whose structure is smooth and amorphous, without spiral arms and ellipsoidal in shape. Ellipticals are redder than spirals of similar mass. Giant ellipticals contain over 10^{12} $M_\odot$ in stars, whereas dwarf ellipticals have masses as low as $10^7 M_\odot$.

emission The process of transition of an electron from an outer orbit to an inner orbit around the nucleus results in a characteristic amount of energy being radiated (as line emission) that corresponds to the lost energy of the electron.

energy density The amount of energy in the form of radiation per unit volume, expressed in ergs cm^{-3}. The energy density of blackbody radiation at temperature T is aT^4, where the radiation constant $a = 7.56 \times 10^{-15}$ erg cm^{-3} (K)$^{-4}$.

entropy A measure of the amount of disorder in a system.

epicycle theory A means of accounting for the apparent motions of the planets in terms of circular motions in a geocentric cosmology. Each planet moves in a circle, the center of which moves in a circle of larger radius, and so on, the largest circles being centered on the earth.

explosive amplification A theory of galaxy formation according to which gigantic explosions, driven by a quasar or multiple supernova events or still more exotic phenomena, drive shells of intergalactic gas. The shells sweep up more and more gas and eventually become unstable, fragmenting into galaxies that in turn seed further explosions.

explosive nucleosynthesis   The nucleosynthesis processes that are believed to occur in supernovae. Explosive carbon burning occurs at a temperature of about 2×10^9 degrees kelvin and produces the nuclei from neon to silicon. Explosive oxygen burning occurs near 4×10^9 degrees kelvin and produces nuclei between silicon and calcium in atomic weight. At higher temperatures, still heavier nuclei, up to and beyond iron, are produced.

fermion   An elementary particle with half integral units of the basic unit of spin, $h/2\pi$; examples are electrons, protons, and neutrinos. No two fermions can exist with identical values of the quantum numbers characterizing them.

fine-structure constant   A coupling constant $e^2/\hbar c$, approximately equal to 1/137 (where e is the electron charge, $\hbar$ is Planck's constant, and c is the speed of light), that measures the strength of the interaction between a charged particle and the electromagnetic field.

flat space   A synonym for ordinary euclidean space.

flux   A measure of the amount of power or radiation received per unit time per unit area measured in erg $cm^{-2}s^{-1}$.

Fraunhofer lines   Absorption lines in the spectrum of the sun were first studied by Joseph Fraunhofer in 1914.

free-free radiation   The acceleration of an unbound, or free, electron by a proton or atomic nucleus results in the emission of electromagnetic radiation.

Friedmann equation   An equation that expresses energy conservation in an expanding universe. It is formally derived from Einstein's field equations of general relativity by requiring the universe to be everywhere homogeneous and isotropic.

Friedmann-Lemaître universes   The three standard big bang models that were formulated by Friedmann and Lemaître.

galactic nucleus   In the innermost region of a galaxy, there is often a concentration of stars and gas, sometimes extending over thousands of light-years from the center of the galaxy.

galaxy correlation function   A measure of the degree of galaxy clustering in a large sample of galaxies. The 2-point correlation function is the probability that there will be a second galaxy at a certain distance from any one galaxy after subtracting the probability for this to occur is a random distribution.

gamma rays   Photons of very high frequency and very short wavelength; the most energetic and penetrating form of electromagnetic radiation.

gaseous fragmentation   The systematic breakup of a gas cloud into smaller and smaller subunits as the gas cools and continues to collapse. The gravitational forces continually overtake the opposing pressure gradients as long as the cloud is able to radiate freely; consequently, the Jeans mass decreases, and fragments divide into smaller subfragments. The process stops only when opacity intervenes to inhibit the cooling and radiation.

geocentric cosmology A model of the universe in which the earth is centrally located, and the sun, planets, and stars revolve around the earth.

global inertial frame A coordinate system or frame of reference anchored with respect to the overall distribution of matter in the universe.

grand unification/unified theories (GUTs) A class of gauge theories that unite the strong, electromagnetic, and weak interactions at high energy. Ultimately, it is hoped that they can be extended to incorporate gravity.

gravitational instability The process by which fluctuations in an infinite medium of size greater than a certain length scale (the Jeans length) grow by self-gravitation. In the expanding universe, the growth is slowed by the expansion, but the unstable fluctuations eventually collapse into stable, gravitationally bound systems.

gravitational lens A concentration of matter, such as a galaxy or a cluster of galaxies, that bends light rays from a background object, resulting in production of multiple images.

gravitational potential energy Energy that a body can acquire by falling through a gravitational field and that decreases as the kinetic energy increases. There is no general reference level (analogous to the state of rest of a body in defining kinetic energy), and so we customarily define the change in gravitational potential energy as the negative of the work done by the gravitational forces during the body's change of position.

gravitational redshift Light is emitted at a lower frequency and longer, or redder, wavelength in a gravitational field than in the absence of a gravitational field. The gravitational redshift of the radiation from a white dwarf star amounts to about 1 part in 10^4; this effect has also been measured for the earth's gravitational field, where it amounts to 1 part in 10^9. Near a black hole, the effect can be much larger, being proportional to the mass of the body divided by its radius.

graviton A quantum of gravitational radiation.

HII region A region of glowing ionized gas that surrounds and is ionized by radiation from massive young hot stars.

hadron Any elementary particle (including mesons and baryons) that interacts strongly in nuclei.

hadronic era The interval lasting until some 10^{-4} seconds after the big bang when the universe was radiation-dominated, containing many hadrons in equilibrium with the radiation field. The hadronic era ended when the characteristic photon energy fell below the rest mass of a pion or π-meson (270 electron masses), and very few hadrons remained (about one hadron for every 10^8 photons).

half-life The length of time required for half the weight of any given sample of radioactive substance to disintegrate.

halo The diffuse, nearly spherical cloud of old stars and globular clusters that surrounds a spiral galaxy.

Heisenberg uncertainty principle The uncertainty in the measurement of the position of an elementary particle varies inversely as the uncertainty in measuring its momentum. Thus, it is impossible to make a measurement of an atomic or nuclear process of arbitrary accuracy; the process will be disturbed by the act of measurement.

heliocentric cosmology A model of the universe in which the sun is centrally located, and the planets and stars are at successively increasing distances from the sun.

Hertzsprung-Russell diagram A plot of stellar luminosity (or absolute magnitude) versus effective temperature (or color). Hydrogen burning stars define a unique locus (the main sequence) in this diagram, which is used to study stellar evolution.

hierarchical clustering The process by which a system of self-gravitating particles will gradually aggregate into larger and larger gravitationally bound groups and clusters. Small clusters merge into larger clusters, which retain little trace of the subunits from which they formed. Elliptical galaxies may have formed in this way from mergers of globular-cluster-sized star clusters; clusters of galaxies may have formed by a similar process.

horizon The observable region of the universe, limited in extent by the distance light has traveled during the time elapsed since the singularity.

hot dark matter A generic name for any species of weakly interacting particle that constitutes the dark matter of the universe and is hot, i.e., that has a significant temperature or velocity dispersion (comparable to the speed of light), when the universe first becomes matter-dominated. This means that only the very largest fluctuations (comparable in size to the horizon at this era) survive to form large-scale structure. An example of a hot dark matter candidate is the massive neutrino, with mass of about 30 electron volts.

Hubble constant The constant (H) of proportionality in the relation between distance of galaxies from us and their recession velocity. A value for H of 50 km s^{-1} Mpc^{-1} or 15 km s^{-1} (million light-years) is currently favored by many astronomers, although a value that is twice as large is also consistent with the astronomical data.

Hubble's law The distance of galaxies from us is linearly proportional to their redshift and therefore linearly proportional to their relative velocity of recession for velocities that are small compared to the speed of light.

hydrostatic equilibrium A balance between the inward gravitational force and the outward gas and radiation pressure in a star.

hyperbolic space A three-dimensional space whose geometry resembles that of a saddle-shaped surface and is said to have negative curvature.

hypergalaxy A system consisting of a dominant spiral galaxy surrounded by a cloud of dwarf satellite galaxies, often ellipticals. Our galaxy and the Andromeda galaxy are hypergalaxies.

inflation, inflationary universe The period during the first 10^{-35} second of the universe when a phase transition accelerates the expansion rate according to some grand unified gauge theories.

intergalactic gas Matter that is present in the region between the galaxies. It has been detected in considerable amounts in great clusters of galaxies, where the intergalactic gas is so hot that it emits copious amounts of x-radiation. In several groups of galaxies, including the Local Group, the evidence for the presence of intergalactic gas is controversial; clouds of atomic hydrogen may be present.

interstellar grains Small needle-shaped particles in the interstellar gas with dimensions from 10^{-6} to 10^{-5} cm. They are primarily composed of silicates and strongly absorb, scatter, and polarize visible light at wavelengths comparable to their size, reemitting the light in the far-infrared region of the spectrum. The amount of visual extinction is wavelength-dependent and leads to a dimming and reddening of starlight.

inverse square law A force law that applies to the gravitational and electromagnetic forces in which the magnitude of the force decreases in proportion to the inverse of the square of the distance.

irradiation Exposure to an intense flux of fast neutrons or to ionizing radiation. The outer layers of a supernova are irradiated by neutrons produced as the core collapses to form a neutron star. The neutrons react with atomic nuclei to form heavier elements.

irregular galaxy A galaxy without spiral structure or smooth, spheroidal shape, often filamentary or very clumpy, and generally of low mass (10^7 to $10^{10}M_\odot$).

isothermal fluctuations Fluctuations in the matter density, without any associated perturbation of the radiation density. The radiation temperature therefore remains uniform. Prior to the decoupling era, isothermal fluctuations were frozen, neither growing nor decaying. After decoupling, isothermal fluctuations became gravitationally unstable, if greater than the Jeans mass, about $10^6M_\odot$.

isotropy No dependence on direction.

Jeans length The critical wavelength above which oscillations (or sound waves) in an infinite homogeneous medium become gravitationally unstable. A density fluctuation of size greater than the Jeans length will decouple by its own self-gravitation from the rest of the medium to form a stable, bound system.

Jeans mass The mass enclosed within a sphere of diameter equal to the Jeans length.

kinetic energy The energy associated with motion; the work that must be done to change a body from a state of rest to a state of motion, equal to $\frac{1}{2}mv^2$ for a body of mass m moving at velocity v.

lepton   Elementary particles that do not participate in the strong interactions, including electrons, muons, and neutrinos.

leptonic era   The era following the hadronic era, when the universe consisted mainly of leptons and photons. It began when the temperature dropped below 10^{12} degrees kelvin some 10^{-4} seconds after the big bang, and it lasted until the temperature fell below 10^{10} degrees kelvin, at an era of about 1 second. At this stage, the characteristic photon energy fell below the rest mass energy of an electron, and the abundance of electron-positron pairs fell by many orders of magnitude. Only 1 electron survived for every 10^8 photons. The universe was subsequently radiation-dominated (substantial numbers of neutrinos were also present, but they did not interact directly with the matter or the radiation).

light-year   The distance traveled in a vacuum by light in 1 year, equal to 9.46×10^{17} centimeters.

Local Group   The system of galaxies to which our Milky Way galaxy belongs is a small group, containing only two large spirals (our galaxy and the Andromeda galaxy, Messier 31) and twenty or more smaller systems.

local inertial frame   A coordinate system or frame of reference defined in the vicinity of the earth in which Newton's first law of motion is valid; that is, a nonrotating and nonaccelerating reference frame.

luminosity distance   If the intrinsic luminosity of a distant object is known, then measurement of its apparent brightness and application of the inverse square law enables the luminosity distance to be calculated. For very distant galaxies, spatial curvature is important, and the luminosity distance differs from other measures of distance.

luminosity function   The distribution of galaxies with respect to their luminosities.

Lyman alpha line   The characteristic spectral line of atomic hydrogen associated with its lowest excited state. The corresponding wavelength is 1216 Ångstroms in the far ultraviolet, and so the Lyman alpha line can only be studied from spacecraft or in the spectra of highly redshifted quasars.

Mach's principle   The hypothesis that the local inertial frame and the inertia of any body is determined by the distribution of all the matter in the universe.

magnetic monopole   A massive particle predicted to exist by grand unified theories. It possesses internal structure and a mass near 10^{-8} gram along with the long-range magnetic field of a single isolated magnetic pole.

magnitude   A measure, on a logarithmic scale, used to indicate the brightness of a celestial object. A 1-magnitude difference in brightness between two stars corresponds to a difference in luminosity by $10^{0.4}$ or 2.51; 5 magnitudes corresponds to a factor of 100 in luminosity.

main sequence   The principal sequence of stars on the graph of luminosity versus effective temperature (the Hertzsprung-Russell diagram) for a group of stars. More than 90 percent of the stars we observe lie on the main sequence, where they remain throughout their hydrogen-burning phase. The observed upper mass limit of the main sequence is about 60 solar masses, more massive stars being unstable, and the calculated lower limit is 0.085 solar mass, less massive stars being incapable of igniting their hydrogen.

massive black hole Utilized in a theoretical model for quasars and active galactic nuclei, according to which the energy source is due to infall (and resultant heating) of gas and stars onto a supermassive central black hole.

mass-luminosity ratio The ratio of the mass of a system, expressed in solar masses, to its visual luminosity, expressed in solar luminosities. The Milky Way has a mass-luminosity ratio in its inner regions of 5, indicating that the typical star is a dwarf of mass about half that of the sun. A rich cluster of galaxies such as the Coma cluster has a mass-luminosity ratio of about 200, indicating the presence of a considerable amount of dark matter.

mean free path The mean distance traversed by a particle before undergoing a significant deflection or collision.

metagalaxy A synonym for the universe.

meteor A "shooting star," or streak of light in the night sky produced by the transit of a meteoroid through the earth's atmosphere. The glowing meteroid is a small rocky body that burns up before it reaches the earth's surface.

meteorites A solid portion of a meteoroid that has reached the earth's surface. The three types of meteorites—stony, iron, and stony-iron—range in size from microscopic to the size of an asteroid. Several hundred tons of meteoric material are estimated to fall on the earth each day.

meteor shower A great number of meteors that appear within a period of a few hours and appear to radiate from a common point in the sky. Meteor showers are often associated with comets.

microwave background anisotropy experiment An experiment designed to measure the intensity of the cosmic microwave background radiation in different directions. A fundamental prediction of the cosmological origin of this radiation is that the earth's motion relative to the distant regions of the universe should be detectable. The effect amounts to an increase of about 10^{-3} degree kelvin in brightness in the direction we are traveling and a similar decrease in the opposite direction.

mini-black holes In a chaotic early universe, black holes may form at eras as early as the Planck time. The characteristic size of these mini-black holes is 10^{-6} gram, the minimum mass of a collapsing inhomogeneity at that time. Larger mini-black holes may form at later eras. Since conventional theories of stellar evolution show that only very massive stars can form black holes, the possible formation of mini-black holes is a unique characteristic of the very early universe.

mixmaster model A cosmological model in which the early universe was homogeneous but highly anisotropic; the unverse expanded at different rates in two directions while collapsing in a third direction and then reversed itself. Any given amount of matter would alternately go through pancakelike and cigarlike configurations that gradually increase in volume as the universe expands.

molecular cloud An interstellar cloud consisting predominantly of molecular hydrogen, with trace amounts of other molecules such as carbon monoxide and ammonia.

molecular cloud complex A region of extensive emission of molecular line radiation by dense, cold interstellar gas. There are often several distinct intensity peaks, each representing individual clumps or clouds of gas and dust in a region that characteristically extends for 50 light-years and is often associated with T-Tauri stars—young, pre-main-sequence stars—and also hot massive stars and the ionized gas around them.

monotonic Either continuously increasing or decreasing.

nebula A term previously applied indiscriminately to all types of fuzzy patches of nebulosity in the sky. Many of these are now recognized to be remote galaxies; others are planetary nebulae, nova shells, supernova remnants, and clouds of ionized gas in our own galaxy.

nebula theory According to this theory of the origin of the solar system, a slowly rotating diffuse gas cloud contracted and cooled, collapsing to a thin disk, which fragmented and eventually formed the planets. At the same time, the material with low angular momentum collected in the core to form the sun.

neutrino A particle with zero rest mass, no charge, but with a characteristic spin, which carries away energy in nuclear reactions. Neutrinos interact very weakly with matter, and provide a potential means of examining conditions in the central core of the sun and other stars.

neutron star A star whose core is composed primarily of neutrons, as is expected to occur at a density above 10^{14} g cm^{-3}. Such a state is reached by a star at the endpoint of its evolution, when its nuclear fuel is exhausted and the star is too massive to become a white dwarf. Stars of 8 to 30 or more solar masses are believed to form neutron stars. The maximum mass of a neutron star is about 4 solar masses, and the remaining material is ejected either during prior evolution in steady mass loss (as in the planetary nebula phase) or in a supernova explosion when the neutron star forms. Pulsars are rotating magnetized neutron stars of about 1 or 2 solar masses with typical radii of 10 kilometers.

newtonian cosmology The simplest cosmological models, including the standard Big Bang models, can be derived in the framework of Newton's classical theory of gravitation, although Birkhoff's theorem from the theory of general relativity is needed to justify the use of newtonian theory in an infinite medium.

nova A star that exhibits a sudden increase in luminosity, by as much as 10^6. The explosion ejects a shell of matter but does not disrupt the star; the nova outburst often repeats in the same star. Novae are found in close binary systems, one component of which is a red dwarf that appears to be losing matter to its white-dwarf companion. The buildup of hydrogen-rich matter on the white dwarf's surface is unstable, resulting eventually in a thermonuclear explosion.

nucleosynthesis   Formation of atomic nuclei by nuclear reactions either in the Big Bang or in stellar interiors. Only the light elements (notably helium and deuterium) are made in significant amounts in the early universe; heavier elements are produced in stars.

nucleosynthetic era   The era following the leptonic era, between 1 second and 1000 seconds after the big bang, when neutrons were abundant and helium and deuterium were synthesized.

number-count test   A cosmological test that involves counting all galaxies down to a certain limiting magnitude and repeating this procedure for fainter and fainter limiting magnitudes. Deviations from the relation expected in euclidean space can help ascertain whether the universe is open or closed. In practice, this test provides a strong constraint on models of galactic evolution and luminosity at past eras.

number-density-redshift test   Measurement of the number density of galaxies simultaneously with determination of their redshift; this enables one to measure the volume element at a remote distance. This means that one should be able to infer the curvature of space, if evolutionary effects that may bias the determination of number density can be understood.

observable universe   The extent of the universe that we can see with the aid of the largest telescopes. Its ultimate boundary is determined by the horizon size.

Olbers' paradox   A paradox formulated by the German astronomer Heinrich Olbers in 1826 that can be traced back to the writings of others, such as de Cheseaux, a century or more earlier. The paradox is: Why is the sky dark at night, if the universe is infinite? We now know that several of the assumptions made by Olbers (explicitly or implicitly) are incorrect.

open space   A space of infinite volume and without any boundary (in the cosmological context).

open universe   A Friedmann-Lemaître cosmological model in which space is infinite and of negative curvature or Euclidean, and which will expand forever.

parallax   The angle subtended by 1 astronomical unit (the mean earth-sun distance) at the distance of a nearby star.

parametric representation   An indirect means of expressiong the solution to a differential equation in terms of an arbitrary parameter. As the parameter is allowed to vary, the parametric expression takes on the various values that the actual solution would have.

parsec (pc)   The distance at which one astronomical unit subtends an angle of 1 second of arc (1 pc = 3.086×10^{18} cm = 3.26 lt-yr.)

perfect cosmological principle The hypothesis, central to the steady state theory, that the universe presents the same aspect to all observers everywhere at all times. These observers would always view approximately the same large-scale distribution of matter in the universe in all regions and in every direction.

period-luminosity relation A correlation between the luminosities and periods of Cepheid variable stars.

photino A hypothetical fermion arising in some supersymmetric theories. Its mass is nonzero, but not precisely known.

photon A quantum of light; a discrete unit of electromagnetic energy. Photons have zero rest mass.

photosphere The visible surface of the sun. In general, the layer of a star that gives rise to the continuum (as opposed to spectral-line) radiation emitted by the star.

Planck distribution The distribution with respect to wavelength of the frequency of the intensity of blackbody radiation can be expressed as

$$B_v = 2\hbar v^3 c^{-2} \exp[\hbar v / kT - 1]^{-1}$$

where $\hbar$ is Planck's constant, T is the radiation temperature, and v is the frequency.

Planck time An instant in the big bang, prior to which Einstein's theory of gravitation breaks down and a quantized theory of gravity is needed. It can be expressed as $(G \hbar/c^5)^{1/2}$, where G is Newton's constant of gravitation, $\hbar$ is Planck's constant, and c is the speed of light, and it equals 10^{-43} second.

planetary nebula A tenuous expanding envelope of ionized gas surrounding a hot, highly evolved star.

planetesimals Asteroid-sized solid bodies that are hypothesized to form when the protosolar nebula collapsed into a disk and fragmented. Most of the planetesimals subsequently accumulated into planets.

pressure gradient A pressure difference between two adjacent regions of fluid results in a force being exerted from the high pressure region toward the low pressure region. In a star, the hot, dense interior and the cooler, more tenuous surface layers supply an outward pressure gradient, which balances the inward attractive force of gravity and stabilizes the star.

primeval fireball The hot, dense, early stage of the universe (predicted by the Big Bang theory) when the universe was predominantly filled with highly energetic radiation, which subsequently expanded and cooled and is now observed as the cosmic microwave background radiation.

positron The antiparticle of an electron, having positive charge but being otherwise similar.

primordial chaos The concept that the early universe might have been highly irregular and inhomogeneous. It could enable us to understand the origin of structure in the universe and why the universe is homogeneous and isotropic on the very largest scales.

primordial quarks All baryons and mesons are believed to be composed of quarks, which are elementary particles of fractional charge. In the high-density, hot-temperature phase of the very early universe, prolific numbers of quarks would have been present in equilibrium with the other elementary particles. As the universe expanded and cooled, some of these quarks may have been frozen out. To what extent independent free quarks could survive is an unresolved issue of elementary particle physics.

proper motion The apparent systematic angular displacement of a star across the celestial sphere, due to its motion transverse to the line of sight.

protogalactic gas cloud A massive gas cloud that collapsed to form a galaxy. Such clouds were produced as a result of the continued growth of density fluctuations after the decoupling era.

protogalaxy A galaxy during the early phase, before it has developed its present shape and mix of stars.

protosolar nebula The slowly rotating cloud of gas and dust from which the solar system formed.

protostellar core The smallest opaque clumps into which a collapsing interstellar gas cloud fragments. The characteristic mass of a protostellar core is only 0.01 solar mass. It grows by accretion as the surrounding matter falls toward it, attaining stellar mass within 10^5 years after it forms. At this stage, it is a protostar—a large cocoon of contracting matter that is radiating predominantly in the far infrared.

pulsar A rotating, magnetized neutron star. Radio pulsars were discovered in 1967, and they emit radio pulses with a very high degree of regularity. More than 100 have been discovered in our galaxy, with periods ranging from 0.001 second to about 3 seconds. The surface magnetic fields are believed to be as high as 10^{12} gauss.

quantum mechanics The theory of the interaction of matter and radiation, which rests on the original idea of Planck that radiating bodies emit energy in discrete units or quanta of radiation whose energy is proportional to the frequency of the light.

quasar An object that appears starlike but whose emission line spectrum reveals a large redshift. Quasars are the most luminous objects known in the universe.

radian A measure of angular distance; 2π radians equals 360 degrees.

radiation era The era from about 1 second to 3×10^5 years after the big bang, when radiation was the dominant constituent of the universe.

radiation temperature The temperature of a blackbody that radiates with the same intensity at the same frequency as a radiation source.

radio galaxy A galaxy that is extremely luminous at radio wavelengths.

radio interferometer Two or more separate radio antennas, each simultaneously receiving radiation from the same source. The different signals traverse slightly different path lengths to the respective antennas, are slightly out of phase, and, when added together, produce a characteristic interference pattern that enables the angular structure of compact radio sources to be mapped.

radio lobes Extended regions of diffuse radio emission, often dumbbell shaped, that surround a radio galaxy.

ram pressure Motion of a blunt body at supersonic velocity through an ambient gaseous medium causes a strong drag or ram pressure to be exerted on the body. In the case of a galaxy moving through the intergalactic gas, the ram pressure is capable of stripping the galaxy of much of its interstellar gas.

Rayleigh-Jeans limit An approximation valid at sufficiently long wavelengths (longward of the peak intensity) to the energy distribution of a blackbody.

recombination The capture of an electron by a positive ion. The dominant process in the early universe was hydrogen recombination. The electrons almost all recombined with protons to form hydrogen atoms, and the matter and radiation consequently decoupled from one another, because no further scattering of the radiation occurs.

red giant A luminous red star that has exhausted the supply of hydrogen in its core and evolved off the main sequence. Its core has contracted and become hotter, which enables helium burning to occur, and an extensive envelope develops as the outer regions of the star expand.

redshift The shift of spectral lines toward longer wavelengths in the spectrum of a receding source of radiation.

redshift-magnitude test A cosmological test involving the plotting of redshifts and apparent magnitudes of distant galaxies. Deviations from the relation expected in euclidean space can help determine whether the universe is open or closed. The redshift-magnitude test is very sensitive to evolutionary effects (whether galaxies were brighter or dimmer in the past).

relativistic cosmology Cosmological applications of Einstein's theory of general relativity. The big bang cosmological models were first derived according to the equations of relativistic cosmology, which are needed for many applications of these models.

relativistic particles Particles whose velocities approach the speed of light.

relaxation The process of gravitational interaction (in the case of a cluster of stars or galaxies) whereby a random distribution of motions is eventually established. The system is said to relax to a state of thermal equilibrium.

rest mass energy The energy that is computed according to the equation

$$E = Mc^2$$

where M is the mass of the particle and c is the speed of light. The rest mass energy is ordinarily liberated only when a particle annihilates with an anti-particle.

retrograde motion   A motion opposite to the prevailing direction of motion.

ring galaxy   A galaxy with a ringlike appearance. The ring contains luminous blue stars, but relatively little luminous matter is present in the central regions. It is believed that such a system was an ordinary galaxy that recently suffered a head-on collision with another galaxy.

Saha equation   An equation that determines the number of atoms of a given species in various stages of ionization that exist in a gas in thermal equilibrium at some specified temperature and total density.

Salpeter function   A simple functional interpolation for the distribution by mass of newly formed stars. Also referred to as the *initial mass function* of stars, the Salpeter function (the number of stars formed per unit mass range) is proportional to $m^{-2.35}$, where m is the mass of a star.

scattering   The process whereby light is absorbed and reemitted in all directions, with essentially no change in frequency. Scattering by free electrons was the dominant source of opacity in the early universe.

Schwarzschild radius   The event horizon of a spherical black hole; the critical radius from which light is unable to escape to infinity. The Schwarzschild radius of a star of solar mass is 2.5 kilometers.

Seyfert galaxy   One of a small class (comprising about 1 percent) of spiral galaxies, whose nuclei are characterized by high luminosity and blue color and whose spectra show intense broad emission lines.

singularity   A region in space-time where the known laws of physics break down and the curvature of space becomes infinite.

solar wind   A radial outflow of hot plasma from the solar corona. The sun ejects about 10^{-13} of its mass per year in the solar wind, which carries both mass and angular momentum away from the sun.

solid angle   A measure of the angular size of an extended object, equal to the area it subtends on the surface of a sphere of unit radius.

space-time   The three physical dimensions of space are combined with time, treated as a fourth dimension, to constitute the space-time continuum that is used as the fundamental framework of the theory of relativity.

spectral lines   Emission or absorption at a discrete wavelength or frequency, caused by atomic or molecular transitions. In the case of atoms, the transitions involve the jump of an electron from one orbit to another; a quantum of light is emitted if the electron jumps toward the nucleus and absorbed if it jumps outward.

spectroscopy   The technique of splitting light into its constituent colors, or obtaining spectra.

spectrum   Light spread out according to wavelength. The spectrum of a given element or compound contains characteristic spectral lines.

spherical space   A three-dimensional space whose geometry resembles that of the surface of a sphere and is said to have positive curvature.

spiral density wave   A wave, due to a local increase in the gravitational field, that produces a series of alternate compressions and rarefactions as it propagates with fixed angular velocity in a rotating galaxy. The compression also acts on interstellar gas in the galaxy, which is triggered to form stars on the leading edges of the spiral arms. The large-scale structure of spiral galaxies can be understood in this way.

spiral galaxy   A galaxy with a prominent nuclear bulge and luminous spiral arms of gas, dust, and young stars that wind out from the nucleus. Masses span the range from 10^{10} to 10^{12} $M_{\odot}$.

standard big bang model   The Friedmann-Lemaître cosmological models of an isotropic and homogeneous universe composed of expanding matter and radiation. There are three possible choices for the geometry of space in a standard big bang model: space can be positively curved, like the surface of a sphere, in which case the universe is finite, closed, and will eventually recollapse; or, space can either be euclidean or have negative curvature (like a saddle-shaped surface), in which the universe is infinite, open, and will expand forever. In all three models, space is unbounded.

standard candle   In astronomy, a class of stars or galaxies whose luminosities are similar and do not vary with age.

statistical parallax   The mean parallax for a group of stars all approximately at the same distance, as determined from radial velocities and proper motions.

steady state theory   The cosmological theory of Bondi, Gold, and Hoyle in which matter is continuously created to fill the voids left as the universe expands. Consequently, the universe should have no beginning and no end and should always maintain the same mean density. The required creation rate amounts to replenishing one atom per cubic meter each 10 billion years, or about 1 atom per 10 cubic kilometers per year. The steady state theory has been discredited and largely abandoned by cosmologists in recent years.

strong interactions   The short-range nuclear interactions responsible for holding nuclei together. The characteristic range of the strong interaction is 10^{-13} cm, and the time scale over which it operates is 10^{-33} second.

supercluster   A cluster of clusters of galaxies, about 10^8 light-years in extent.

supergalactic plane   An apparent plane of symmetry, passing through the Virgo cluster of galaxies, about which many of the brightest galaxies in the sky are concentrated. These galaxies form the Local Supercluster.

supernova   A prodigious increase in luminosity by up to 10^8, resulting in ejection of the outer envelope of the star and collapse of the inner core to form a neutron star. A star of more than about 8 solar masses becomes a supernova when it has exhausted its internal supply of nuclear fuel.

supernova remnant   The expanding shell of gas ejected at a speed of about 10,000 km s^{-1} by a supernova explosion, observed as an expanding diffuse gaseous nebula, often with a shell-like structure. Supernova remnants are generally powerful radio sources.

supersymmetry   A property demanded of certain unified gauge theories which creates a symmetry between fermions and bosons. If this symmetry is a local gauge symmetry, it can include the properties of the gravitational interaction and the resultant theory is called supergravity. These theories are now being actively investigated by theorists.

superstring A 26-dimensional configuration that has a linelike topology in which the fundamental forces, including gravity, can all be unified. Superstrings represent one of the best hopes that cosmologists have for developing a theory of quantum cosmology. The collapse of the extra dimensions to our three-dimensional universe is necessary to make a realistic theory; such a compactification has not yet been successfully achieved. Superstrings also exist in 10 dimensions.

synchrotron radiation The radiation emitted by charged relativistic particles spiraling in magnetic fields. The acceleration of the moving charges causes the particles to emit radiation. Radio galaxies and supernova remnants are intense sources of synchrotron radiation. Characteristics of synchrotron radiation are its high degree of polarization and nonthermal spectrum.

thermal equilibrium The equilibrium attained by a system that is in intimate contact with a thermal reservoir at some specified temperature.

tidal force The differential gravitational pull exerted on any extended body in the gravitational field of another body.

tidal theory The theory of the origin of the solar system involving the near collision of a massive body with the sun. The original version of a tidal theory, due to Buffon (1785), considered passage of a comet, but modern versions of this theory invoke a passing star. The gaseous debris torn from the sun by tidal forces is supposed to have condensed into the planets; however, this theory has been replaced by the nebular theory.

tired light The hypothesis that light may be degraded in energy, thereby increasing in wavelength and becoming redshifted, during its passage through intergalactic space. This would provide an alternative to the big bang model in accounting for the redshifts of distant galaxies. However, there is no evidence for any such tired-light effect.

Titius-Bode law A prescription for calculating planetary distances: the distance to the $(n+2)$th planet is $0.4 + (0.3)(2)^n$ astronomical units. This law works surprisingly well out to Uranus but then breaks down.

top-down theory A scenario for large-scale structure formation in which the largest objects (superclusters of galaxies) form first, and structure subsequently develops by fragmentation until the smallest galaxies have formed.

tritium A radioactively unstable heavy isotope of hydrogen. The tritium nucleus consists of two neutrons and one proton. The half-life of tritium is twelve years.

T-Tauri stars Luminous variable stars with low effective temperatures and strong emission lines, associated with interstellar gas clouds and found in very young clusters. They are believed to be still in the process of gravitational contraction from their protostellar phase and have not yet arrived at the main sequence and begun to burn hydrogen.

twin-exhaust model A theoretical model for radio galaxies, in which a compact source in the galactic nucleus is assumed to emit twin beams of rapidly moving plasma that traverse hundreds of thousands of light-years, eventually splattering to a halt in the ambient intergalactic gas, where the resulting dissipation accumulates in and energizes the radio lobes.

variable-mass theory A theory of Hoyle and Narlikar in which the masses of fundamental particles are assumed to vary with time in a manner that precisely accounts for the Hubble redshift law.

very long baseline interferometry Radio antennas that are several thousands of miles apart are used as separate elements of a radio interferometer to measure extremely fine-scale angular structure of radio sources.

virial theorem For a bound gravitational system, the time-averaged kinetic energy equals one-half of the (negative) gravitational potential energy.

virtual pairs Particles and antiparticles that exist for an extremely short time, over which Heisenberg's uncertainty principle allows the law of mass-energy conservation to be violated. According to Dirac's theory, the vacuum consists of a sea of virtual electron-positron pairs that can only be released or separated when sufficient energy is made available.

viscosity The internal friction of a fluid or liquid that tends to resist and dissipate its flow.

vortex theory The cosmogony of the solar system by Descartes (1644), who argued that the planets and sun accumulated from matter that moved in a system of vortices extending over all scales.

wavelength Electromagnetic radiation is characterized by wavelength or frequency, the product of which equals the speed of light. The wavelength is the distance between successive wavefronts (where the intensity peaks), and the frequency is the number of wavefronts passing a given point per second. In the visual region, optical light has a wavelength of from 4000 to 7000 Ångstroms, and a frequency of from 7×10^{14} to 4×10^{14} hertz.

weak interactions The short-range nuclear forces responsible for radioactivity and for the decay of certain unstable nuclei, such as neutrons. They occur at a rate slower than the strong interactions by a factor of about 10^{13}. The neutrino is the elementary particle primarily associated with the weak interactions.

white dwarf A dense, hot star of low luminosity that has exhausted its supply of nuclear fuel and is at the final stage of its evolution. Stars of less than 1.4 solar mass eventually become white dwarfs, with typical radii about 1 percent that of the sun and density in the range from 10^5 to 10^8 g cm^{-3}.

white noise Completely random and uncorrelated noise, with equal power at all frequencies.

ILLUSTRATION CREDITS

Frontispiece: National Radio Astronomy Observatory, operated by Associated Universities, Inc., under contract with the National Science Foundation (VLA observations by C. P. O'Dea and F. N. Owen).

2.4: Courtesy of Muzeum Okregowego W Toruniu. *2.5:* Copernicus' own drawing as it appears in his original manuscript. The figure in the printed book is different and inferior and was probably changed without reference to Copernicus. *2.7:* Yerkes Observatory photograph. *2.9:* Photographie Cramponi Brussels. *2.10:* This item is reproduced by permission of The Huntington Library, San Marino, California. *2.11:* American Institute of Physics, Niels Bohr Library. *3.5:* From "Pulsating Stars and Cosmic Distances" by Robert P. Kraft. Copyright © 1959 by Scientific American, Inc. All rights reserved. *3.6:* Data from P. Schechter, *Astrophysical Journal*, vol. 203, p. 297 (1976). *3.7:* S. Strom, Kitt Peak National Observatory. *3.9, 3.10, 3.11, 10.10, 10.13, 10.14, 10.15, 11.8(a), 12.6(a), 13.1:* Hale Observatories. *3.13:* P. J. E. Peebles. *4.2(a,b):* C. B. Moore. *4.2(c):* Courtesy of American Museum of Natural History. *4.4, 13.7, 13.9, 14.5, 15.2, 16.3:* Lick Observatory. *4.6:* The Mullard Radio Astronomy Observatory, Cambridge, England. *4.7:* Bell Laboratories. *4.8:* Data from D. Woody and P. Richards, *Physical Review Letters*, vol. 42, p. 925 (1979). *6.4:* D. Bennett and F. Bouchet. *7.1, 7.2:* Data from R. V. Wagoner, *Astrophysical Journal*, vol. 179, p. 343 (1973). *7.3:* Data from D. York and J. Rogerson, *Astrophysical Journal*, vol. 203, p. 378 (1976). *10.5:* Data from E. J. Groth and P. J. E. Peebles, *Astrophysical Journal*, vol. 217, p. 385 (1977). *10.6, 10.7, 14.8:* M. Davis, G. Efstathiou, C. Frenk, and S. D. M. White. *10.11:* Data from T. van Albada and J. van Gorkom, *Astronomy and Astrophysics*, vol. 54, p. 121 (1977). *11.1:* Data from E. Turner and J. Gott, *Astrophysical Journal Supplement*, vol. 32, p. 411 (1976). *11.3:* From V. de Lapparent, M. Geller, and J. Huchra, *Astrophysical Journal*, Letters, vol. 302, p. L1 (1986). *11.4:* P. Gorenstein. *11.5:* Data from R. J. Mitchell, J. L. Culhane, P. J. N. Davison, and J. C. Ives, *Monthly Notices of the Royal Astronomical Society*, vol. 176, p. 29P (1976). *11.6:* D. Malin and D. Carter. *11.7:* L. Hernquist and P. Quinn. *11.8:* Data from M. S. Roberts and R. N. Whitehurst, *Astrophysical Journal*, vol. 201, p. 343 (1975). *12.1:* From "The Cosmic Background Radiation" by Adrian Webster. Copyright © 1974 by Scientific American, Inc. All rights reserved. *12.2(a):* Cerro Tololo Inter-American Observatory. *12.2(b):* Data from J. Grindley, *Astrophysical Journal*, vol. 199, p. 49 (1975). *12.3(a):* G. K. Miller, K. J. Wellington, H. van der Laan, *Astronomy and Astrophysics*, vol. 38, p. 381 (1985). *12.3(b):* C. P. O'Dea and F. N. Owen, *Astrophysical Journal*, vol. 301, p. 841 (1986). *12.6(b):* Data from B. N. Swanenburg et al., *Nature*, vol. 275, p. 298 (1978). *12.7:* From "The Evolution of Quasars" by Maarten Schmidt and Francis Bello; copyright © 1971 by Scientific American, Inc.; all rights reserved. *12.8:* A. M. Wolfe. *12.10:* V. Petrosian and R. Lynds. *13.6:* H. Ungerechts and P. Thaddeus. *13.8:* Data from J. Scalo, in *Protostars and Planets,* edited by T. Gehrels (University of Arizona Press), p. 265 (1978). *13.10:* B. J. Bok. *15.4:* J. Kristian, Hale Observatories. *15.8:* Ellis Miller. *15.9(a):* C. B. Moore. *15.9(b):* Courtesy of American Museum of Natural History. *16.2:* From "The Discovery of Icarus" by Robert S. Richardson. Copyright © 1965 by Scientific American, Inc. All rights reserved. *16.4:* Courtesy of American Museum of Natural History. *16.10:* J. Tarter. *17.1:* Data from J. Kristian et al., *Astrophysical Journal,* vol. 221, p. 383 (1978). *17.2:* Data from M. Ryle. Reproduced, with permission, from *Annual Review of Astronomy and Astrophysics,* vol. 6, p. 2490. Copyright © 1968 by Annual Reviews, Inc. *17.3:* Data from J. A. Tyson. *17.4:* W. Harris and D. Vanden Berg. *17.5:* O. Lahav. *17.6:* NASA.

Cartoons on pages 57 and 368: Sidney Harris.

Color plates: 1: P. Lubin. *2:* National Radio Astronomy Observatory, operated by Associated Universities, Inc., under contract with the National Science Foundation (VLA observations by C. P. O'Dea and F. N. Owen). *3, 4, 6, 10, and 11:* European Southern Observatory. *5:* T. Dane and P. Thaddeus. *7 and 9:* Infrared Astronomy Satellite, NASA. *8:* Canada-France-Hawaii Telescope. *12:* R. B. Partridge.

INDEX